Sustainable Energy Systems and Technology

Sustainable Energy Systems and Technology

Edited by **Kurt Marcel**

R CALLISTO REFERENCE

New York

Published by Callisto Reference,
106 Park Avenue, Suite 200,
New York, NY 10016, USA
www.callistoreference.com

Sustainable Energy Systems and Technology
Edited by Kurt Marcel

International Standard Book Number: 978-1-63239-763-8 (Hardback)

The publisher's policy is to use permanent paper from mills that operate a sustainable forestry policy. Furthermore, the publisher ensures that the text paper and cover boards used have met acceptable environmental accreditation standards.

Trademark Notice: Registered trademark of products or corporate names are used only for explanation and identification without intent to infringe.

Printed in the United States of America.

Contents

Preface

This book provides comprehensive insights into the field of sustainable energy systems and technology. It elucidates new techniques and their applications in a multidisciplinary approach. Sustainable energy refers to the form of energy which can be used without causing harm to nature. The objective of sustainable energy systems is to find efficient ways to use resources so that they fulfill the need of the present generation while being available for the future as well. It aims to protect environmental and natural resources. The different types of sustainable energies currently available are solar energy, hydropower, biofuels, green power, etc. From theories to research to practical applications, case studies related to all contemporary topics of relevance to sustainable energy have been included in this book. The objective of this text is to provide a comprehensive overview of the different areas of this field along with technological progress that has future implications. It will serve as a valuable source of reference for engineers, environmentalists, ecologists, researchers, students and for anyone who wants to delve deeper into the study of sustainable energy.

This book is a result of research of several months to collate the most relevant data in the field.

When I was approached with the idea of this book and the proposal to edit it, I was overwhelmed. It gave me an opportunity to reach out to all those who share a common interest with me in this field. I had 3 main parameters for editing this text:

1. Accuracy – The data and information provided in this book should be up-to-date and valuable to the readers.

2. Structure – The data must be presented in a structured format for easy understanding and better grasping of the readers.

3. Universal Approach – This book not only targets students but also experts and innovators in the field, thus my aim was to present topics which are of use to all.

Thus, it took me a couple of months to finish the editing of this book.

I would like to make a special mention of my publisher who considered me worthy of this opportunity and also supported me throughout the editing process. I would also like to thank the editing team at the back-end who extended their help whenever required.

Editor

A New Approach for Converting Renewable Energy to Stable Energy

Mohamed Talaat[1], Reda Edris[2], Naglaa Ibrahim[2], Fatma Omar[2], Mohamed Ibrahim[2]
[1]Electrical Power and Machines, Faculty of Engineering, Zagazig University, Zagazig, Egypt
[2]Electrical and Computer Engineering, Higher Technological Institute, 10th of Ramadan City, Egypt
Email: m_mtalaat@zu.edu.eg, Reda.Edris@gmail.com

ABSTRACT

A renewable energy plant which relies on wind speed or solar insolation is unreliable because of the stochastic nature of weather patterns. It is theorized that by using multiple renewable energy plants in separate areas of a region, the different weather conditions might approach a probabilistically independent relationship. The goal of this paper is to utilize the power system technology to help disseminate wind and solar power systems to get a stable energy. A new approach to get appropriate stable energy is achieved by using the interrupted energy that obtained from wind farm and solar insolation. This is achieved by lifting water to a higher level with appropriate pumps and storing it in the form of potential energy. Then a stable energy is obtained by reliving water to the lower level. In this paper, the efficiency obtained from the renewable energy is compared with that obtained from traditional ones. An experimental model to simulate the process of converting the renewable energy to a stable energy is presented. The obtained results from experimental model explained that the renewable energy can be converted to a stable one with high efficiency.

Keywords: Renewable Energy; Stable Energy; Energy Stored; Wind Farm; Solar Energy; Application of Renewable Energy

1. Introduction

Renewable energy sources, such as wind and solar, have vast potential to reduce dependence on fossil fuels and greenhouse gas emissions in the electric sector. Climate change concerns, state initiatives including renewable portfolio standards, and consumer efforts are resulting in increased deployments of both technologies. Both solar PV and wind energy have variable and uncertain sometimes referred to as "intermittent" output, which are unlike the dispatchable sources used for the majority of electricity generation [1-2].

The variability of these sources has led to concerns regarding the reliability of an electric grid that derives a large fraction of its energy from these sources as well as the cost of reliably integrating large amounts of variable generation into the electric grid [3-5]. Because the wind doesn't always blow and the sun doesn't always shine at any given location, there has been an increased call for the deployment of energy storage as an essential component of future energy systems that use large amounts of variable renewable resources. However, this often-characterized "need" for energy storage to enable renewable integration is actually an economic question [2,4].

To determine the potential role of storage in the grid of the future, it is important to examine the technical and economic impacts of variable renewable energy sources. It is also important to examine the economics of a variety of potentially competing technologies including demand response, transmission, flexible generation, and improved operational practices. While there are clear benefits of using energy storage to enable greater penetration of wind and solar, it is important to consider the potential role of energy storage in relation to the needs of the electric power system as a whole [2,5].

In this paper, the role of energy storage in the electricity grid has been explored, focusing on the effects of variable renewable sources (primarily wind and solar energy), a new technique of simulation like FEM and CSM are used to simulate the field and energy [6-7]. The goal is to utilize and forecast the power system technology [8] to help disseminate wind and solar power systems to get a stable energy. This is achieved by lifting water and storing electrical energy in the image of potential energy to lift water to a higher level with appropriate pumps then obtain a stable energy by releasing electricity impulsively water to the lower level. The current role that energy storage plays in meeting the varying electricity demand is important. The impact of variable renewable on the grid is then discussed, including how these energy sources will require a variety of enabling techniques and technologies to reach their full potential. Finally, the potential role of several forms of enabling technologies including energy storage has been evaluated.

2. Energy Storage Technologies

A variety of technologies are available for storage of energy in the power system [9,10]. When identifying the most relevant storage solutions it is necessary to include considerations on many relevant parameters, such as: cost, lifetime, reliability, size, storage capacity and environmental impact. All these parameters should be evaluated against the potential benefits of adding storage in order to reach a decision on which type of storage should be added. There may also be cases where the value of adding storage is not large enough to justify such an investment.

Energy storage technologies for power applications can be divided according to the form of energy stored: Mechanical, electro-chemical, electromagnetic, or thermal storage.

Mechanical storage includes pumped hydro storage, CAES and flywheels. Electrochemical storage includes all types of batteries and fuel cells, and electromagnetic storage includes super capacitors and superconducting magnetic energy storage. Each technology has certain attributes with regard to for example storage capacity, power, reaction time and cost [10].

Figure 1 shows some of the most relevant storage technologies, grouped according to the form of stored energy as well as energy storage capacity. The medium capacity storage technologies seem very relevant for storage in relation to wind power. The medium capacity storage technologies are primarily batteries and flow batteries, which all have the advantage in relation to wind power plants that they are modular and scalable.

2.1. High-Energy Batteries

For many batteries, there is considerable overlap between energy management and the shorter-term applications discussed previously. Furthermore, batteries can generally provide rapid response, which means that batteries "designed" for energy management can potentially provide services over all the applications and timescales discussed.

Several battery technologies have been demonstrated or deployed for energy management applications. In addition to the chemistries discussed previously, the commercially available batteries targeted to energy management include two general types: high-temperature batteries and liquid electrolyte flow batteries.

2.2. Pumped Hydro Storage (PHS)

Pumped hydro is the only energy storage technology deployed on a gigawatt scale in the United States and worldwide. Many of the sites store 10 hours or more, make the technology useful for load leveling. PHS is also used for ancillary services. PHS uses conventional pumps and turbines and requires a significant amount of land and water for the upper and lower reservoirs. PHS plants can achieve round-trip efficiencies that exceed 75% and may have capacities that exceed 20 hours of discharge capacity. Environmental regulations may limit large-scale above-ground PHS development. However, given the high round-trip efficiencies, proven technology, and low cost compared to most alternatives, conventional PHS is still being pursued in a number of locations.

2.3. Compressed Air Energy Storage (CAES)

CAES technology is based on conventional gas turbine

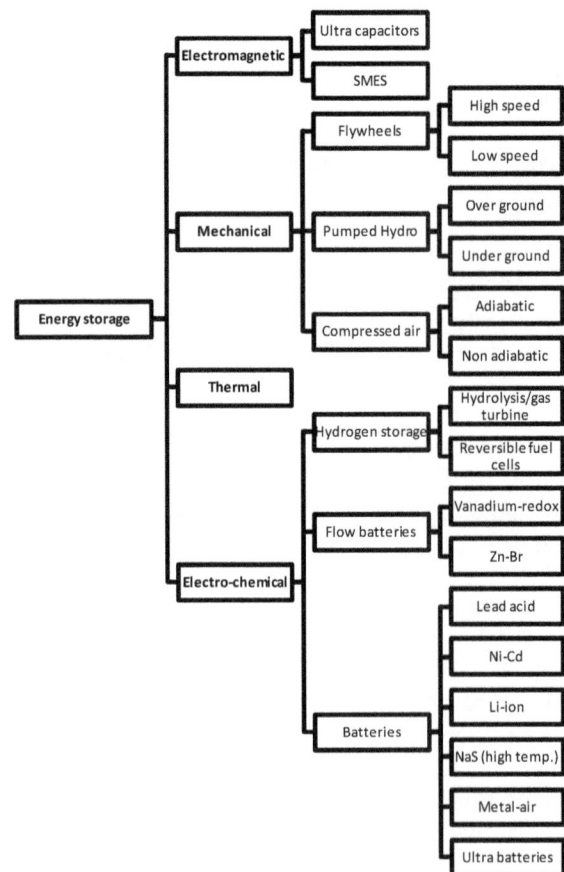

Figure 1. Energy storage technologies grouped according to form of energy as well as energy storage capacity. Typical timescales have been indicated [10].

technology and uses the elastic potential energy of compressed air. Energy is stored by compressing air in an airtight underground storage cavern. To extract the stored energy, compressed air is drawn from the storage vessel, heated, and then expanded through a high-pressure turbine that captures some of the energy in the compressed air. The air is then mixed with fuel and combusted, with the exhaust expanded through a low-pressure gas turbine. The turbines are connected to an electrical generator.

The primary disadvantages of CAES are the need for an underground cavern and its reliance on fossil fuels. Alternative configurations for CAES have been proposed using manufactured above-ground vessels, new turbine designs to reduce fossil fuel use, or designs that re-use the heat of compression and avoid fuel use altogether.

2.4. Thermal Energy Storage

Thermal energy storage is sometimes ignored as an electricity storage technology because it typically is not used to store and then discharge electricity directly. However, in some applications, thermal storage can be functionally equivalent to electricity storage. One example is storing thermal energy from the sun that is later converted into electricity in a conventional thermal generator. Another example is converting electricity into a form of thermal energy that later substitutes for electricity use such as electric cooling or heating.

In the low capacity end, ultra capacitors may be of relevance in relation to wind power conditioning. Hydrogen fuel cells may also be relevant, both as medium- and high capacity storage.

Due to short-term energy market closure delay, dispatch levels of wind farm-energy storage system should be decided according to wind power LF [8]. The forecast accuracy and dispatch strategy are both of importance to energy storage sizing. Next, wind power forecast method and the dispatch strategy for minimizing size of energy storage system will be studied.

The low capacity storage technologies seem less relevant in relation to overall improvement of wind energy quality because of high cost pr. unit stored energy and relatively short storage time scale. The very high capacity technologies, PH storage and CAES, involve large investments and civil engineering efforts, as well as special requirements with regard to placement. But these technologies may well be the best solution in relation to large scale storage, where a few major energy storage facilities ensure overall power system stability. Especially in cases where hydro-electric plants with unused storage capacity already exist, large benefits may be obtained by combining them with wind power plants.

3. Mathematical Model

The PM DC motor is the one that is most commonly used in PVPS. It does not need field current and the magnetic field is provided separately by the permanent magnet. When a voltage V_t is applied at time t across the armature winding including an inductance L_a and resistance R_a. Then the relationship between V_b L_a and R_a can be expressed as.

$$V_t(t) = E_b(t) + R_a I_a(t) + L_a \frac{dI_a(t)}{dt} \quad (1)$$

where; I_a is the motor current and E_b is the internally induced voltage by the motor. Which is proportional to motor rotational speed ω (red/sec) such that,

$$E_b(t) = K_E \omega(t) \quad (2)$$

where; K_E is the voltage constant in $V.sec/red$. Combining Equations (1) and (2) gives

$$V_t(t) = K_E \omega(t) + R_a I_a(t) + L_a \frac{dI_a(t)}{dt} \quad (3)$$

The motor develops an electric torque T_d which basically depends on the motor current

$$T_d(t) = K_t I_a(t) \quad (4)$$

where; K_t is the torque constant (N·m/A).

The lead torque is the pump torque T_p which depends on the pump type, as will be seen later .If the motor pump coupling losses are neglected, the dynamic equation of the motor and load is,

$$T_d(t) = T_p + J \frac{d\omega(t)}{dt} \quad (5)$$

where; J is the moment of inertia of motor-pump system (N·m·sec²/red). Combining Equations (4) and (5) gives,

$$K_t I_a(t) = T_p + J \frac{d\omega(t)}{dt} \quad (6)$$

The two first order differential Equations (3) and (6) represent the dynamic model of the system. The dynamic performance can be obtained by solving these equations numerically to obtain the instantaneous values of the motor current and speed.

Under steady sate conditions Equations (3) and (6) become.

$$V_t(t) = K_E \omega(t) + R_a I_a \quad (7)$$

$$K_t I_a(t) = T_p \quad (8)$$

Equations (7) and (8) represent the mathematical model of the system under steady conditions. The solution of these equations at any climatic conditions gives the operating parameters of the motor current and speed.

$$V_t = K_E \omega + R_a \frac{T_p}{K_t} \qquad (9)$$

The torque constant

$$k = \frac{V_t}{\text{speed}} \qquad (10)$$

Hence the torque is $T_d(t) = kI_a(t)$ and the output power is given as

$$P_{\text{out}} = T_d(t) \times \frac{2\pi \times \text{speed}}{60} \qquad (11)$$

4. Experimental Model

An experimental model was established to simulate the mechanism of utilize the energy from wind farm and solar energy and convert this energy to a stable energy using a water pump see **Figure 2**.

A model of wind turbine or photovoltaic cell was connected to a DC pump through a PLC device. The DC pump was 24 V and 400 mA. The pump received the interrupted energy from wind if it was active or PV cell if it was active, the activation decided through the PLC sensor device. The water pump using these interrupted energy to left the water to a high tank, which use this water as a hydraulic high dam to convert this energy to a stable energy.

The performance of the wind turbine observation depends on the amount of *emf* produced with variable speed of wind. Also the characteristics of the DC pump depend on the amount of current taken from pump at each speed to lift the water to high tank.

A new microcontroller circuit is designed for measuring the values of current, voltage, and speeds, see **Figure**

3.

5. Results

The torque constant according to Equation (10) is given as

$$k = \frac{17}{2105 * \frac{2\pi}{60}} = 0.0771 \frac{\text{Volts}}{\text{rad/sec}} = 0.0771 \frac{\text{N} \cdot \text{m}}{\text{A}}.$$

where, $V_t = emf$ – armature drop = 24 – 7 = 17 Volts, and the corresponding angular speed is given as 2105 rpm, as given from **Table 1** at no load of pump.

The voltage output from wind turbine, and the current delivered to the pump at different speed of wind are given in **Table 2**. Also the torque calculated from Equation (4), and the calculated powers from Equation (11) are calculated in the same table.

Figure 4 shows the variation of output voltage from wind turbine with wind speed at no load. Also **Figure 5** shows the variation of delivered current to the DC pump with each wind turbine speed.

The variation of output torque with output current of wind turbine is shown in **Figure 6**. Also **Figure 7** shows the variation of DC pump current with wind turbine speed. **Figure 8** gives the variation of wind efficiency with wind speed.

Finally, the overall efficiency of the wind and pump is calculated and shown in **Figure 9**, which gives a stable power.

6. Discussion

Tables 1 and **2**, also **Figures 4-7** show that the energy obtained from the wind turbine with no load or to the DC pump is not stable and interrupted, so a curve fitting is

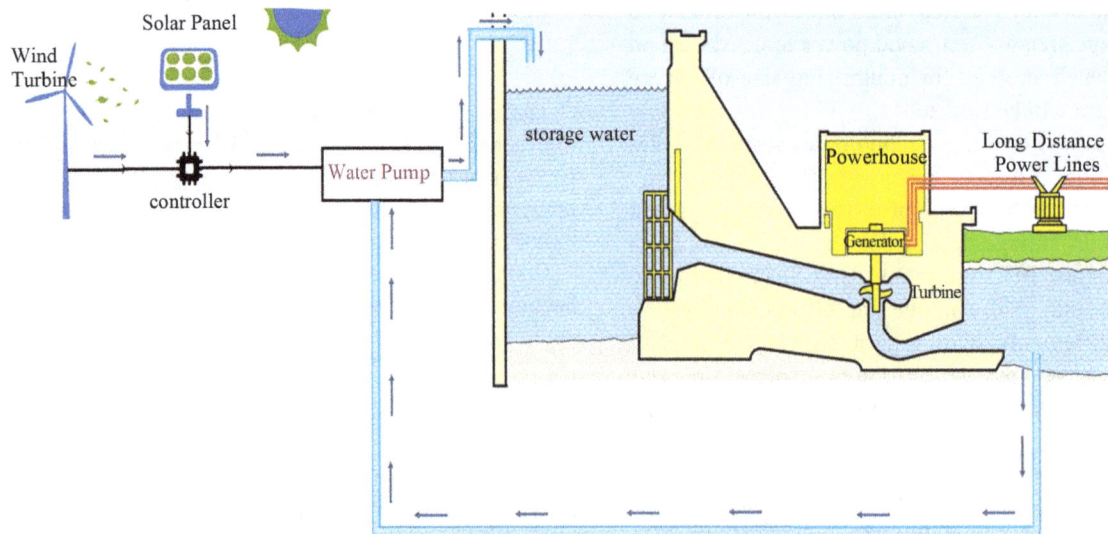

Figure 2. Energy storage simulation from wind and solar.

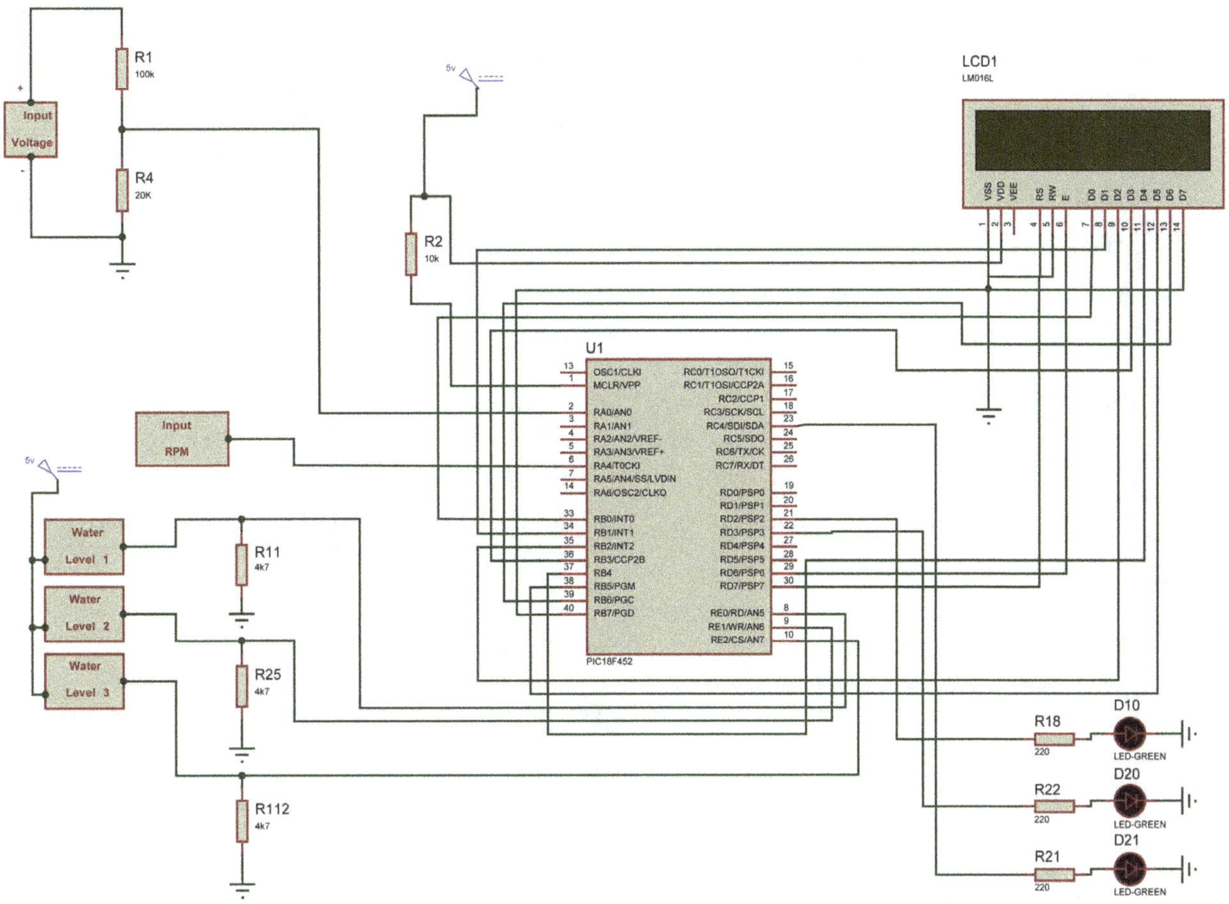

Figure 3. Microcontroller circuit used for measuring voltage, current, and speed.

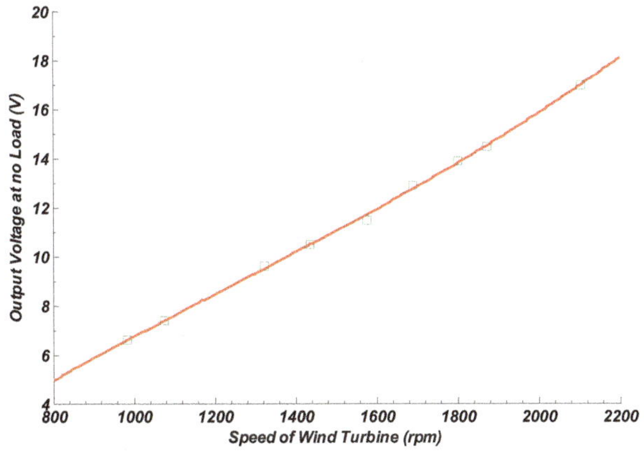

Figure 4. Variation of output no load voltage with wind turbine speed.

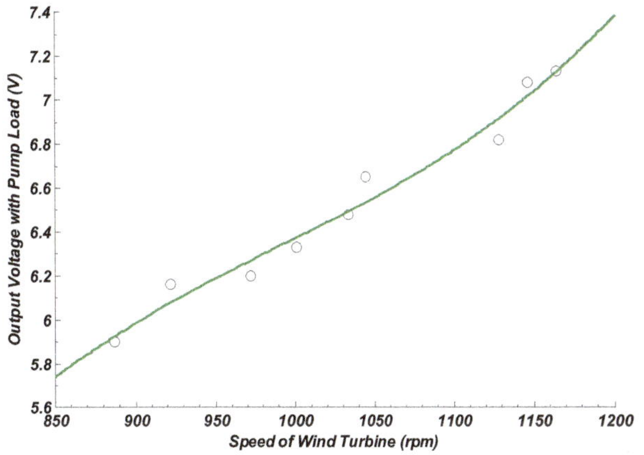

Figure 5. Variation of output voltage with wind turbine speed.

used to get a stable curve. Also **Figure 8** explained that the variation of efficiency of wind with wind speed is not stable. So instead to use this unstable energy for power generation which will be unstable also, the new technique uses this unstable energy for lifting the water to high tank to generate a stable energy, according to **Figure 9** the overall efficiency gives a stable energy.

7. Conclusion

An experimental model to simulate the process of converting the renewable energy to stable energy is presented. A new approach to get appropriate stable energy is achieved by using the interrupted energy that obtained from wind farm and solar insolation. This is achieved by

Table 1. Variation of output voltage with wind turbine speed at no load.

Wind Turbine Speed (rpm)	Output Voltage (V)
985	6.6
1076	7.4
1323	9.6
1435	10.5
1577	11.5
1690	12.9
1800	13.9
1874	14.5
2105	17

Table 2. Variation of delivered current to DC pump with wind turbine speed.

Wind Turbine Speed (rpm)	Output Voltage (V)	Current (A)	Output Power (W)	Torque (Nm)
887	5.9	0.01	0.059	0.0008
922	6.16	0.05	0.308	0.0039
972	6.20	0.08	0.496	0.0062
1001	6.33	0.12	0.7596	0.0093
1033	6.48	0.18	1.1664	0.0139
1044	6.65	0.24	1.596	0.0185
1128	6.82	0.32	2.1824	0.0247
1146	7.08	0.40	2.832	0.0308
1164	7.13	0.42	2.9946	0.0324

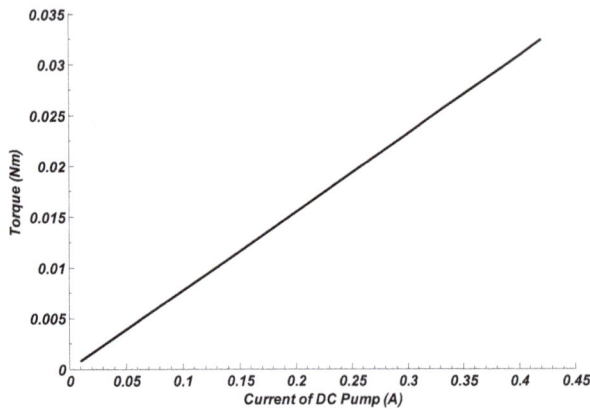

Figure 6. Variation of output torque with output current.

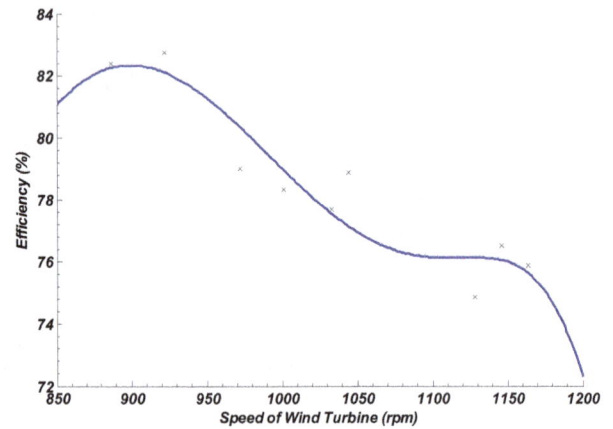

Figure 8. Variation of wind efficiency with wind turbine speed.

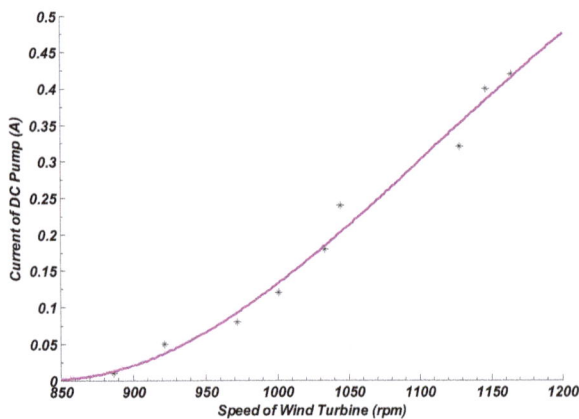

Figure 7. Variation of DC pump current with wind turbine speed.

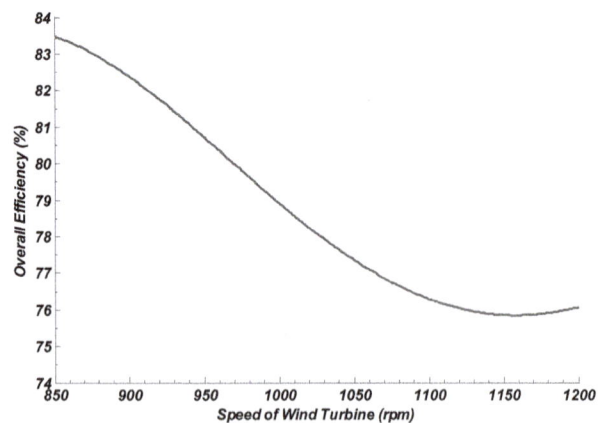

Figure 9. Variation of overall efficiency with wind turbine speed.

lifting water and store electrical energy in the image of potential energy to lift water to a higher level with appropriate pumps then obtain a stable energy by releasing

electricity impulsively water to the lower level. The obtained results from experimental model explained that the renewable energy can be converted to a stable one with high efficiency.

REFERENCES

[1] World Wind Energy Association, "Highlights of the World Wind Energy Report," 2009. http://www.wwindea.org/home/index.php

[2] "The Role of Energy Storage with Renewable Electricity Generation," Technical Report NREL/TP-6A2-47187, 2010. http://www.osti.gov/bridge

[3] L. L. Freris, "Wind Energy Conversion Systems," Prentice Hall, Upper Saddle River, 1990.

[4] B. C. Ummels, E. Pelgrum and W. L. Kling: "Integration of Large-Scale Wind Power and Use of Energy Storage in the Netherlands' Electricity Supply," *IET Renewable Power Generation*, Vol. 2. No. 1, 2008, pp. 34-46.

[5] E. Spahic, G. Balzer, B. Hellmich and W. Münch, "Wind Energy Storages—Possibilities," IEEE PowerTech, 2007.

[6] M. Talaat and A. El-Zein, "A Numerical Model of Streamlines in Coplanar Electrodes Induced by Non-Uniform Electric Field," *Journal of Electrostatics*, Vol. 71, No. 3, 2013, pp. 312-318. http://dx.doi.org/10.1016/j.elstat.2012.12.034

[7] M. Talaat, "Charge Simulation Modeling for Calculation of Electrically Induced Human Body Currents," *IEEE Annual Report Conference on Electrical Insulation and Dielectric Phenomena CEIDP*, West Lafayette, 17-20 October 2010, pp. 644-647.

[8] M. A. Farahat and M. Talaat, "The Using of Curve Fitting Prediction Optimized by Genetic Algorithms for Short-Term Load Forecasting," *International Review of Electrical Engineering (IREE)*, Vol. 7, No. 6, 2012, pp. 6209-6215.

[9] M. Swierczynsky, R. Teodorescu, C. N. Rasmussen, P. Rodriguez and H. Vikelgaard, "Storage Possibilities for Enabling Higher Wind Energy Penetration," *EPE Wind Energy Chapter Symposium*, Stafford, 15-16 April 2010.

[10] C. N. Rasmussen, "Energy Storage for Improvement of Wind Power Characteristics," IEEE PowerTech, Trondheim, 2011.

Nomenclature A

PV: Photovoltaic
FEM: Finite Element Methods
CSM: Charge Simulation Method
LF: Load Forecasting
CAES: Compressed Air Energy Storage
PH: Pumped Hydro
PM: Permanent Magnet
PHS: Pumped Hydro Storage
PVPS: Photovoltaic Pump Storage

2

Prime Energy Challenges for Operating Power Plants in the GCC

Mohamed Darwish, Rabi Mohtar

Qatar Environment and Energy Research Institute, Qatar Foundation, Doha, Qatar

Email: madarwish@qf.org.qa

ABSTRACT

There is a false notion of existing available, abundant, and long lasting fuel energy in the Gulf Cooperation Council (GCC) Countries; with continual income return from its exports. This is not true as the sustainability of this income is questionable. Energy problems started to appear, and can be intensified in coming years due to continuous growth of energy demands and consumptions. The demands already consume all produced Natural Gas (NG) in all GCC, except Qatar; and the NG is the needed fuel for Electric Power (EP) production. These countries have to import NG to run their EP plants. Fuel oil production can be locally consumed within two to three decades if the current rate of consumed energy prevails. The returns from selling the oil and natural gas are the main income to most of the GCC. While NG and oil can be used in EP plants, NG is cheaper, cleaner, and has less negative effects on the environment than fuel oil. Moreover, oil has much better usage than being burned in steam generators of steam power plants or combustion chambers of gas turbines. Introducing renewable energy or nuclear energy may be a necessity for the GCC to keep the flow of their main income from exporting oil. This paper reviews the GCC productions and consumptions of the prime energy (fuel oil and NG) and their role in electric power production. The paper shows that, NG should be the only fossil fuel used to run the power plants in the GCC. It also shows that the all GCC except Qatar, have to import NG. They should diversify the prime energy used in power plants; and consider alternative energy such as nuclear and renewable energy, (solar and wind) energy.

Keywords: Gulf Co-Operation Council (GCC); Electric Power; Natural Gas; Crude Oil; Renewable Energy; Gas Turbine Combined Cycle; Integrated Solar Combined Cycle; Oil and Natural Gas Reserves

1. Introduction

The Arab Gulf Co-operation Countries (GCC) includes Qatar, Saudi Arabia (SA), United Arab Emirates (UAE), Kuwait, Bahrain, and Oman. They have about 57% of world petroleum oil reserves and 28% of world Natural Gas (NG) reserves, [1].

The GCC are the main producers and suppliers of oil and NG to the world, see **Figures 1-5**, [1,2], and **Table 1**, [3]. The returns from exporting oil and NG represent the primary income to most of the GCC. For example, SA is the world's largest producer and exporter of total petroleum liquids in 2010, and the world's second largest crude oil producer behind Russia. SA's economy depends heavily on crude oil. In 2010, SA oil export revenues have accounted for 80% - 90% of total revenues and more than 40% of gross domestic product (GDP). Kuwait Petroleum export revenues account for half of the GDP, 95% of total export earning, and 95% of government revenues in 2010. Although Bahrain has a minor role in oil production, its economy depends heavily on hydrocarbon exports, mostly refined products, which account for 70% of the government revenues. Qatar's oil and gas sectors accounted for over half of its GDP in 2010. Qatar is wholly dependent on oil and NG for all of its primary energy demands. In 2009, the NG consumption in some of the GCC was very high. For example, SA consumed 68.3 billion cubic meters BCM (2.3% of total worldwide consumption), and UAE consumed 39.6 BCM (1.3% of world consumption). Also, SA has the highest crude oil consumption per capita in the world. Its crude oil consumption in 2009 was 2.376 Million (M) barrels per day (Mbbl/d) or 0.3778 M ton/d. For 25 M population, this gives 33.7 bbl per year per capita (bbl/y.ca), or 5.36 ton/y.ca.

The plentiful reserves and high productions of prime energy (*i.e.* NG and fuel oil) give the notion of existing available, abundant, and long lasting fuel energy in GCC; with continual income of its export. This is not true; and the sustainability of this income is questionable.

[4] pointed that there are major challenges that necessitate significant changes in the GCC energy policy such

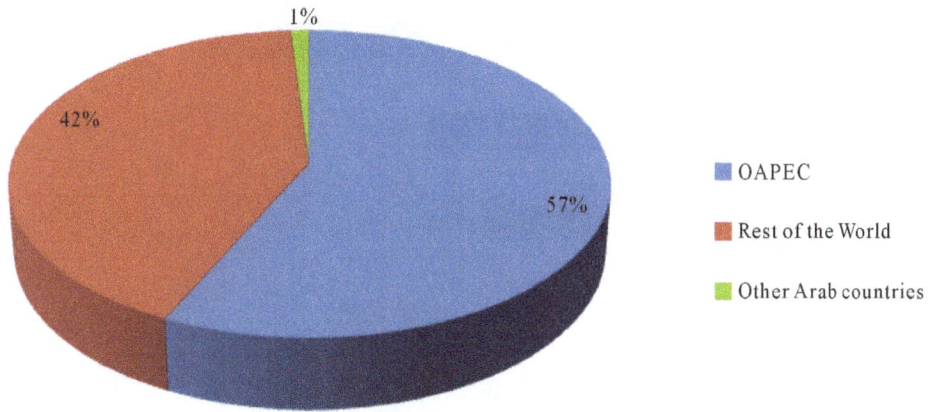

Figure 1. Share of Organization of Arab Petroleum Exporting Countries (OAPEC) in world fuel oil proven reserve percentage 2009 [1].

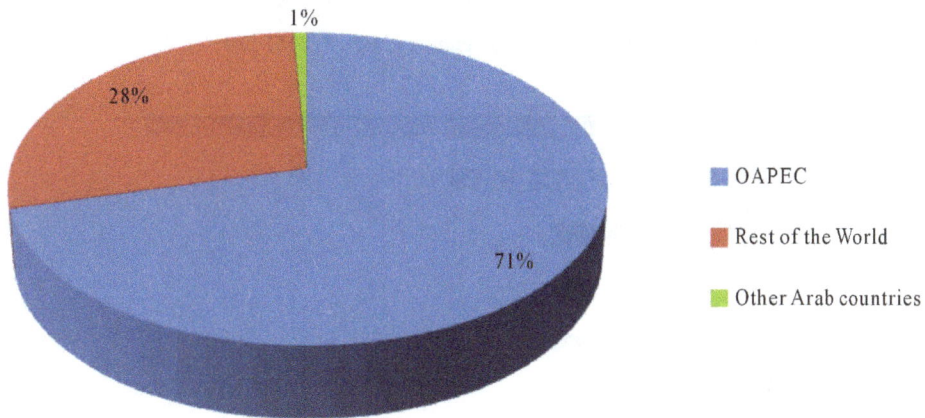

Figure 2. Share of OAPEC in world natural gas proven reserve percentage, 2009 [1].

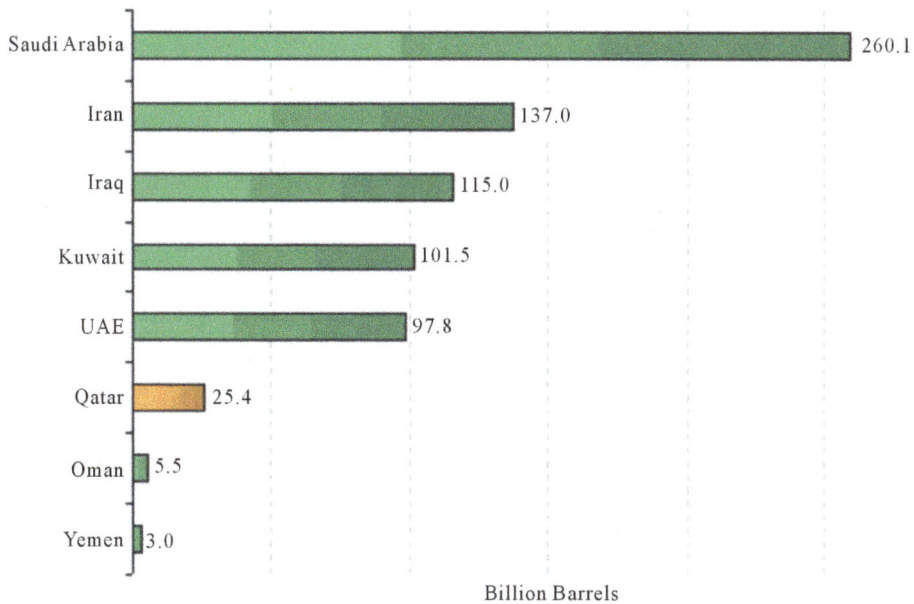

Source: Oil & Gas Journal, Jan. 1, 2011

Figure 3. Proven fuel oil reserve by country in Middle East as in Jan. 2011 [2].

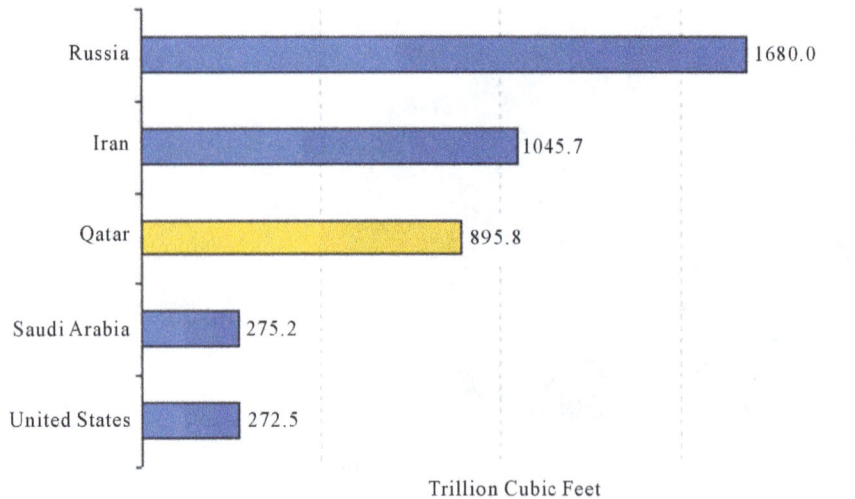

Source: Oil & Gas Journal, Jan. 1, 2011 and EIA Natural Gas Navigator (U.S. Only), Dec. 31, 2009.

Figure 4. Natural gas reserve by country, Jan. 1, 2011 [2].

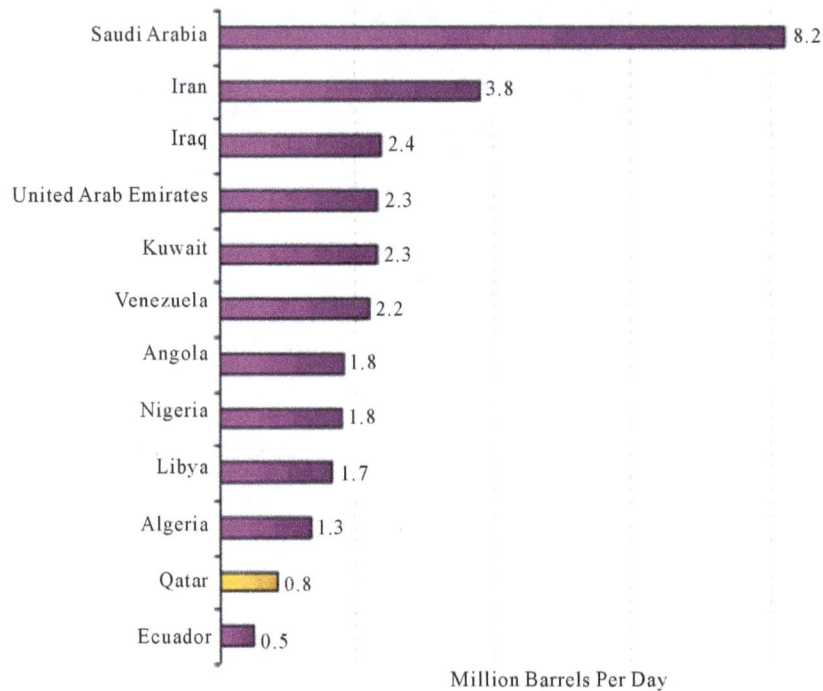

Source: EIA Short Term Energy Outlook, December 2010, Production values do not include lease condensate.

Figure 5. OPEC crude oil production by country, [2].

as the rising consumption of petroleum products, the water shortages and desertification, and the climate change.

The UAE, is already on the verge of NG crisis; by 2015-2016. The NG production in the UAE is predicted to increase significantly, by an estimated 14.4 BCM/y when several non-associated fields start producing. But this increase will be accompanied by a significant increase in gas demands over the same period. In spite of

the gas allocation issues, the UAE's state-owned companies are pressing ahead with diversification into petrochemicals, building additional plants that demand large quantities of gas [5].

2. NG Fuel Shortage in GCC

Although the GCC has huge NG resources (about 28% of the world), there is a serious shortage in the needed NG

Table 1. NG produced by some Arab countries and other and their rank worldwide, in BCM in 2009, [3].

Rank	Country	Production (BCM)
1	United States	593.400
2	Russia	546.8
5	Iran	116.3
7	Algeria	86.500
10	Saudi Arabia	77.1
11	Qatar	76.98
14	Egypt	62.7
18	United Arab Emirates	50.24
28	Oman	24
33	Libya	15.9
37	Kuwait	12.7
38	Bahrain	12.64
47	Syria	6.04
58	Iraq	1.88
93	Yemen	0.455

to run their Power Plants (PP), due to continually rising EP consumption. The subject was recently raised in few publications, (e.g. [6-8]). The NG demand is much more than the region's gas exploration and production, and NG have to be imported, [6]. [6] mentioned that although the GCC began in 1977 exporting gas as the UAE built the region's first Liquified Natural Gas (LNG) liquefaction terminal and began sending LNG to Japan; the GCC began importing gas, as Kuwait received its first LNG cargo from Russia at its fast-track LNG receiving terminal at Mina Al-Ahmadi Gas Port (MAAGP). Other GCC, with the exception of Qatar, are considering importing NG to meet rapidly rising demands. Construction of Kuwait's MAAGP commenced in January 2008, and in April 2008 the Emirate of Dubai, UAE appointed Shell Company as advisor for building a fast-track LNG receiving terminal and expects to receive its first gas in 2010. The growing shortage in gas is due to, [6]: increasing power consumption and the high share of gas in power generation; depleting oil fields and the gas use for oil recovery enhancement, increasing economic emphasis on the steel, aluminum, and petrochemicals sectors; gas exploration and production challenges; and long-term gas export commitments limit local supply.

The gas supply outlook for the GCC remains bleak. Even if the prolonged recession slows down the gas demand increase, the gas shortage is expected to increase from about 19 BCM in 2009 to about 31 BM in 2015 in

the GCC. If the growth returns to historical levels, the shortage is expected to increase to more than 50 BCM in 2015, see **Figures 6(a)** and **(b)**.

[7] indicated that it seems counterintuitive that the GCC need to import gas (including from outside the region). The reasons for this need include the fact that production has not kept pace with demand for gas in the region, but also relate to the development of the gas industry in the Middle East. Another reason is much of the gas in the Middle East is sour or tight gas. This is particularly the case in SA, the UAE and Oman. Recovery of such gas is time-consuming, technologically difficult and expensive. Sing mentioned that importing gas or LNG is the most economically and environmentally sound solution to the GCC's problem, especially for the power sector

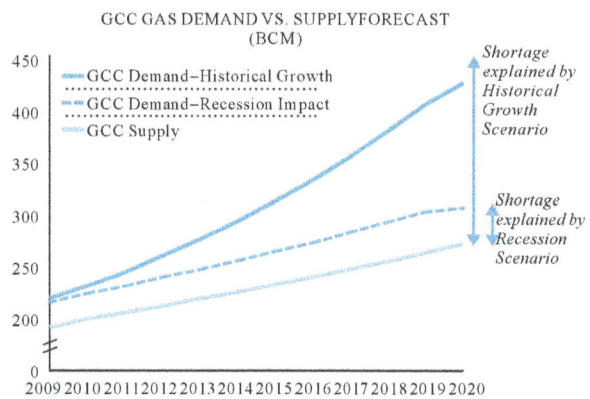

(a)

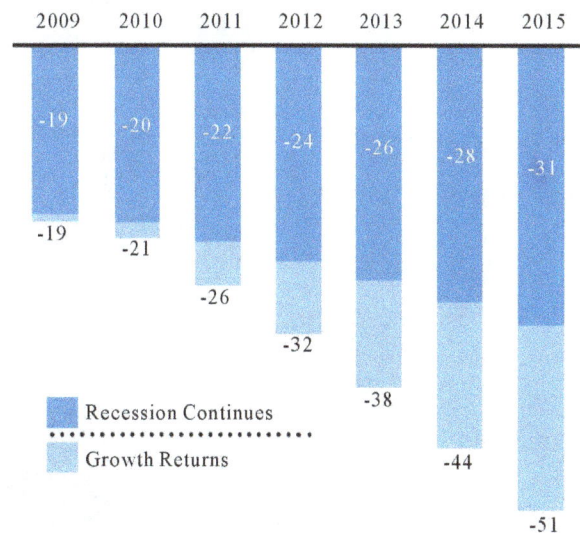

(b)

Figure 6. (a) Forecast NG demand in GCC with and without slow economic growth (recession), [6]; (b) NG shortage in GCC with and without slow economic growth (recession), [6].

[8]. Importing gas or LNG will enable GCC countries to continue their economic diversification efforts using local gas as well as allow them to continue to export crude oil, sending value-added refined products to foreign markets instead of burning them as fuel for their own power needs. [9], and [10] urge the use of renewable energy as alternatives to NG.

Saudi Arabia's place in the world oil market is threatened by unrestrained domestic fuel consumption. In an economy dominated by fossil fuels, current patterns of energy demand are not only wasting valuable resources and causing excessive pollution, but also rendering the country vulnerable to economic and social crises. Glada and Stevens identified the insufficiency of the current political approaches, and the need for change emerges. The authors highlight as a dominating measure for the consolidation of consumption and the avoidance of a possible crisis, the increase in the fuel's price [11].

3. Escalation of Energy Consumption in GCC

Powerful long run trends continue to shape the modern energy economy: industrialization, urbanization and motorization. These trends are associated with increasing, [12]:

- Quantities of energy consumption.
- Efficiency of energy use, in production and consumption.
- Diversification of sources of energy.
- Demand for clean and convenient energy at the point of use.

According to US Energy Information Administration (EIA) predicts that 50% increase in power generation in the GCC from 2010 to 2030; and more than 90 percent of this incremental will be fulfilled by gas, significantly increasing the GCC power sector's reliance on gas, see **Figure 7(a)**.

The GCC energy problems started and are expected to be intensified in near future due to high and continuous growth in energy demands and consumptions, especially in electric power (EP) generation. Most of EP plants are using either gas turbines (GT) or GT Combined Cycle (GTCC) with steam turbines as shown in **Figure 7(a)** for the whole GCC, and in **Figure 7(b)** for the case of SA as

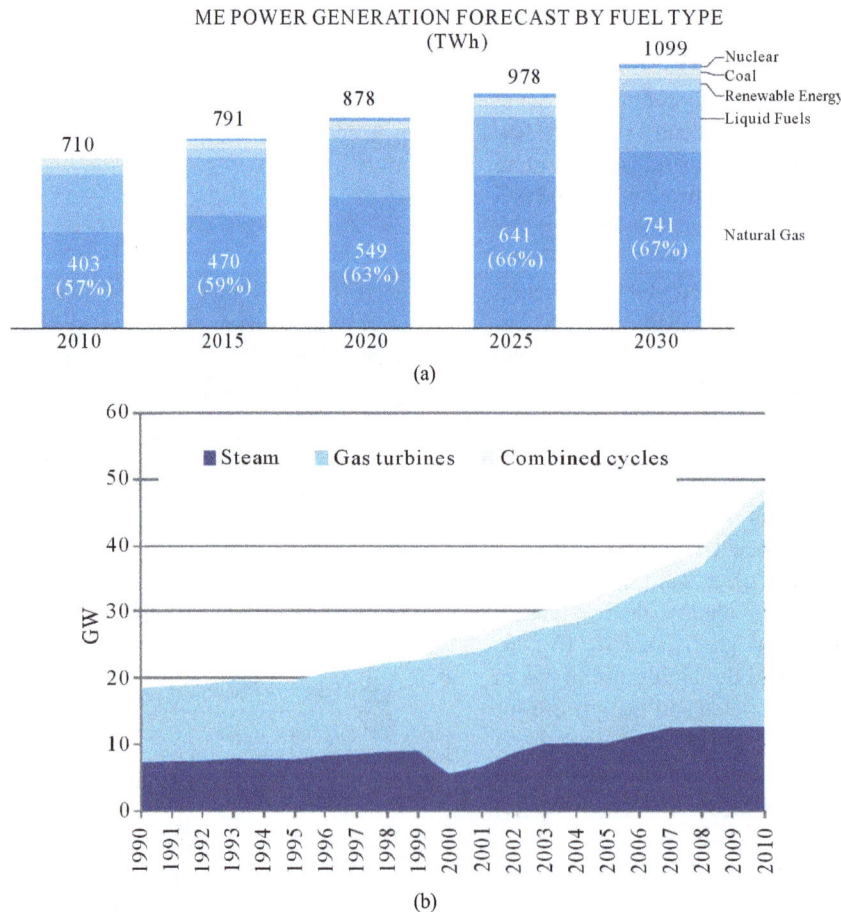

Figure 7. (a) Projection of fuel used for the power generation in the GCC, [6]; (b) Electric Power capacity by steam, gas turbines, and gas turbines combined cycle, [13].

example. These PP are mainly using Gas Turbines (GT) and GTCC with recommended NG as fuel, as shown in **Figure 7(b)**.

One of the main reasons of high consumption is the politically motivated low prices for both electricity and fuel. Other reasons include the high fuel consumptions by systems used for desalting seawater, which is main potable water resource in most GCC. The full production of the GCC petroleum products can be locally consumed within two to three decades if the current rate of fuel consumption prevails.

In stark contrast to energy, water is an extremely scarce resource in the GCC, which is one of the world's most arid regions. With only limited groundwater resources, and amid growing signs that groundwater is being depleted by over-use, the GCC is facing potential water shortages.

The GCC economies account for more than 40% of the world's water desalination capacity, and much of that capacity is energy-intensive. To meet demand, governments continue to build new desalination plants. Since these plants run on fossil fuels, efforts to boost the supply of energy, diversify fuel sources and improve energy efficiency will have a strong impact on the provision of water.

Most of the GCC's desalination plants use thermal sources, mainly natural gas. There is a significant and ongoing investment in dual-purpose co-generation plants, which produce both electricity and desalinated water through a combined thermodynamic cycle, which is more efficient than separate production processes. These are encouraging initiatives, but there is still a lot of room for further energy savings in the desalination process [14].

Escalation in EP consumptions in some of the GCC is illustrated by a few examples from the GCC.

In SA, the EP consumption increased from 163,151 GWh to 193,472 GWh during 2006 to 2009, or 6% annual increase. The daily DW production increased from 1070 MIGD to 1013 MIGD MW during 2006 to 2009, almost steady. The EP generated plants installed capacity increased from 35,885 to 51,195 MW from 2006 to 2009. The peak load increased from 31,240 to 39,900 MW from 2006 to 2009, or 9% annual increase, [15]. In 2007, the EP plants consumed 45.76 M ton equivalent fuel. This consists of 20.5 M metric ton equivalent of NG (22.78 BCM), 9.288 M ton Diesel oil, 7.566 M ton crude oil, and 6.233 M ton heavy fuel oil. The increases in NG and oil consumptions in EP plants are given in **Figure 8**. In 2010, the oil production in SA was 10.21 Mbbl/d; consisting of 8.4 Mbbl/d of crude oil, and 1.8 Mbbl/d of NG liquids (NGLs). Out of the 10.2 Mbbl/d, 7.3 Mbbl/d were exported, and 2.65 Mbbl/d were consumed. The 2009 consumption was 50% more than that of 2000. The consumption growth was due to strong economic and industrial growth and subsidized prices, [17]. The trend of oil consumption increase is shown in **Figure 9**.

According to Saudi Aramco forecasts, NG demand in SA is expected to be more than doubled to 14.5 trillion (T) cubic feet per day (TCF/d) by 2030, up from an estimated demand of 7.1 TCF/d in 2007. Saudi Arabia's energy consumption pattern is unsustainable. The country currently consumes over one-quarter of its total oil production—some 2.8 million barrels a day. This means that on a "business as usual" trajectory it would become a net oil importer in 2038 (see **Figure 10**). No one is suggesting this is the most likely outcome but the possibility does signal the urgency of the need for change. More oil reserves may be discovered and production raised, population growth may decline and new policies and technology may change consumption patterns, but in the absence of such events and with the country's high dependence on oil revenues the economy would collapse before that point. The Saudi Arabia's oil balance on a business-as-usual trajectory is shown in **Figure 10**.

In order to free up petroleum for export, all current and

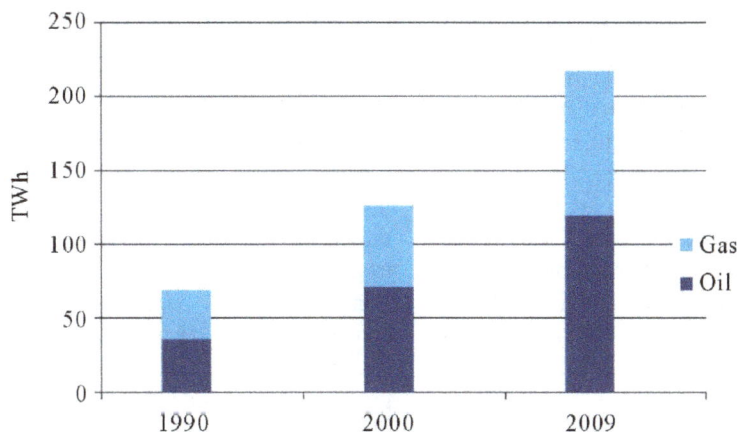

Figure 8. Power generation by source, [16].

Saudi Arabia Oil Consumption

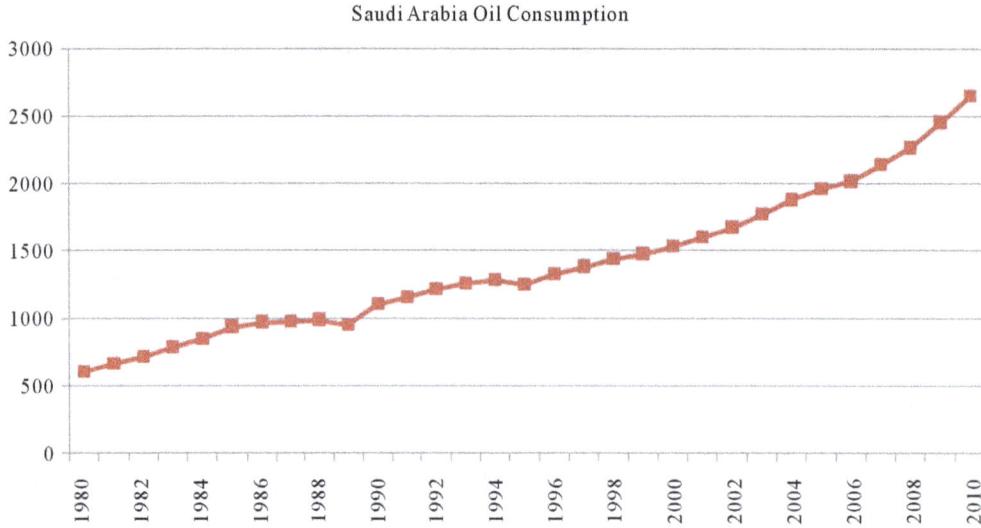

Figure 9. Oil consumption increase in Saudi Arabia, [17].

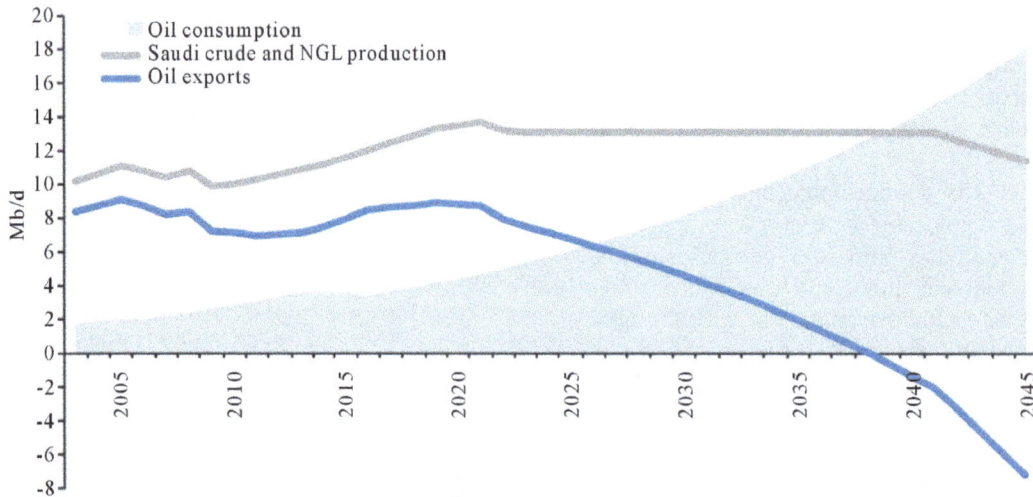

Figure 10. The Saudi Arabia's oil balance on a business-as-usual trajectory, [11].

future gas supplies (except NG liquids) reportedly remain for use in domestic industrial consumption and desalination in SA. However, NG production (estimated at 2.7 TCF in 2007) remains limited, as the soaring costs of production, exploration, processing and distribution of gas have squeezed supply, while an estimated 13% to 14% of total production is lost in venting, flaring, reinjection and natural processes according to OPEC and other sources, [15]. SA has no net imports or exports of NG. The 2009 NG production was estimated by 3.2 TCF; and was totally consumed locally. The total consumed fuels (NG and oil) increased from 1.798 million equivalent barrels per day (Mebbl/d) in 2000 to 3.0322 Mebbl /d in 2009, an annual increase of 6%, [15]. The total produced energy in 2009 was 11.3402 Mebbl /d, and thus the consumed to produced ratio was 26.7% in 2009. The

consumed energy is expected to reach most of the oil production (8.5 M-boe/d) in 2028.

In the UAE, the EP consumption increased in Abu Dhabi from 25,424 to 31,478 GWh from 2003 to 2008, and in Dubai from 16,572 to 23,571 GWh from 2005 in 2010, [9]. The consumed fuel energy increased from 0.686 Mebbl/d in 2000 to 1.4467 Mebbl/d in 2009, annual increase of 8.6%. The UAE total fuel energy production in 2009 was 3.477 Mebbl/d/d, or the consumed to produced ratio was 41.6%. The reported UAE population in 2009 was 5.066 million, and this gives annual consumed Mebbl per year per capita (Mebbl/y.ca) as 104 Mebbl/y.ca, [18]. The UAE is more aware of the energy problem than other GCC. This is clear from starting to build a nuclear power plant of 4 reactors of 1400 MW each, and their active participation in renewable energy program.

In Kuwait, the EP consumption increased from 43,734 to 51,749 GWh from 2005 to 2008, [19]. The daily DW production increased from 317 to 423 MIGD MW from 2004 to 2008. **Figure 11** shows the expected total consumed fuel (given in blue) would meet the total expected production in almost 25 years (y). The fuel used for EP and desalted seawater generation (in red) is about half of the total consumption, [20]. In Bahrain the EP consumption increased from 8267 to 12,224 GWh from 2004 to 2009. The daily DW production increased from 93 to 131 MIGD MW from 2004 to 2009. The oil production in 2010 was 46,000 bbl/d, of which 76% was crude oil. The oil consumption in 2009 was 45,000 bbl/d, see **Figure 12**, [21].

The projection here was based on business-as-usual trend estimate, given known technology used for generating both electric power and desalted seawater by both natural gas and crude oil.

In Qatar, the consumed EP increased from 13,232 GWh in 2004 to 28,144 GWh in 2010, more than doubled in 6 years, significant annual increase of 13.6%, [22, 23]. For 1.7 million (M) population, this gives 1650 kWh per capita per year (kWh/y.ca) of EP in 2009.

In Oman, the EP consumption increased from 13,867 to 19,121 GWh from 2007 to 2010. The daily DW production increased from 84 to 129 MIGD from 2004 to 2008, [24].

While consumed energy is on the rise in all GCC, the produced prime energy in most of GCC is almost constant, or even decreasing. The produced oil was decreasing at the ratios of 8.2% in UAE, 9.6% in SA, and 13% in Kuwait from 2005 to 2009. The exported fuel is the main source of income to the GCC. So, the income from selling energy productions in GCC is not sustainable, [1].

One of the known sustainability conditions requires that the rate of using non-renewable resources such as fossil fuel does not exceed the rate of developing sustainable substitutes such as renewable energy (RE). It is unfortunate that practical application of RE use does not even started yet. Another sustainability condition requires that pollutants emission rate does not exceed the environment's capacity to absorb, or render them harmless. Unfortunately, the GCC has the highest CO_2 per capita with no insight solution to this problem.

Examples of the kWh/y.ca of EP in different countries are given in **Figure 13**, [25]. **Figure 13** shows that Qatar and Kuwait have the highest kWh/y.ca of EP in the world. In Qatar, it was 950, 1500, and 1650 kWh/y.ca in 1990, 2003, and 2010 respectively [23]. As a result, Qatar, Kuwait, and UAE have the highest CO_2 emission in the world, as shown in **Figure 14**, [25]. **Figure 15** shows the drastic increase of the CO_2 emission in Qatar, which has the highest per capita in the world, [26].

Currently, fuel oil and NG are consumed in the GCC's power plants (PP). Both oil and NG have limited supplies, and their costs are continuously rising. Fuel consumed in Saudi Arabia (SA), Kuwait, UAE almost doubled every 10 years or less. The fuel productions in these countries in 2009 are: 11.34 Mbbl/d in SA, 2.55 Mbbl /d in Kuwait, and 3.476 Mbbl/d in UAE. So, its full productions can be consumed locally in almost thirty years if the present consumption rates prevail. Thus, there is an urgent need to change to more sustainable fuel energy.

An important reason for the poor and worsening record of energy inefficiency in the GCC region lies in the domestic pricing policies that keep prices well below international levels, which need to be addressed soon. However, according to Alyousef and Stevens [27], even the methodology used by the International Energy Association (IEA) does not help the situation to improve; the authors claim that the method used grossly overstates the levels of subsidies on oil products in the region. This fact at best creates negative attitudes towards energy issues at a time when they need to attract serious discussion and debate while at worse undermines and discredits the very real practical concerns over domestic pricing derived from a careful economic analysis.

Securing fuel supply for an EP generation in the GCC is a major challenge in the coming decades; as it consumes about half of the total consumed fuel. This can be achieved through: reducing dependence on one source of energy, exploiting more fuel or renewable energy, and lowering the demand by energy conservation. There is real need for alternative energy use to generate EP generation, and to keep the income from selling oil. There are available prime energy alternatives to generate EP, such as renewable sources (e.g. solar and wind); and nuclear energy.

To a similar direction, Dehen, the CEO of Energy Sector and Member of the Managing Board of Siemens AG has proposed in the World Energy Insight 2010 a

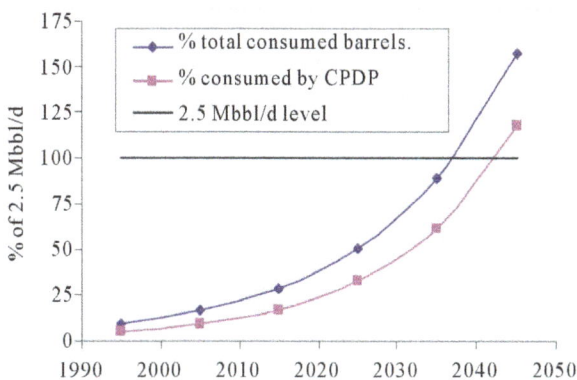

Figure 11. Percentage of expected fuel consumption by all sectors and by CPDP and their percentage of total fuel oil production of 2.5 M-bbl/d., [20].

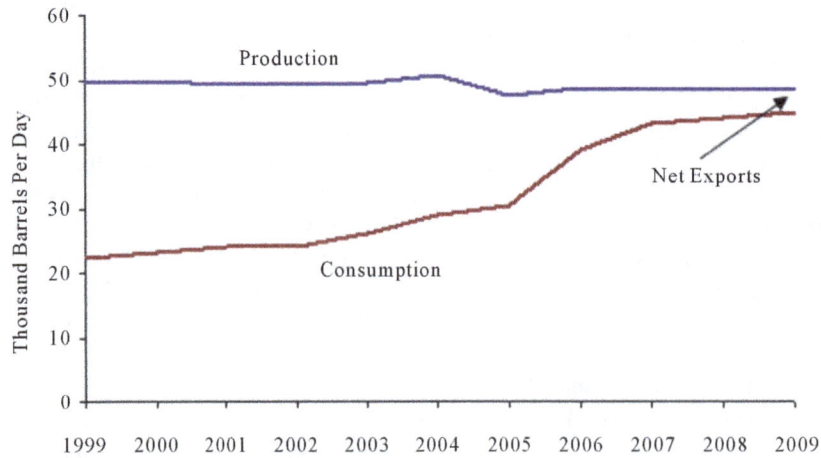

Source: Energy Information Administration

Figure 12. Bahrain total oil production and consumption, 1999-2009, [21].

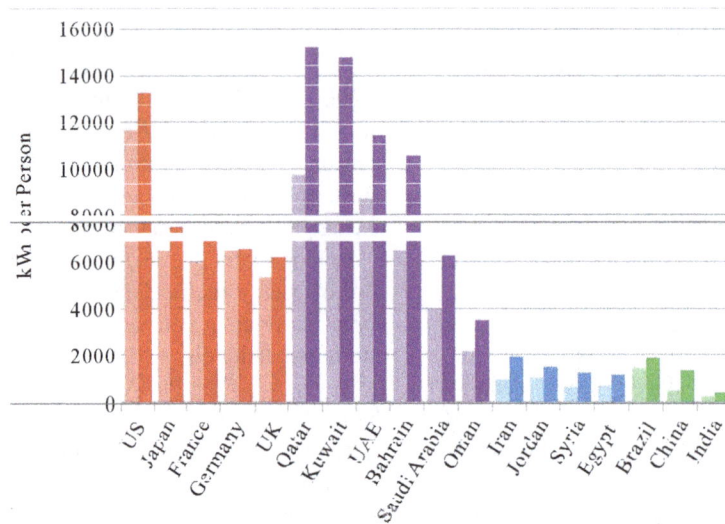

Figure 13. Per-capita electricity consumption of selected major developed (red), GCC (purple), regional (blue), and major developing nations (green) for 1990 and 2003. Lighter, left-hand bars, are for 1990; darker, right-hand bars, are for 2003, [25].

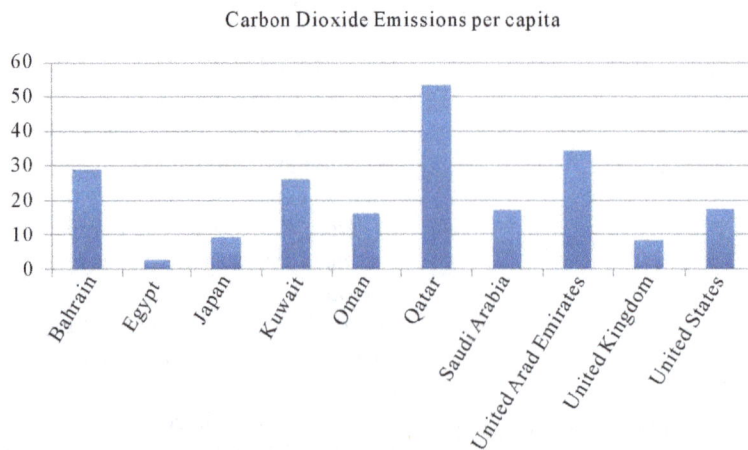

Figure 14. Representative of CO_2 per capita emission in some countries, in 2008, [25].

CO₂ Emission from consumption of fossil fuels in Qatar

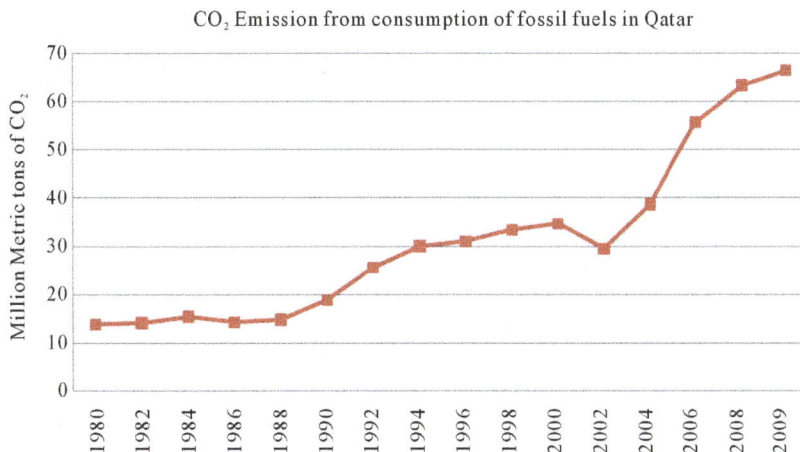

Figure 15. History of the CO₂ emission in Qatar, [26].

three-step strategy to formulate an efficient energy system; the proposed strategy highlights the optimization of the energy mix, technical improvements to achieve greater efficiency and reduce fossil fuel consumption and systemic optimization of the energy system in order to transform today's passive consumers of the energy system into interactive "procumers"—that produce the energy they consume [28].

4. Power Plants Fossil Fuel Choices

The NG is the preferred Fossil Fuels (FF) used in PP worldwide. **Figure 16** shows that the FF used in power plants are coal, NG, and oil, with negligible share of oil. In the GCC, only NG and oil are used, [29].

The projection in **Figure 16** is based on the International Energy Outlook Reference case in 2011 (*IEO*2011, [29]). This Reference case projection is a business-as-usual trend estimate, given known technology and technological and demographic trends. The *IEO*2011 cases generally assume that current laws and regulations are maintained throughout the projections. Thus, the projections provide policy-neutral baselines that can be used to analyze international energy markets. While energy markets are complex, energy models are simplified representations of energy production and consumption, regulations, and producer and consumer behavior. Projections are highly dependent on the data, methodologies, model structures, and assumptions used in their development. Behavioral characteristics are indicative of real-world tendencies, rather than representations of specific outcomes. Energy market projections are subject to much uncertainty. Many of the events that shape energy markets are random and cannot be anticipated. In addition, future developments in technologies, demographics, and resources cannot be foreseen with certainty. Key uncertainties in the *IEO*2011 projections are addressed

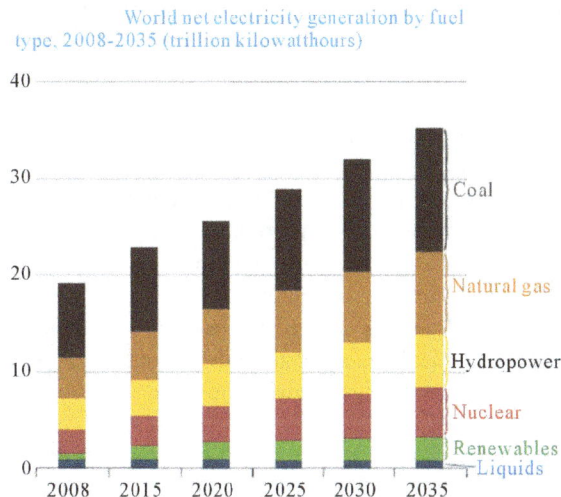

World net electricity generation by fuel type, 2008-2035 (trillion kilowatthours)

Figure 16. Prime energy used in electric power generation, [29].

through alternative cases. EIA has endeavored to make these projections as objective, reliable, and useful as possible. They should, however, serve as an adjunct to, not a substitute for, a complete and focused analysis of public policy initiatives.

The non-Organization for Economic Cooperation and Development (OECD) countries account, in general, for 80% of the global rise in gas consumption, with annual growth averaging 3% per year (y) to 2030. Gas use is driven mostly by economic growth, accompanying industrialisation, industrial policy, the power sector and the development of domestic resources. The gas consumption in the Middle East grows 3.9%/y over 2010-2030. The power sector accounts for 44% of this growth as domestic gas and imports in some countries displace oil burning. Petrochemical industries are contributing to the projected 3.2%/y growth in industrial gas use [12].

The idea of using coal in the GCC is dismissed as it is

not practical for several reasons. Coal combustion affects the environment more badly than both NG and oil. Compared with oil and gas, coal emits more greenhouse gases (GHG), mainly CO_2, causing global warming; and more air polluting gases such as sulfur oxides and nitrogen oxides. The infrastructure for transporting coal is not available in GCC, such as railways. A coal PP of say 3000 MW, a typical capacity of a power plant in the GCC, would consume about 500 kg of coal per second, or 43,200 tons daily. This requires 4320 full trucks of 10 tons load daily to supply the coal to the power plant. Large power plants using coal is usually located close to coal mines, or served by railways to transport the coal. Moreover, supplying coal from a foreign country is expensive and unsecured.

Crude or heavy oil is used in boilers of steam Power Plants (PP). This practice of using heavy oil is slowly dying out in favor of NG because lot of pollutants is released into the air due to burning of the oil compared to NG. In the US, the residual oil share in generating EP decreased from 16.8% in 1973 to 6.2% in 1983, [30]. In 2005, electricity generation from all forms of petroleum, including diesel and residual fuel was only 3% of total production. The decline is the result of price competition with NG and environmental restrictions on emissions. For PP, the costs of heating heavy oil, extra pollution control and additional maintenance required after burning are often outweighed the low cost of the heavy (or residual) fuel. Burning residual fuel oil also produces much darker smoke and uniformly higher carbon dioxide emissions than NG, [30].

There are no prospects of using oil in running PP worldwide in the future, as shown in **Figure 17**, [29]. This figure shows that the share of oil in power generation is either kept the same by the already oil operated PP or even decreasing in the future. The oil has much better usage than being burned in PP such as steam generator of steam PP or combustion chambers in PP using GT, as shown in **Figure 17**, [29].

Most PPs in the GCC is using gas turbines (GT) or GT combined cycle (GTCC), see **Figure 7** for the case of Saudi Arabia. In GCC, the hot gases discharged from the GT are used in heat recovery steam generator (HRSG) to produce steam. The steam is then used to drive steam turbine to generate more EP without adding more fuel. NG or light fuel oil is burned in the combustion chambers of the GT. The light oil is more expensive than heavy oil, and much higher than NG. Oil has much better uses for transportation and petrochemical industry than being burned in boilers of steam turbines or in combustion chambers. **Figures 18** and **19** show the cost of different fuels (coal, NG, and heavy oil) used for EP generation; and the cost ratio of NG to oil. **Figure 20** shows the operating cost of generating EP when different fuels (nuclear, coal, NG, and oil) are used. In conclusion, the use of fuel oil in EP production is much more costly and more polluting to the environment compared to NG. *So, oil burning in PP's is decreasing and should be completely stopped in the GCC countries.*

The remaining and more economical FF used for PP is NG. This also came as a result of power plant industries, where the GTCC using NG is the most preferred PP worldwide.

The power plants using GT are the easiest and cheapest PP to build. The GTCC have the highest efficiency. The GTCC is the first best choice for EP generation in the GCC. Besides, the transportation of gas is more expensive than that of oil; and exporting oil return is much higher than that of NG, even in liquefied form. Qatar and the UAE use only NG for EP production. Other GCC, such as SA and Kuwait, augment their NG production with fuel oil. It is unfortunate that Kuwait and SA still make extensive use of crude oil for power generation because of limited gas availability. In 2010, the only 34% of electricity was produced from gas in Saudi Arabia, see **Figure 21**, [34].

5. Natural Gas in the GCC and EP Generation

Conventional PP and (or CPDP) utilizing GT and GTCC should use only NG. The NG is clean fuel with less polluting gases when combusted than oil. Moreover, the price of oil is much higher than the price of NG, almost twice [32].

Figures 22(a)-(f), show the NG history of production and consumption in the GCC, [35-40]. **Figure 22(a)** shows that the production is more than the consumption in the GCC countries as a whole. However, with respect

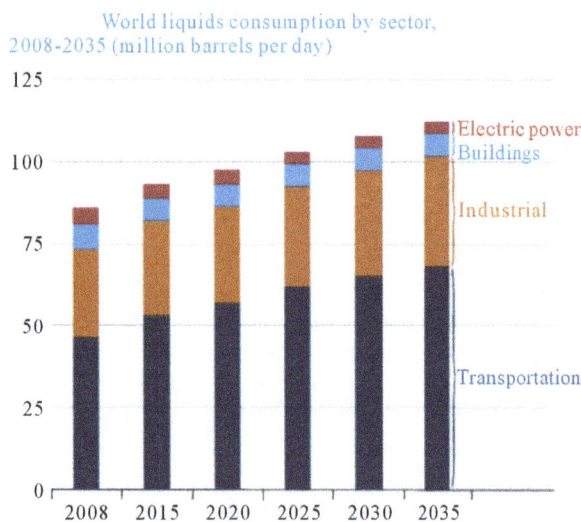

Figure 17. Prospects of fuel oil usages, [29].

Cost of Fossil-Fuel Receipts at Generating Plants

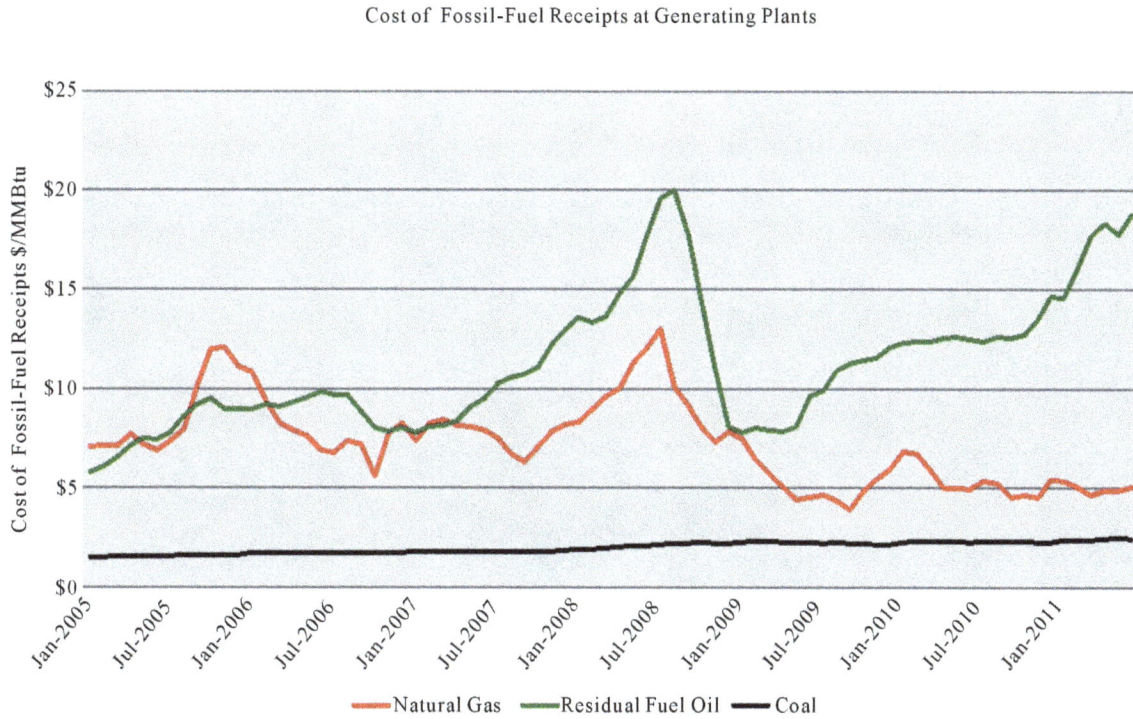

Figure 18. Cost comparison of different fuels used in EP generation in last six years, [31].

Delivered Gas to Oil Price Ratio

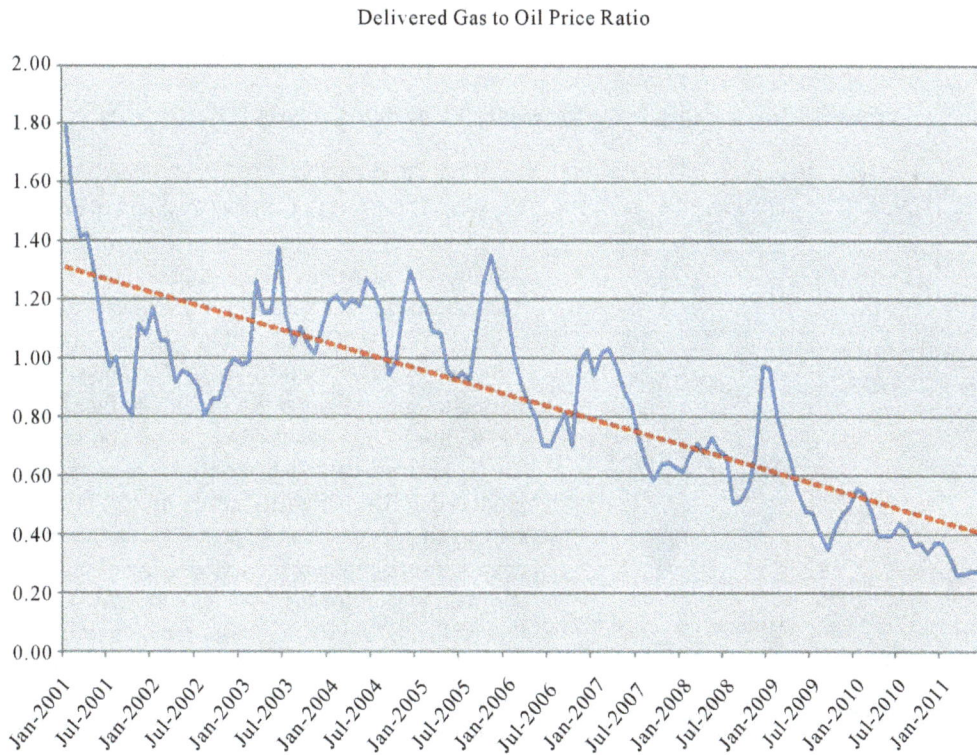

Gas prices have declined in value to oil prices since2001,
http://www.petrostrategies.org/Graphs/gas_and_residual_fuel_c,
[12]omparison.htm

Figure 19. The ratio of natural gas to residual oil costs, [32].

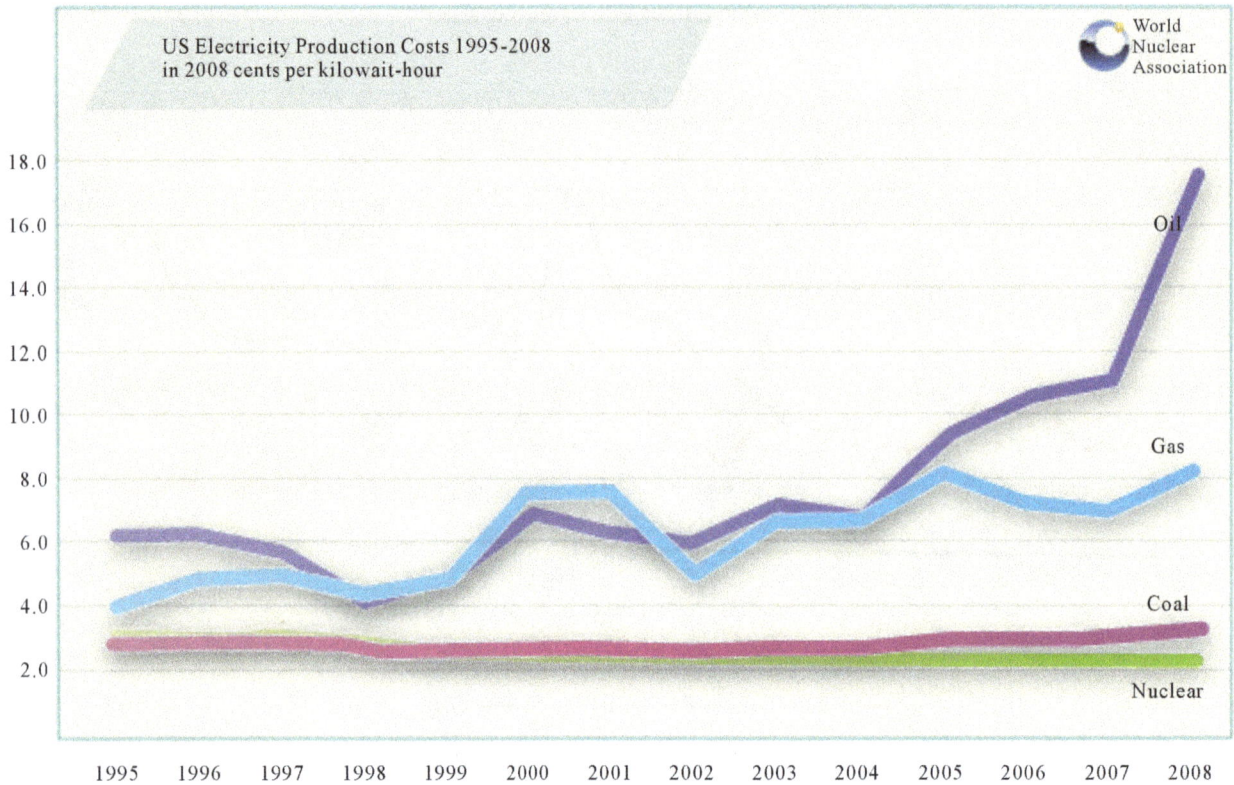

US Electricity Production Costs 1995-2008
in 2008 cents per kilowait-hour

World
Nuclear
Association

Production Costs = Operations & Maintenance + Fuel. Production costs do not include indirect costs or capital.
Source: Ventyx Velocity Suitem via NEI

Figure 20. The operating cost of generating EP by different fuels, [33].

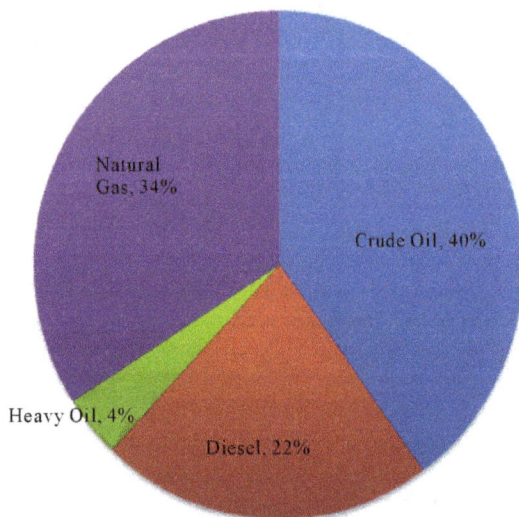

Figure 21. Share of different fuels in producing EP in SA, year 2010, [34].

to each country, this is right only for Qatar, **Figure 22(b)**, but not for the all other five GCC. Qatar NG production in 2009 was 3145 BCF, while consumption was 745 BCF. The difference between the productions and consumptions

in all the GCC represents the amount exported by Qatar. **Figures 22(c)-(e)** and **Table 2** shows that all the produced natural gas is completely consumed for domestic purposes in Kuwait, Saudi Arabia, and Bahrain. This does not mean that the NG produced satisfies all the demands. In 2010, Kuwait natural gas production and consumption were 414 and 446 BCF respectively. Kuwait started to import natural gas in 2007 to face the decline in its NG production and to lower its oil consumption in PP. The UAE uses natural gas only in power plant, and the produced NG satisfied its needs up to 2007, when they start to import gas from Qatar. This is the result of NG continuous rise in demand and decline of production; and the main reason for the UAE to adopt the use of Nuclear Power Plants (NPP). In 2009, Oman's natural gas production and consumption (**Figure 22(f)**) were 875 BCF and 520 BCF respectively. Much of the remaining natural gas reserves are locked in geological formations that are smaller and more difficult to access. For example, the concession of the Khazzan and Makarem natural gas fields operated by BP highlight the technical difficulties facing development of natural gas in Oman. BP has recently increased its estimation for these fields to between

Natural Gas Production and Consumption

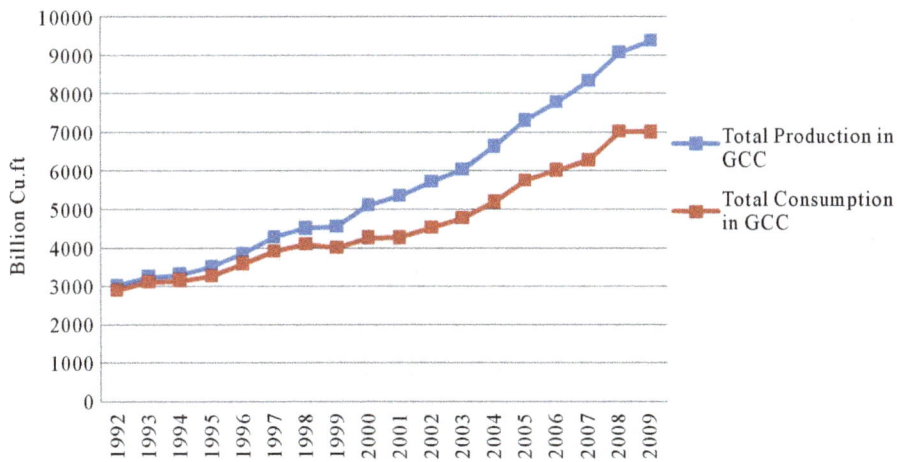

(a)

Qatar NG Production and Consumption

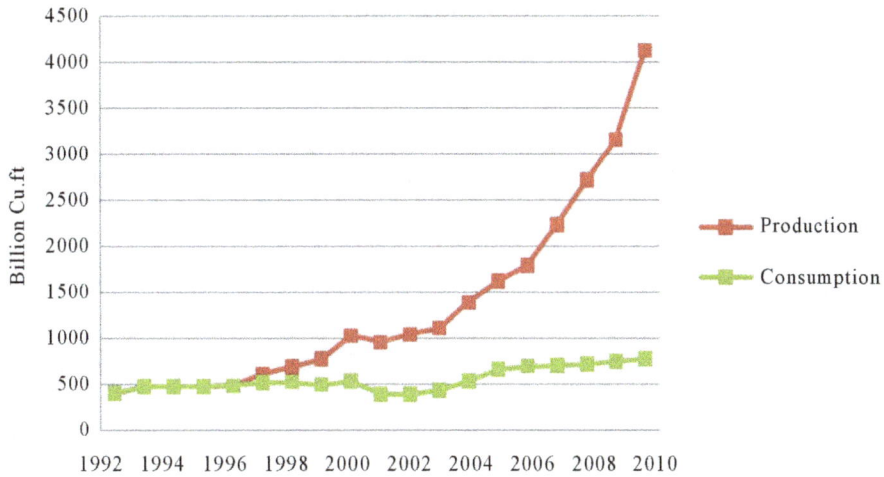

(b)

Kuwait NG Production and Consumption

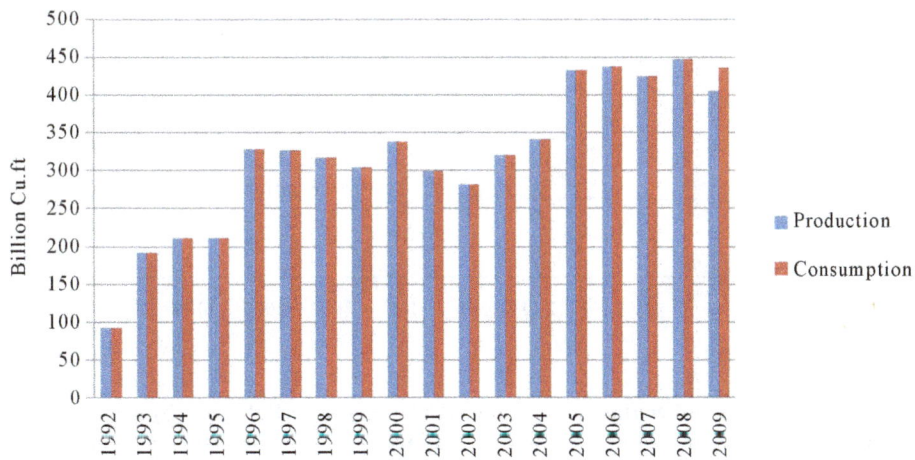

(c)

Saudi Arabia NG Production and Consumption

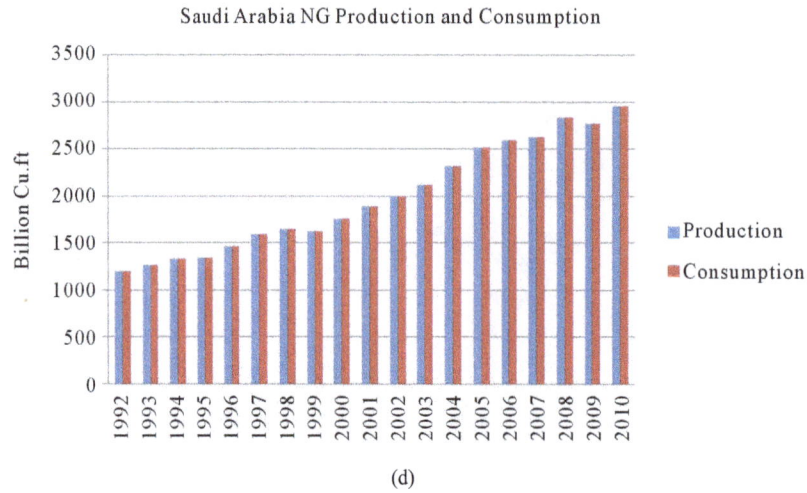

(d)

UAE Natural Gas Production and Consumption

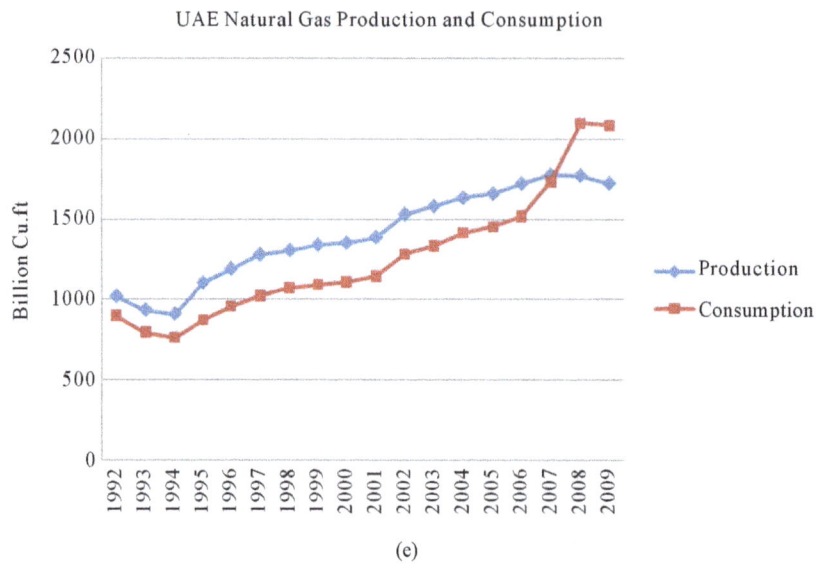

(e)

Omani Natural Gas Production and Consumption,
1999-2009

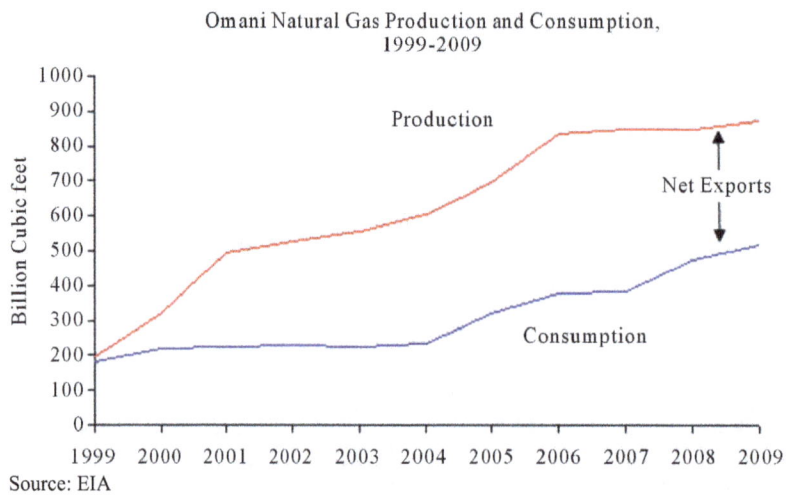

Source: EIA

(f)

Figure 22. (a) History of NG production and consumption in the GCC, (data from Ref. [1]); (b) History of NG production and consumption in Qatar, [37]; (c) Kuwait NG production and consumption, [38]; (d) Saudi Arabia NG production and consumption, [39]; (e) UAE natural gas production and consumption, [35,36]; (f) Oman history of NG consumption and production, [40].

50 and 100 TCf of reserves in-place, of which only 10 TCf are recoverable. Oman would consume all its NG production and start importing NG within few years. So, while the demands are always on the rise, the NG productions do not match this demand. This happened in UAE, SA, Kuwait, and Bahrain; and soon will happen in Oman. Part of its produced NG is re-injected in oil fields to increase oil production, and takes up rising proportion of domestic production. Although Oman is a net exporter of oil and natural gas, it also imports small volumes of natural gas. The Dolphin pipeline provides Oman's only natural gas imports, providing approximately 200 million cubic feet per day (MCf/d).

The Oman and Qalhat LNG projects are the sole source of natural gas exports from Oman, with a nameplate capacity of 506 BCF/y, (1.388 BCF/d). **Table 2** shows that in 2009, there was 411 Billion Cubic feet (BCF) deficit between the NG production and consumption in Bahrain, Kuwait, UAE, and SA. This deficit in NG is even less than the actual demands, because oil fuel substitutes the difference between the demand and consumption. These countries, and soon Oman, have to import NG from Qatar or other NG producer countries with surplus. The situation in Saudi Arabia and Kuwait are critical since they use heavy and crude oil in huge quantities to augment the natural gas in power plants and CPDP. The fuel oil is relatively expensive and the GHG emissions are relatively higher than natural gas.

So, all the GCC countries are in deficit of natural gas needed to run their power plants or CPDP, except Qatar. The use of oil for burning in PP is a bad practice since it has much better usage as a liquid for transportation, and petrochemical industries. More than that, the use of this oil drains these countries resources.

In 2008, the prime energy consumed in SA was almost 56% petroleum-based and 44% NG, [34]. Saudi Arabia is moving forward with plans to build nuclear power plants

Table 2. Natural gas production and consumption in the GCC countries and Iran.

Country	Production A	Consumption B	A − B =
Qatar	3154	745	2408
Bahrain	444	444	0
Iran	4632	4649	−18
Kuwait	406	437	−31
Oman	875	520	355
Saudi Arabia	2770	2770	0
United Arab Emirates	1725	2086	−362
Total	14,005	11,653	2352
Total without Iran	9373	7004	2370

(NPP) by 2020 to meet domestic power needs and to free up oil for export. Saudi Arabia is also participating in the GCC's efforts to link the power grids of member countries in order to reduce shortages during peak power periods

The upgrading of SA's refineries will reduce the share of undesirable heavy fuel oil, thereby reducing its available supply.

The GCC countries should stop completely burning petroleum oil in PP. They should start using renewable energy to take share in EP production.

6. The Alternative of Nuclear Energy

The GCC countries had a negative attitude towards nuclear energy, viewing it primarily as a potential competitor of the hydrocarbons. This negative attitude, which in essence translates into a denial of access to nuclear energy technology, might have been justified as long as hydrocarbon energy was abundant and cheap. Since the turn of the century, conditions have radically changed and a reconsideration of the case is imperative today [41].

To this direction the UAE tentatively decided to integrate nuclear energy into the electricity scheme to mitigate CO_2 emissions as declared by the government. In [42] an evaluation of the effectiveness of the UAE's nuclear strategy, presents that nuclear energy is more practical and economic viable option in mitigating CO_2 than renewable energy and carbon capture.

Nuclear energy is capital intensive and an excellent store of value for the future: it is the kind of investment that a country with large financial resources and limited investment opportunities would logically find very attractive as a basis for long-term economic diversification and sustainability [41]. However, investments in nuclear energy are perceived of high risk and therefore are treated with suspicion.

The use of nuclear energy to generate electric power EP and desalt seawater D raises many concerns in the GCC. In spite of these serious concerns, the question is not to accept nuclear energy or not, as it may be the only option they have. The real question should be how and when nuclear energy would be inherently safe, not prohibitively expensive, and when it can be applied safely in countries at different development stages. Nuclear energy can present a sustainable way to produce EP and D if its standing problems are resolved. The introduction of NE to UAE (presently) and to Saudi Arabia (as planned) to generate EP (*i.e.* nuclear power plants, NPP) and D (nuclear desalination ND) can encourage other GCC to diffuse some of the public resistance to NE. There are arguments that these countries have enough fuel oil and natural gas reserve to satisfy its present needs. Also, there are fears of large catastrophic accidents like what happened in Chernobyl, Ukraine, and the Three Mile Island in USA. Moreover, there are standing problems of nuclear waste disposal, nuclear plants de-commissioning, possible radioactive contaminations, excessive capital and operating costs, lack of nuclear fuel technology and trained personnel in developing countries. It also imposes dependency on the foreign country supplying the NPP to re-fuel the reactor for entire life of the plant. The supplying country should also have access to the spent (used) fuel to avoid its reprocessing for unlawful uses. It is necessary to have qualified manpower for safe operation and maintenance of the NPP. So, Kuwait's personnel, for example, should acquire training in the country supplying the plant, thus requiring additional time and cost. This training is not limited to scientists but to all levels including engineers and technicians. These factors apply to some GCC like SA, UAE, and Egypt, [43].

7. Activities towards Sustainable Energy in the GCC Countries

The activities of using solar or wind energy are very limited to some scattered research work. The output of this work is very hard to find its application on a large scale to curb the increasingly consumed oil and NG. Although Qatar was the first country in the GCC to form Ministry of Environment, the awareness of environmental issues is low. SA is the most needed country to diversify its fuel; as it is facing shortage in NG needed to operate its PP, and its power demand is growing at an alarming rate, [44]. The increase of SA's population and oil consumption lower the per capita income (mainly depends on oil export) since oil production is almost constant or decreasing.

The UAE is the first country in the GCC to deal the fuel energy problem surrounding the use of oil in EP generation; and tries effectively to reduce consumed fuel energy. All its power plants are operated by NG (part of it is imported). Public transportation metro and efficient bus systems are operating in Dubai and Abu Dhabi to reduce the transportation fuel consumption. The government subsidization of gasoline (for car) in the UAE is the least, compared to other GCC. For example, the 2008 car super gasoline price was $0.45/liter ($l$) in UAE; while it was 0.16/l in SA, $22/l in Qatar, $21/l in Bahrain, $24/l in Kuwait, and $31/l in Oman. The UAE diversifies its PP primary fuel by starting to build Nuclear (N) PP having four reactors of 1400 MW (electric) each. The activities in renewable energy are clear in Masdar project. This project initiated in 2006 in Abu Dhabi, UAE. It is planned to build a city relying entirely on solar energy and other renewable energy sources, with a sustainable, zero-carbon, zero-waste ecology. It will cost US$22 billion and take some eight years to build. Masdar will employ a variety of renewable power resources. Among the first construction projects will be a 40 to 60 MW solar PP, built by the German firm to supply power for all construction activity. Also solar panels will be placed on rooftops to provide supplemental solar energy of 130 MW. Wind farms will be established outside Masdar's, and capable of producing up to 20 MW.

In this context, the number of discussions and research activities relating to the necessity of policies promoting more sustainable forms of energy is recently increasing. In [45,46] the solar and wind energy potentials are highlighted in the GCC. The authors have also listed the major renewable energy (RE) projects (mainly solar and wind) in each of the six GCC and have proposed a mechanism to accelerate the RE utility in these countries.

From a different perspective, particular reference has also been made in the international scientific literature for the social and political specificities of the region regarding the adoption of alternative energy sources; in [47] the authors have composed an analytical review of the current Renewable Energy Sources (RES) and Rational Use of Energy (RUE) development status in the GCC region, giving special emphasis to the business opportunities that the region offers for regional and international companies involved in this market. Patlitzianas et al. (2006), have identified and assessed sustainable energy investments in the framework of the EU-GCC cooperation. Indeed, the smooth cooperation of EU and GCC key energy players is considered important for the challenging objective to engage GCC countries in a more sustainable development path [48-50].

8. Conclusions

The fuel mix changes relatively slowly, due to long asset lifetimes, but gas and non-fossil fuels gain share at the expense of coal and oil. GCC should diversify fuel sources by introducing renewables and nuclear energy and investment in these resources. Energy demand management initiatives, including curtailment of energy subsidies.

The most preferred PP type using fossil fuel is the GTCC; which is the predominantly installed and applied in the GCC. The standard fuel used in the NG; and all GCC countries, except Qatar, have to import NG to run their PP. Although UAE and Kuwait started to import NG, the most needed country, SA, does not start yet. This may be for supply security reasons. Burning the too expensive crude oil in PP is equivalent to burning money. The per capita CO_2 emission in GCC is the highest in the world, and 100% use of fossil fuel should stopped by using renewable energy. The UAE started to build NPP to take share in satisfying its EP needs. SA also has plans to build NPP. The UAE started also to build two large concentrating solar plants (CSP). One plant has 100 MW capacity and using parabolic trough solar collectors. The second is solar tower having 10 MW electric capacity, [8]. A share of solar energy in EP generation should be adopted and fulfilled. The simple way is to augment the GTCC by solar energy (to become integrated solar combined cycle). The realization of electric power grid between the GCC opens the door for better energy co-operation between the GCC. The availability of NG in Qatar can be used to build excess power capacity in Qatar to supply other needed GCC short of NG for EP when needed. EP was exported in last two years from Qatar to Kuwait. Conserving energy must be the highest priority for the GCC countries, which have the highest per capita of energy consumption worldwide.

REFERENCES

[1] Organization of Arab Petroleum Exporting Countries (OAPEC), "2010 Annual Statistical Report, in Arabic," 2010, pp. 10-13.

[2] US Energy Information Administration EIA, Qatar, 2012. http://www.eia.gov/countries/cab.cfm?fips=QA

[3] "The MENA Region and Global Energy Risk," MEES Research Special Report, 2010. http://geology.com/oil-and-gas/natural-gas-production-map/Jallah

[4] J. Dargin, "Addressing the UAE Natural Gas Crisis: Strategies for a Rational Energy Policy," Harvard Belfer Center, Cambridge, 2010.

[5] R. Kombargi, O. Waterlander, G. Sarraf and A. Sastry, "Gas Shortage in the GCC, How to Bridge the Gap," *Booz & Company Analysis*, pp. 3-6. http://www.booz.com/media/uploads/Gas_Shortage_in_the_GCC.pdf

[6] N. Williamson and M. M. Garcia, "Gas Shortages in the Middle East: an Unlikely Paradox, Gas Regulation 2011," Global Legal Group, 2011. http://www.iclg.co.uk/index.php?area=4&show_chapter=4139&ifocus=1&kh_publications_id=175

[7] P. R. Weems and F. Midani, "A Surprising Reality: Middle East Natural Gas Crunch," 2012. http://www.kslaw.com/Library/publication/MiddleEastNaturalGasCrunch.pdf

[8] I. J. Bachellerie," Renewable Energy in the GCC Countries: Resources, Potential, and Prospects," Gulf Research Center Publications, Dubai, 2012.

[9] Oman, "Study on Renewable Energy Resources," 2008. http://www.aer-oman.org/pdf/studyreport.pdf

[10] G. Lahn and P. Stevens, "Burning Oil to Keep Cool: The Hidden Energy Crisis in Saudi Arabia," Chatham House (The Royal Institute of International Affairs), London, 2011, pp. 2-3.

[11] BP Energy Outlook 2030, 2011.

[12] Saudi Arabia Energy Efficiency Report, 2012.

[13] Economist Intelligent Unit, "The GCC in 2020: Resources for the Future," 2010.

[14] US Energy Information Administration EIA, "Analysis, Saudi Arabia." http://www.eia.gov/countries/cab.cfm?fips=SA

[15] "Saudi Arabia, Energy Efficiency Report." http://www05.abb.com/global/scot/scot316.nsf/veritydisplay/f90e53733342b472c125786400519e97/$file/saudi%20arabia.pdf

[16] Saudi Arabia Oil Consumption, United States Energy Information Administration. http://www.indexmundi.com/energy.aspx?country=sa&product=oil&graph=consumption

[17] US Energy Information Administration EIA, "Analysis, United Arab Emirates." http://www.eia.gov/countries/cab.cfm?fips=TC

[18] "US Energy Analysis, Kuwait." http://www.eia.gov/countries/cab.cfm?fips=KU

[19] M. A. Darwish, F. M. Al-Awadhi and A. M. Darwish, "Energy and Water in Kuwait, Part I. A Sustainability View Point," *Desalination*, Vol. 225, No. 1-3, 2008, pp. 341-355. doi:10.1016/j.desal.2007.06.018

[20] US Energy Information Administration EIA, "Analysis, Bahrain." http://www.eia.gov/countries/cab.cfm?fips=BA

[21] US Energy Information Administration EIA, "Analysis, Qatar." http://www.eia.gov/countries/country-data.cfm?fips=QA

[22] Qatar General Electricity and Water Corp, "(KAHRAMAA) Statistical Yearbooks," 2010.

[23] Oman, "Energy Report, Economist Intelligent Unit." http://www.eiu.com/index.asp?layout=ib3Article&article_id=357910420&pubtypeid=1142462499&country_id=430000043&page_title=

[24] S. Hertog and G. Luciani, "Energy and Sustainability Policies in the GCC," Kuwait Programme on Development, Governance and Globalisation in the Gulf States.

[25] "Carbon Dioxide Emissions [tonnes] per Capita," 2012. http://en.wikipedia.org/wiki/List_of_countries_by_carbon_dioxide_emissions_per_capita

[26] Y. Alyousef and P. Stevens, "The Cost of Domestic Energyprices to Saudi Arabia," *Energy Policy*, Vol. 39, No. 11, 2011, pp. 6900-6905. doi:10.1016/j.enpol.2011.08.025

[27] "World Energy Insight 2010," Official Publication of the World Energy Council to Mark the 21st World Energy Congress. http://www.worldenergy.org/documents/wec_combined.pdf

[28] International Energy Outlook, DOE/EIA, 2011.

[29] Emission Estimation Technique Manual for Combustion in Boilers, Version 1.2.

[30] "National Pollutant Inventory." www.npi.gov.au

[31] Petro Strategies, Inc., "US Natural Gas and Residual Fuel Oil Price Comparison."

http://www.petrostrategies.org/Graphs/gas_and_residual_fuel_comparison.htm

[32] D. Bacher, "How Contributor, Price Comparison of Oil vs. Natural Gas." http://www.ehow.com/about_6688369_price-oil-vs_-natural-gas.html

[33] "The Economics of Nuclear Power." http://www.world-nuclear.org/info/inf02.html

[34] "Saudi Arabia Electricity and Co-Generation Regulatory Authority 2010 Report." http://www.ecra.gov.sa/

[35] International Energy Statistics, US Energy Administration Information, Natural Gas Consumption. http://www.eia.gov/cfapps/ipdbproject/iedindex3.cfm?tid=3&pid=26&aid=2&cid=r5,&syid=1992&eyid=2010&unit=BCF

[36] International Energy Statistics, US Energy Administration Information Natural Gas Production. http://www.eia.gov/cfapps/ipdbproject/iedindex3.cfm?tid=3&pid=26&aid=1&cid=regions&syid=1992&eyid=2010&unit=BCF

[37] US EIA, Qatar, 2012. http://www.eia.gov/countries/cab.cfm?fips=QA

[38] US Energy Information Administration (EIA), "Countries, Kuwait." http://www.eia.gov/countries/cab.cfm?fips=KU

[39] US EIA, "Saudi Arabia Analysis." http://www.eia.gov/countries/cab.cfm?fips=SA

[40] US Energy Information Administration EIA, "Analysis, Oman." http://www.eia.gov/countries/cab2.cfm?fips=MU

[41] G. Luciani, "Nuclear Energy Developments in the Mediterranean and the Gulf," *The International Spectator*, Vol. 44, No. 1, 2009, pp. 113-129. doi:10.1080/03932720802692947

[42] H. L. Al Farra and B. Abu-Hijleh, "The Potential Role of Nuclear Energy in Mitigating CO_2 Emissions in the United Arab Emirates," *Energy Policy*, Vol. 42, No. C, 2011, pp.

272-285.

[43] M. A. Darwish, F. M. Al-Awadhi, A. Akbar and A. Darwish, "Alternative Primary Energy for Power Desalting Plants in Kuwait: The Nuclear Option I," *Desalination and Water Treatment*, Vol. 1, No. 1-3, 2009, p. 25. doi:10.5004/dwt.2009.133

[44] http://www.bloomberg.com/news/2011-04-03/solar-nuclear-energy-to-reduce-saudi-oil-demand-official-says.html

[45] W. E. Alnaser and N. W. Alnaser, "The Status of Renewable Energy in the GCC Countries," *Renewable and Sustainable Energy Reviews*, Vol. 15, No. 6, 2011, pp. 3074-3098. doi:10.1016/j.rser.2011.03.021

[46] W. E. Alnaser and N. W. Alnaser, "Solar and Wind Energy Potential in GCC Countries and Some Related Projects," *Journal of Renewable and Sustainable Energy*, Vol. 1, No. 2, 2009, pp. 1-28. doi:10.1063/1.3076058

[47] H. Doukas, K. D. Patlizianas, A. G. Kagiannas and J. Psarras, "Renewable Energy Sources and Rational of the Energy Development in the Countries of GCC: Myth or reality," *Renewable Energy*, Vol. 31, No. 6, 2006, pp. 755-770. doi:10.1016/j.renene.2005.05.010

[48] K. D. Patlitzianas, H. Doukas and J. Psarras, "Enhancing Renewable Energy in the Arab States of the Gulf: Constraints & Efforts," *Energy Policy*, Vol. 34, No. 18, 2006, pp. 3719-3726. doi:10.1016/j.enpol.2005.08.018

[49] H. Doukas, I. Makarouni, C. Karakosta, V. Marinakis and J. Psarras, "EU-GCC Clean Energy Cooperation: From Concept to Action," In: M. Tortora, Ed., *Sustainable Systems and Energy Management at the Regional Level: Comparative Approaches*, University of Florence, Kent State University, Florence Program Abroad, 2012, pp. 288-308.

[50] A. Flamos, K. Ergazakis, D. Moissis, H. Doukas and J. Psarras, "The Challenge of a EU-GCC Clean Energy Network," *International Journal of Global Energy Issues*, Vol. 33, No. 3-4, 2010, pp. 176-188. doi:10.1504/IJGEI.2010.036955

Nomenclature

Bbl	Barrels
BCF	Billion cubic feet
Bm^3	Billion cubic meters
BMC	Billion cubic meters
Ca	Capita (person)
CPDP	Cogeneration power desalting plants
D	Day
DW	Desalted seawater
Ebbl	Equivalent barrel
EIA	Energy information administration, from the US government
EP	Electric power
FF	Fossil fuel
GCC	Gulf co-operation council countries
GDP	Gross domestic product
GHG	Greenhouse gases
GT	Gas turbine
GTCC	Gas turbine/steam turbine combined cycle
GWh	Giga Watt hour = 1000 MWh = 10^6 kWh
H	Hours
kW	Kilo Watt = 1000 kJ/s
kWh	Kilo-Watt hour, 3600 kJ
LNG	Liquefied natural gas
M	Minute, or meter
MAAGP	Mina Al-Ahmadi Gas Port
MIG	Million imperial gallons, 4546 m^3
MIGD	Million imperial gallons per day, 4546 m^3/d, or 52.62 kg/s
MW	Mega Watts = 1000 kW
NG	Natural gas
OAPEC	Organization of Arab petroleum exporting countries
PP	Power plant
SA	Saudi Arabia
TCF	Trillion cubic feet
UAE	United Arab emirates
Y	Year

3

Knowledge and Perceptions of Energy Alternatives, Carbon and Spatial Footprints, and Future Energy Preferences within a University Community in Northeastern US

Joanna Burger[1]*, Michael Gochfeld[2]
[1]Division of Life Sciences, Consortium for Risk Evaluation with Stakeholder Participation,
Environmental and Occupational Health Sciences Institute, Rutgers University, Piscataway, USA
[2]Consortium for Risk Evaluation with Stakeholder Participation, Environmental and Occupational Health Sciences Institute,
Environmental and Occupational Medicine, Rutgers Medical School,
Rutgers University, Piscataway, USA
Email: *burger@biology.rutgers.edu

ABSTRACT

Our overall research aim was to examine whether people distinguished between the spatial footprint and carbon footprint of different energy sources, and whether their overall "worry" about energy types was related to future developed of these types. We surveyed 451 people within a university community regarding knowledge about different energy sources with regard to renewability and spatial and carbon footprints and attitudes about which energy type(s) should be developed further. Findings were: 1) Gas, oil and coal were rated as the least renewable, and wind, solar and hydro as the most renewable; 2) Oil and coal were rated as having the largest carbon footprint, while wind, solar and tidal were rated the lowest; 3) There were smaller differences in ratings for spatial footprints, probably reflecting unfamiliarity with the concept, although oil and gas were rated the highest; 4) Energy sources viewed as renewable were favored for future development compared with non-renewable energy sources, and coal and oil were rated the lowest; 5) Worry-free sources such as solar were favored; and 6) There were some age-related differences, but they were small, and there were no gender-related differences. Overall, subjects knew more about carbon footprints than spatial footprints, generally correctly identified renewable and non-renewable sources, and wanted future energy development for energy sources which were less worried about (e.g. solar, wind). These perceptions require in-depth examination in a large sample from different areas of the country.

Keywords: Energy Sources; Preferences; Survey; Ecological Footprint; Carbon Footprint; Spatial Footprint

1. Introduction

Public knowledge about environmental issues can affect attitudes and beliefs about pollution, development, and environmental protection [1,2]. Recently, many environmental concerns have focused on energy, renewable energy options, and the environmental costs of different energy options [3-5]. Carbon footprints have received great attention, but ecological footprints have received less. The calculation of ecological footprint of fuel types is complicated and consists of three main components: area needed for energy production (including mining and processing), area needed to sequester emissions of greenhouse gases, and the area needed for safe deposition of nitrogen, sulphur and other waste products [6,7]. These usually translate into carbon equivalent emissions, using global warming potential recommended by the International Panel on Climate Change [8]. Calculations of carbon equivalent emissions quickly lead to discussions of sustainability, production capabilities, and alternative fuels [7,9-11].

These discussions have involved the public, and there are assessments of how the public views energy sources and renewable energy [2,12-15]. Many papers examine one type of energy or another, or report support for renewable energy in general [2,13]. Nuclear power has received much attention because of controversy surrounding safety, environmental risk and public opposition [16,17]. With nuclear, siting issues, population density, accidents and emergency routes are concerns [18-22], as large as concerns about proximity to nuclear facilities [16,23-24]. While the carbon footprint of different energy sources has figured prominently in these discussions [3,5,8], spatial footprint has not. That is, perceptions of the actual size required for different energy sources have not been examined.

Our overall aim was to explore whether people understood the relative size of the spatial footprint (and carbon footprint) of different energy sources, how much they worried about different energy sources, and whether their worry was related to which energy sources they thought should be further developed. We define spatial footprint as the actual physical space needed to support a given energy type—how much land is required for a wind or solar farm, or how much land is required for a nuclear power plant or a hydroelectric plant? This paper also examines the hypothesis that there is a relationship between perceptions of possible harm (personal worry) and the energy sources favored for development. Six questions are addressed: 1) What are perceptions of the relative size of the spatial footprint of different energy sources; 2) What are perceptions of the carbon footprint of different energy sources; 3) Which energy sources are renewable; 4) What is their overall worry rating for each energy source; 5) Which energy sources would they like to see developed; and 6) Are there any age-related differences in these perceptions?

We surveyed 451 students and non-students in a university community in central New Jersey in 2011. The energy sources listed in the survey were natural gas, nuclear, coal, solar, wind, tidal, hydro, oil, and geothermal, although worry was not addressed for the last two. We test the null hypotheses that: 1) there are no significant differences in perceptions about spatial and carbon footprints among energy sources; 2) there are no significant differences in ratings for energy sources to be further developed; 3) there are no age-related differences in these perceptions; and 4) there is no relationship between overall worry and energy sources to be further developed. Any findings apply to the study population, sampled at one time, and are meant to serve as a basis for further study in other communities and countries. Our data thus reflect local, rather than global perceptions, and thus can be related to local development or lack thereof. Even when people support particular technologies, they often do not accept them within their own community [e.g. 23,25], although Greenberg [2] reported that people living near nuclear facilities favored more development of nuclear than the general population. Greenberg also found age-related differences in that older respondents

were more likely to support increasing reliance on coal, gas, oil and nuclear power than younger respondents. For this reason, we examined age in our study.

Information on the public's views deal with perceptions or worries about renewable and non-renewable energy, rather than on their knowledge base [2,16]. Dalton et al. [26] surveyed tourist attitudes about renewable energy use in a hotel, and found that about 50% favored renewable energy, such as wind, but wanted to see onshore rather than offshore development. There is often a gap between perceptions of preferred energy types, and siting acceptance [23,27]. Others have focused on economic valuation of land for sustainable development [28].

Social trust is critical in risk/benefit decisions about environmental safety and health [20,29], but so is knowledge. Based on 239 published studies, Beierle [30] found that involving stakeholders in decisions resulted in higher-quality decisions, but only if the public had a sufficient information base about alternatives. Reversing public opposition requires both understanding of public views and knowledge about the issues, as well as appropriate steps to obtain public approval [24,30], although knowledge does not always change attitudes [15,23,25]. Understanding public perceptions and knowledge about energy sources is a first step in involving the public and other stakeholders in decision-making, leading to better environmental decisions [30,31]. While positive perceptions of energy types may not lead to acceptance of facilities at a local level [25,32], information on future energy type preferences and perceptions of worry can inform decision-makers. Other concerns, such as housing values, noise, and unsightliness, also influence personal decisions [13, 33,34].

The concept of ecological footprints is older than that of carbon footprints, and deals with a resource accounting tool that measures how much productive land and sea is appropriated for a given human use (e.g. the footprint) [35]. In this paper spatial footprint is used to denote the physical space needed to operate a given energy source.

2. Methods

The overall protocol was to interview students (aged 18 - 22; $N = 196$) and others (over age 22; $N = 255$) living and working in a university and surrounding community (restaurants, bus stops) in central New Jersey, to examine knowledge and views of different energy options. Interviews took place from 1 April to 15 May 2011. Subjects were selected by approaching the first person encountered, and then approaching the third or fourth person encountered thereafter. Although this approach is not completely random, there is no reason to assume biases. Interviewers identified themselves as from Rutgers University, gave a brief description of the study, and answered all questions following the interview. Refusal rate was less than 5%, and people refused because they were late for class or other appointments, had small children, or were rushing to board a bus. The interviews required about 20 minutes; many were longer due to subject's questions or comments about energy and politics which were allowed after the survey was completed. The protocol was approved as exempt by the Rutgers Institutional Review Board.

The questionnaire contained 4 parts concerning: 1) the relative size of the carbon footprint and the spatial footprint [on a scale of 1-5]; 2) whether each energy source was renewable or not; 3) how much they worried about different energy sources and favored further development of each energy source for the United States; and 4) demographics. Carbon footprint was defined as the relative amount of carbon emissions per kilowatt hour (kwh) of electric output, and spatial footprint was defined as how much land was required per kwh. Demographics included gender and age. A pilot survey of 10 students indicated that their ratings were not significantly different on two different days. There were no clear gender differences using Kruskal-Wallis non-parametric Analysis of Variance ($P > 0.10$ for comparisons), so gender is not discussed further.

We focused on nine energy sources: coal, oil, natural gas, solar, wind, tidal, nuclear, geothermal and hydroelectric. On a scale of 1 to 5, respondents were asked in a forced choice manner to rate from "low" to "high" the size of the carbon footprint and spatial footprint separately for each energy source. Respondents were also asked separately whether each energy source should be developed more in the US from "not at all" to "a lot" , and were asked "How worried are you about....," emphasizing individual rather than societal concern.

The "worry" question covered six of the energy types (oil, geothermal, and tidal were not on the worry question). Geothermal and tidal are not used in New Jersey and surrounding states, and oil accounts for only 1% of US electricity production [27]. The "worry" questions explored direct individual concerns including impacts on food, water, exposure of workers in the source facility, and exposure of wildlife from the facility. Exposure referred to radiation or radionuclides from nuclear facilities, mercury from coal-fired plants, carbon dioxide and sulphur emissions, and noise. Ratings were on a Likert Scale of 1 (no concern) to 5 (great concern). A composite worry index (mean score of the different worries) was computed from each energy source. On the pilot study, there was not a significant difference (Kruskal-Wallis tests, $P > 0.05$) on ratings for two different days.

Kruskal-Wallis non-parametric Analysis of Variance was used to compare estimates of renewability, spatial footprint, carbon footprint, worry and desirability across energy types, and also to analyze by age (up to age 22 versus 23 or older). A $P < 0.05$ was considered statistically significant, but readers should keep in mind the multiple comparisons inherent in this design.

3. Results

3.1. Carbon and Spatial Footprint

Across the nine energy types there were significant differences in the estimates of the carbon and spatial footprint ratings (**Figure 1**, statistics given on figure). The differences were dramatic for carbon footprint, and subtle for spatial footprint. For carbon footprint, coal and oil were rated highest and hydro, wind, tidal and solar were rated the lowest, and results generally matched our own understanding. There were no age-related differences in the ratings. We concluded that most respondents had a basic understanding of carbon footprint.

For spatial footprint, there was little variation in the scores, hovering around 3 - 3.5, probably reflecting a

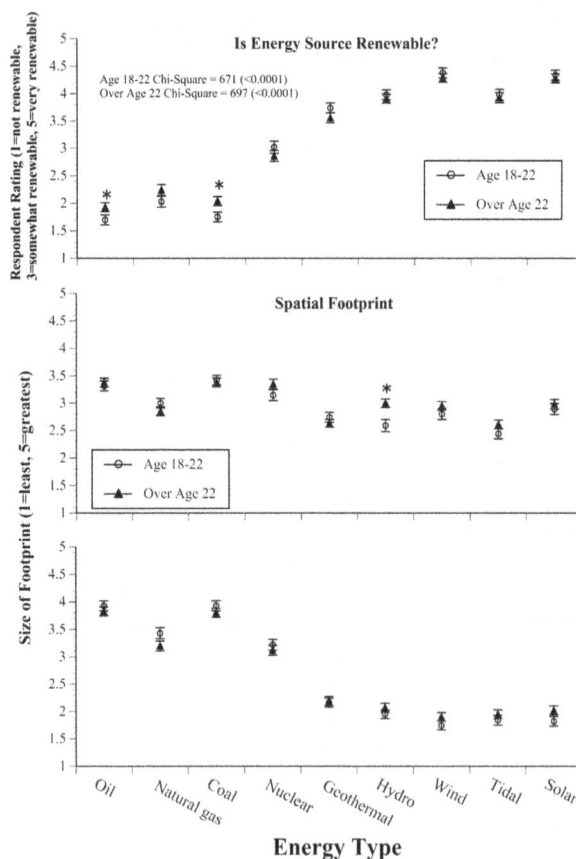

Figure 1. Ratings of people in a university community in central New Jersey about the relative size of the spatial and carbon footprints of different energy types. Shown also (bottom panel) are the ratings for whether an energy source is renewable or not. For the footprints, 1 = smallest, and 2 = largest. Shown are means ± standard error. Star equals significant age-related difference.

lack of knowledge about spatial footprint—a concept generally ignored in media coverage. Indeed only 32% of responses gave a 1 or 5 compared with 54% for carbon footprint ($\chi^2 = 7.12$; $P < 0.01$). In general, subjects thought coal, nuclear, and oil had the largest spatial footprint, and geothermal, hydroelectric, and tidal had the smallest. The only age-related difference was for hydro, where older people thought hydro had a larger footprint than did younger people (**Figure 1**).

3.2. Renewability of Resources

There were significant differences in whether subjects rated energy sources as renewable or not (**Figure 1**, top panel). Generally, oil, gas and coal were rated as non-renewable, with nuclear and geothermal in the middle. However, there was not a clear dichotomy; some respondents provided intermediate ratings. There were only two significant age-related differences. Older subjects rated oil and coal as more renewable than did students.

3.3. Energy Sources for Future Development

Subjects were asked to rate their views about which sources should be developed more in the United States. On **Figure 2**, we illustrate the percentage of subjects that rated each energy type a 1 ("do not develop" bottom panel) or a 5 (develop further, top panel). The energy source that had the lowest mean rating for future development was coal, followed by oil and nuclear, then by natural gas. The energy source with the highest rating was solar, followed by wind. Thus, people felt most strongly about developing solar, and not developing coal, than for the other energy sources. There were few age related differences. More young people were negative about oil and natural gas than were older people (**Figure 2**).

3.4. Overall Worry Rating

There were significant differences among energy type in the overall rating of worry (**Figure 3**, top panel). Nuclear had the highest worry index, followed by natural gas, and coal. People were less worried about wind, and solar (**Figure 3**). Generally, older people were more worried about more forms of energy (5 out of 6 categories, binomial 2-tailed $p = 0.125$) than were younger people, although the differences were significant only for nuclear. Although older subjects were more worried about nuclear, this was not reflected in the future development question.

The factors in the worry index included transportation risks, exposures from the plant or facility, exposure from food or water, exposure of workers, and exposure of wildlife. We also computed the mean worry score for each type of worry for all energy types combined. There

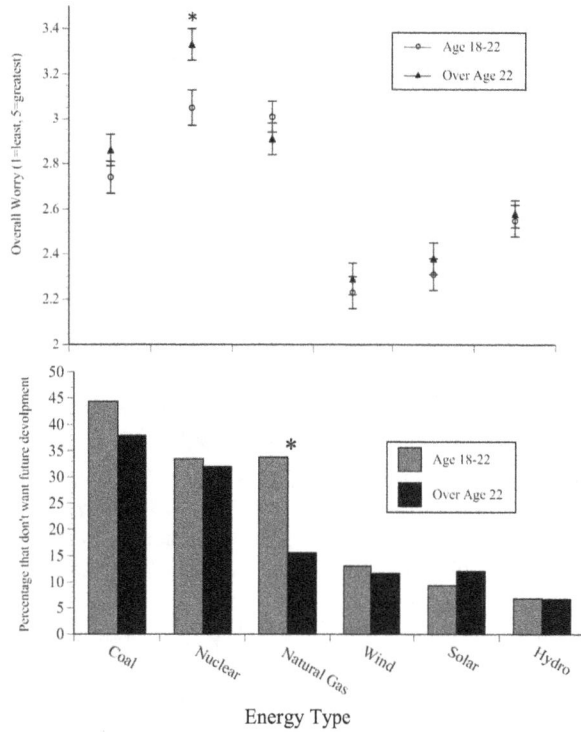

Figure 2. Percent of people wanting to see a particular energy type developed more in the future (rating of 5 of 5) and percent not wanting to see any future development (rating of 1 of 5). Star indicates significant age-related difference.

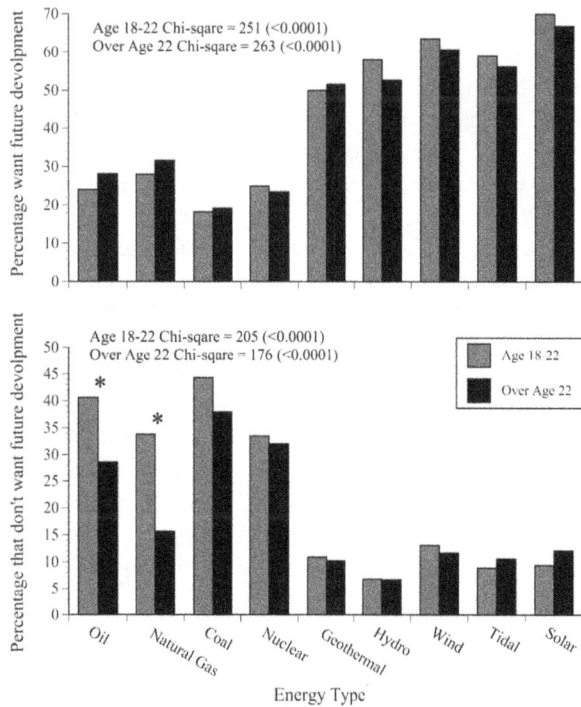

Figure 3. Overall worry score or index for different energy sources for college students (open circles) and people over 22 (black triangle). Shown are means ± standard error. Star equals significant age-related difference.

were significant differences, with people being more worried about risks/exposures to food (mean of 3.12 ± 0.05) and wildlife (mean of 3.03 ± 0.05) than they were for workers (2.67 ± 0.05) and from the facility itself (2.49 ± 0.04). Other components scored even lower. Older respondents were significantly more worried about exposure from the plant itself, and from transportation, than were younger people (X^2 tests, $P < 0.05$).

The relationship between overall worry and energy types subjects did not want to develop, is also shown in **Figure 3**. Although generally related, people did not want to see coal developed further, even though they were less worried about coal than for either nuclear or natural gas. In contrast, people were more worried about hydro than their response on future development would suggest.

In summary, the null hypotheses were rejected with respect to all questions. There were significant differences as a function of energy type in rating of relative spatial and carbon footprint size, understanding of renewable energy, perceptions of which energy source to develop, and in overall worry about different energy forms. The relationship between overall worry and the percentage of people who do not want future development were generally related, except for coal (**Figure 3**). There were few age-related differences.

4. Discussion

This study tested both knowledge and perceptions, and inevitably revealed some misconceptions regarding energy and climate. A 2011 report from the Yale University Center on Climate Policy reported that 90% of respondents identified development of clean energy and 70% rated global warming as medium, high, or very high priority for the country, even to the point that 65% supported a "carbon tax" [36]. A 2012 report for the World Energy Summit found that American concerns for energy security and the economic impacts of energy choices ran high, accompanying an interest in renewable and alternative energy sources [37]. The report reflected a strong interest in incentivizing renewable energy, based on both environmental and economic concerns [37]. Thus it is reasonable to predict that students and residents in a University town would have at least a basic understanding of these issues and their relationships. Further, one might expect younger people to have a greater understanding because they are still in school and exposed to some of these issues [see 2].

4.1. Carbon and Spatial Footprints

In this survey we asked specifically about "carbon footprint" and "spatial footprint" rather than ecological footprint, a more complex and controversial construct [8]. Understanding carbon footprints, reducing carbon emissions, and reversing global climate change is one of the foremost current ecological and media issues. Considerable attention has been given to examining global drivers, and to the need to reduce emissions from fossil-burning fuel (for electricity and transport) and industrial processes that have been accelerating rapidly [3,20]. The currency for these discussions is "carbon footprint', which relates to the amount of carbon released as carbon dioxide per unit (typically per kilowatt hour). Carbon is released by the burning of all types of fossil fuel and the carbon/kwh depends on the thermal density of the fuel, the efficiency of the combustion process, and air pollution control devices (although the latter only redirect the carbon from air to some other disposal process).

We expected respondents to have at least a basic grasp of the general issue of carbon associated with familiar energy types, since this has received extensive media attention. Respondent rankings correspond well to our own expectations, including the recognition that geothermal emits more carbon (in addition to sulfur) than other "renewable" sources [38]. Their ratings were generally correct despite the fact that neither hydrothermal nor hydro are used in New Jersey and thus the ratings do not reflect local experience.

Determining spatial footprints is difficult because of differences in the physical environment. This topic gets little media coverage, and not surprisingly respondents seemed unfamiliar with the concept, judging by their middle-of-the-road responses. A few examples of the complexities of spatial footprint will suffice: 1) slope, updrafts and local geography influence how many wind towers can be efficiently placed on a given amount of land, 2) the size and depth of the thermal field determines how much electricity can be generated from the field, and 3) weather patterns and latitude influence solar capacities, and how many solar cells are needed, facing in what directions, and how much energy is required to rotate them. For solar, the spatial footprint on a roof can be discounted compared to the usurpation of otherwise productive agricultural acreage or natural landscape. In the production of energy from biomass (not examined in this study), a water footprint must be considered since different plants (crops) require different amounts of water to produce a unit of energy [39], and the release of carbon from biomass burning varies by crop type [40].

Further, determination of both carbon and spatial footprints depend upon whether only direct footprints are considered, or indirect are as well. For example, the direct footprint of a hydropower generation plant includes the occupied area of the dam and plant, the build-up of land surrounding the facility, and the flooding of land behind a dam. Indirect effects include machinery production, building materials, what workers require to run the plant and the energy (either from hydro or fossil fuel) that is required for all machinery and materials to run to hydropower plant. Similarly, the direct footprint of a nuclear plant is the land area occupied by reactors, other buildings, storage pools and pads, as well as buffer areas. But the area involved in mining and processing, and ultimately the off-site disposal or reprocessing of spent fuel rods must be considered. Surface mining versus underground mining, and surface disposal (currently on site) versus repository storage of fuel, would provide different footprints.

Although there were differences among energy types in the spatial footprint responses, the responses reflect unfamiliarity rather than knowledge. That is, the ratings by subjects did not reflect current science. Few calculations have been made to compare with the perceptions reported in this paper. Stöglehner [9], however, provided some comparisons, and found that spatial footprints per energy produced decreased as follows: coal (highest spatial footprint, relative value of 20), oil (12), gas (10), biofuels, hydropower, solar, and wind (all less than 1). Geothermal, nuclear, and tidal were not examined. Huijbregts [41] provided another accounting of ecological footprint (in decreasing order) of biomass, hydro, wind and solar, and fossil and nuclear energy, but did not examine indirect footprints. Using energy chains for cars, Holden and Høyer [7] came up with a ecological footprint ranking of biomass > oil > natural gas > hydro. The most inclusive ranking is from Sovacool [42]: coal > oil > natural gas >> nuclear > geothermal = biomass > solar = hydro = wind. The estimated release ranges from 1000 g CO_2/kWh for coal to about 10 g/kWh for solar, wind and hydro. There are several discrepancies depending on assumptions. For example, the full nuclear cycle includes substantial carbon emission in the front end, although negligible carbon is released during the reactor operations [42]. These discordant analyses illustrate the importance of scientists deciding on a uniform method of calculating spatial or ecological footprints.

In the present study of respondents from a university community in central New Jersey, the ranking of spatial footprint (in decreasing order) was: coal/oil/nuclear/ geothermal > hydro/wind/solar > tidal, but the differences were small with mean ratings for spatial footprint between 2.5 and 3.5, while they rated carbon footprints as varying from 1.8 to 4. Moreover, there were few age-related differences in knowledge about carbon and spatial footprints (refer to **Figure 1**). Older people thought the spatial footprint for hydro was larger than did younger people, but the differences were not great and may not be meaningful.

4.2. Renewability of Resources

Much of the public debate, public-policy decisions, and

international agreements concern the dichotomy between renewable and non-renewable resources. Renewable resources are those that are naturally renewed, such as solar, wind, tidal, geothermal, and to varying extents biomass [43,44]. Our definition of "renewable" is an energy source that is not depleted by use. In our view, there is not a perfect dichotomy, but instead there are intermediate stages. Geothermal, for example, in not completely renewable because it requires recharge to maintain the steam source [38].

Various polls have shown wide-spread support for the concept of renewable energy for environmental, economic, and security reasons [37]. Some people, however, have questioned whether the current high material living standards in developed nations; can be maintained using only renewable energy [45]. Several agencies and governments have addressed the development of energy plans [46] and systems that are 100% renewable [47,48], acknowledging that these would involve major societal changes in farming practices (if biofuels are key), use of land (if solar and wind), and possible offshore effects (if offshore wind), not to mention the direct environmental effects.

A reasonable public debate that leads to public policy decisions and the siting of energy facilities, however, requires an understanding of which sources are renewable, as well as the relative spatial footprint each requires. Clearly the most renewable energy source is solar, since the sun's energy striking the earth is relatively constant, taking into account latitude/season and atmospheric clarity, and wind energy which results from the sun's differential heating of the earth's surface. In the present study, subjects rated both solar and wind as the most renewable, although the average rating was less than 5, meaning that some people did not consider it completely renewable. At present, New Jersey has little solar or wind energy, although these are being encouraged by State government and the media.

Improved technology aims at increasing the efficiency of energy conversion for solar as well as for other forms of energy considered renewable (wind, tidal, geothermal, hydro), which requires energy-dependent generators to convert the renewable energy into electricity [49]. Geothermal, companies, for example, developed methods of powering the generators from geothermal energy rather than depending upon oil, but were slow in becoming independent [50]. Geothermal has the clear advantage of not influencing global warming [51].

We suggest that there are other distinctions that are rarely made when considering renewable resources—the degree of renewability and the predictability of the resource. For example, the sun will continue to shine, but wind is much less reliable, and geothermal is reliable but it can be overexploited. That is, if too much water is withdrawn from the geothermal field, the water table can drop (E. Gunniaugsson, Reykjavik Energy, Iceland, Pers. Comm.). Thus, there are complexities to the term "renewable" that require exploration. Further, methods of energy storage are critical for many forms of energy; the sun doesn't shine at night, and wind is not always strong enough to turn turbines [52].

Subjects in this survey correctly recognized solar and wind as renewable, and rated natural gas, coal and oil as non-renewable. Even so, however, everyone did not rate them a 1 (not renewable). Nuclear energy, usually considered non-renewable, but advantageous because of its low carbon emission, was rated as intermediate with a wide range of scores from 1 to 5. Thus, there seems to be less understanding among the respondents regarding the renewability status of nuclear. There were few age-related differences, and those that were significant were not great and may not be meaningful.

Greenberg and Truelove [53], in a survey of 3200 US residents, showed that there are multiple publics with respect to energy preferences and risk benefits. In our study, with a relatively homogeneous population within a university community, there was a wide difference in knowledge about the renewability of energy sources under discussion. It suggests public forums on energy resources and sustainability need to clearly define renewable, and identify the resources being discussed.

4.3. Worry, Knowledge and Energy Sources for Development

There is a very large literature on public preferences for, and worries about, different energy sources, with literally hundreds of opinion polls. Overall these polls show a clear preference for renewable sources of energy, and major reservations about coal and nuclear fuel [reviewed in 2]. Greenberg's national survey of 2701 US residents showed that over 90% wanted greater reliance on solar and wind, and over 70% wanted more reliance on hydroelectric sources. There is still concern, however, about the effect of wind on global climate [54].

In the present study, there was also a preference for wind and solar, followed by tidal, hydroelectric, and geothermal. Nuclear was more preferred for future development than natural gas, oil, and coal, which was surprising, given that the survey was conducted only weeks after the Fukushima nuclear event (March 2011), when the story was still receiving daily coverage in the media. In another series of questions, about 55% said that the Fukushima event and the Deep Water Horizon Gulf oil spill influenced their views about energy use (Burger, unpubl. data). Thus, it is likely that the Fukushima accident influenced the ratings, making it more surprising that nuclear was rated higher for future development than

natural gas, oil, or coal.

The mean worry score for different energy sources was generally related to the percent of subjects who did not want that form of energy developed (**Figure 3**). However, this was only generally true. A higher percentage of subjects were opposed to further development of coal than their worry score would indicate. Generally subjects were not very worried about renewable energy forms (hydro, solar, wind), and few people opposed further development. Some people did, however, feel the renewables should not be further developed, and this bears further study.

4.4. Implications and Conclusions

Overall, subjects in this study had a reasonable understanding of the relative size of the carbon footprint, but less of an understanding of spatial footprints. The implications of this are that people may not be aware of the ecological consequences, in terms of physical space and the amount of ecosystems that would be disrupted, of different energy sources. It also suggests the importance of examining the relative physical impact of different energy sources on natural ecosystems. Understanding spatial footprint is particularly important for the state of New Jersey because it is a small state, with the highest population density in the US, where land is at a premium. We also suggest that permanence should be examined. That is, if a particular energy source is developed, can the ecosystem it replaces ever be restored once the energy source is developed? For example, could an ecosystem be restored if a wind farm or solar panels are placed there? This is an especially important question for New Jersey, where some farmland is being covered with solar panels.

Further, subjects correctly knew which energy sources were renewable and which were not, and they wanted to see more development of renewable resources, and less of non-renewable resources. Younger people wanted to see less future development of oil and natural gas than did older people, and a conclusion which agrees with the findings of Greenberg [2]. Thus, these results suggest that older people are less reluctant to move away from oil and natural gas, toward other forms of energy. However, when the data on the percentage of people who wish to see future development were examined (refer to **Figure 2**), there were no significant differences as a function of age. All age groups wanted to see future development of renewable energy sources, suggesting further support by people of efforts to develop renewable resources.

Worry can be used by managers to understand educational needs, and discrepancies between worry and their desire to forego further development of some energy sources. For example, people were less worried about coal than their preference for no more development would suggest. It seems people are not worried about it, but do not want to see further coal development. This also suggests that there is another reason for they wish not to see coal development that is not captured by their "worry" scores. The combination of preferences (or lack thereof) for future development, in conjunction with worry scores, may provide another way to examine personal perceptions of energy development.

Finally, this survey clearly indicated that people worry about the development of some energy sources (gas, oil, coal), and worry much less about others (wind and solar, followed by tidal, hydroelectric, and geothermal). The subjects interviewed generally wanted to see more future development of the energy sources that they were less worried about. The only energy source which did not fit this was nuclear (nuclear was more preferred for future development than natural gas, oil, and coal). Thus, overall, surveys can provide information on different aspects of future energy development, such as the public's rating of which sources to develop, their worry about different energy sources, and their knowledge (and worry) about carbon and ecological footprints.

5. Acknowledgements

This research was partly funded by the Consortium for Risk Evaluation with Stakeholder participation (DE-FC01-06EW07053), and NIEHS Center Grant (P30ES005022), and EOHSI. The views expressed herein are solely those of the authors, and do not represent any of the funding agencies.

REFERENCES

[1]　M. Greenberg and K. Crossney, "The Changing Face of Public Concern about Pollution in the United States: A Case Study of New Jersey," *Environment*, Vol. 26, No. 4, 2006, pp. 255-268.

[2]　M. R. Greenberg, "Energy Sources, Public Policy, and Public Preferences: Analysis of US National and Site-Specific Data," *Energy Policy*, Vol. 37, No. 8, 2009, pp. 3242-3249. doi:10.1016/j.enpol.2009.04.020

[3]　M. R. Raupach, G. Marland, P. Ciais, C. LeQuere, J. G. Canadell, G. Klepper and C. B. Field, "Global and Regional Drivers of Accelerating Cos Emissions," *Proceedings of the National Academy of Sciences*, Vol. 104, No. 24, 2007, pp. 10288-10293. doi:10.1073/pnas.0700609104

[4]　B. J. M deVries, D. P vanVuuren and M. M. Hoogwijk, "Renewable Energy Sources: Their Global Potential for the First Half of the 21st Century at the Global Level: An Integrated Approach," *Energy Policy*, Vol. 35, No. 4, 2007, pp. 2590-2610. doi:10.1016/j.enpol.2006.09.002

[5]　EIA (Energy Information Administration), Energy-Related Carbon Dioxide Emissions, US Energy Information Administration, 2010. http://www.eia.gov/oiaf/ieo/emissions.html

[6]　N. Chambers, C. Simmons and M. Wackernagel, "Sharing Nature's Interest: Ecological Footprints as an Indicator of Sustainability," Earthscan, London, 2000.

[7]　E. Holden and K. G. Høyer, "The Ecological Footprints of Fuels," *Transportation Research, Part D*, Vol. 10, No. 5, 2005, pp. 395-403.

[8]　IPCC (International Panel on Climate Change), Climate Change 2007, Contribution of Working Group to the Fourth Assesment Report of the Intergovernmental Panel on Climage Change, Cambridge University Press, Cambridge, 2007.

[9]　G. Stoglehner, "Ecological Footpring—A Tool for Assessing Sustainable Energy Supplies," *Journal of Cleaner Production*, Vol. 11, No. 3, 2003, pp. 267-277. doi:10.1016/S0959-6526(02)00046-X

[10]　S. M. Benson and F. M. Orr Jr., "Sustainability and Energy Conversion" *MRS Bulletin*, Vol. 33, No. 4, 2008, pp. 297-302. doi:10.1557/mrs2008.257

[11]　C. J. Bromley, M. Mongillo, G. Hiriart, B. Goldstein, R. Bertani, E. Huenges, A. Ragnarsson, J. Tester, H. Muraoka and V. Zui, V, "Contribution of Geothermal Energy to Climate Change Mitigation: The IPCC Renewable Energy Report," *Proceedings of the World Geothermal Congress*, Bali, 25-29 April 2010, pp. 1-5.

[12]　P. Upham, L. Whitmarsh, W. Poortinga, K. Purdam, A. Darnton, C. McLachlan and P. Devine-Wright, "Public Attitudes to Environmental Change: A Selective Review of Theory and Practice," Research Councils, Swindon, 2009. www.lwec.org.uk

[13]　J. Zoellner, P. Schweizer-Ries and C. Wemheurer, "Public Acceptance of Renewable Energies: Results from Case Studies in Germany," *Energy Policy*, Vol. 36, No. 11, 2008, pp. 4136-4141. doi:10.1016/j.enpol.2008.06.026

[14]　A. Spence, W. Poortinga, C. Butler and N. F. Pidgeon, "Perceptions of Climate Change and Willingness to Save Energy Related to Flood Experience," *Nature*, Vol. 1, 2011, pp. 46-49.

[15]　G. Ellis, J. Barry and C. Robinson, "Many Ways to Say 'No'-Different Ways to Say 'Yes'; Applying Q-Methodology to Understand Public Acceptance of Wind Farm Proposals," *Journal of Environmental Planning and Management*, Vol. 50, No. 4, 2007, pp. 517-551. doi:10.1080/09640560701402075

[16]　M. R. Greenberg, "How Much Do People Who Live near Major Nuclear Facilities Worry about Those Facilities: Analysis of National and Site-Specific Data," *Journal of Environmental Planning and Management*, Vol. 52, No. 7, 2009, pp. 919-937. doi:10.1080/09640560903181063

[17]　H. C. Hung and T. W. Wang, "Determinants and Mapping of Collective Perceptions of Technological Risk: The Case of the Second Nuclear Power Plant in Taiwan," *Risk Analysis*, Vol. 31, No. 4, 2011, pp. 668-682. doi:10.1111/j.1539-6924.2010.01539.x

[18]　R. R. Kasperson, O. Renn, P. Slovic, H. S. Brown, J. Emel, R. Goble, J. X. Kasperson and S. Ratick, "The Social Amplification of Risk: A Conceptual Framework,"

Risk Analysis, Vol. 8, No. 2, 1988, pp. 177-187. doi:10.1111/j.1539-6924.1988.tb01168.x

[19] M. Wolsink, "Entanglement of Interests and Motives: Assumptions behind the NIMBY Theory on the Facility Sitting," *Urban Studies*, Vol. 31, No. 6, 1994, pp. 851-866. doi:10.1080/00420989420080711

[20] G. O. Rogers, "Siting Potentially Hazardous Facilities: What Factors Impact Perceived and Acceptable Risk?" *Landscape and Urban Planning*, Vol. 39, No. 4, 1998, pp. 265-281. doi:10.1016/S0169-2046(97)00087-X

[21] P. Slovic, "The Perceptions of Risk," In: J. Slovic, Ed., *The Perception of Risk*, Earthscan, London, 2000, pp. 221-230.

[22] D. L. Feldman and R. A. Hanahan, "Public Perceptions of a Radioactively Contaminated Site: Concerns, Remediation Preferences, and Desired Involvement," *Environmental Health Perspectives*, Vol. 104, No. 12, 1996, pp. 1344-1352. doi:10.1289/ehp.961041344

[23] M. R. Greenberg, "NIMBY, CLAMP, and the Location of New Nuclear-Related Facilities: US National and 11 Site Specific Surveys," *Risk Analysis*, Vol. 29, No. 9, 2009b, pp. 1242-1245. doi:10.1111/j.1539-6924.2009.01262.x

[24] H. C. Jenkins-Smith, C. L. Silva, M. C. Nowlin and G. deLozier, "Reversing Nuclear Opposition: Evolving Public Acceptance of a Permanent Nuclear Waste Disposal Facility," *Risk Analysis*, Vol. 31, No. 4, 2011, pp. 629-644. doi:10.1111/j.1539-6924.2010.01543.x

[25] D. Bell, T. Gray and C. Haggett, "The 'Social' Gap in Wind Farm Sitting Decisions; Explanations and Policy Responses," *Environmental Policy*, 2005, pp. 49-64.

[26] G. J. Dalton, D. A. Lockington and T. E. Baldock, "A Survey of Tourist Attitudes to Renewable Energy Supply in Australian Hotel Accommodation," *Renewable Energy*, Vol. 33, No. 10, 2008, pp. 2174-2185. doi:10.1016/j.renene.2007.12.016

[27] United States Energy Information Administration (USEIA), US Carbon Dioxide Emissions from Energy Sources, EIA of DOE, 2010.

[28] T. Soderqvist, H. Eggert, B. Olsson and A. Soutukorva, "Economic Valuation for Sustainable Development in the Swedish Coastal Zone," *Ambio*, Vol. 34, No. 2, 2005, pp. 169-175.

[29] M. Siegrist, G. Cvetkovih and C. Roth, "Salient Value Similarity, Social Trust, and Risk/Benefit Perception," *Risk Analysis*, Vol. 20, No. 3, 2000, pp. 353-362. doi:10.1111/0272-4332.203034

[30] T. C Beierle, "The Quality of Stakeholder-Based Decisions," *Risk Analysis*, Vol. 22, No. 4, 2002, pp. 739-749. doi:10.1111/0272-4332.00065

[31] T. Dietz and P. C. Stern, "Public Participation in Environmental Assessment and Decision-Making," National Academy Press, Washington DC, 2008.

[32] P. Devine-Wright, "Local Aspects of UK Renewable Energy Development; Exploring Public Beliefs and Policy Implications," *Local Environment*, Vol. 10, No. 1, 2005, pp. 57-69. doi:10.1080/1354983042000309315

[33] J. Blake, "Overcoming the 'Value-Action Gap' in Environmental Policy: Tensions between National Policy and Local Experience," *Local Environ*, Vol. 4, No. 3, 1999, pp. 257-278. doi:10.1080/13549839908725599

[34] R. Kahn, "Siting Struggles; the Unique Challenge of Permitting Renewable Energy Power Plants," *The Electric Journal*, Vol. 13, No. 2, 2000, pp. 21-33. doi:10.1016/S1040-6190(00)00085-3

[35] J. Kitzes, A. Peller, S. Goldfinger and M. Wachernagel, "Current Methods for Calculating National Ecological Footprint Accounts," *Scientific Environmental Sustainability Society*, Vol. 4, 2007, pp. 1-9.

[36] A. Leiserowitz, E. Maibach, C. Roser-Renouf, N. Smith and J. D. Hmielowski, "Climate Change in the American Mind: Public Support for Climate & Energy Policies in November 2011," Yale University and George Mason University. New Haven, 2011.

[37] Council on Foreign Relations, US Opinion on Energy Security, 2012.

[38] J. L. Renner, "Geothermal Energy," In: T. M. Letcher, Ed., *Future Energy: Improved, Sustainable and Clean Options for Our Planet*, Elsevier, New York, 2008, pp. 211-224.

[39] P. W. Gerbens-Leenes, A. Y. Hoekstra and T. H. van der Meer, "The Water Footprint from Biomass: A Quantitative Assessment and Consequences of an Increasing Share of Bio-Energy in Energy Supply," *Ecological Economy*, Vol. 68, No. 4, 2009, pp.1032-1060. doi:10.1016/j.ecolecon.2008.07.013

[40] P. Champagne, "Biomass," In: T. M. Letcher, Ed., *Future Energy: Improved, Sustainable and Clean Options for Our Planet*, Elsevier, New York, 2008, pp. 151-170.

[41] M. A. J. Huijbregt, S. Hellweg, R. Frischknecht, K. Hungerbuhler and A. J. Hendriks, "Ecological Footprint Accounting in the Life Cycle Assessment of Products," *Ecology Economics*, Vol. 64, No. 4, 2008, pp. 798-807. doi:10.1016/j.ecolecon.2007.04.017

[42] B. Sovacool, "Valuing the Greenhouse Gas Emissions from Nuclear Power: A Critical Survey," *Energy Policy*, Vol. 36, No. 8, 2008, pp. 2940-2953. doi:10.1016/j.enpol.2008.04.017

[43] D. M. Berman and J. T. O'Connor, "Who Owns the Sun? People, Politics and the Struggle for a Solar Economy," Chelsea Green Publishing Co., White River Junction, VT, 1996.

[44] W. Shi, "Renewable Energy: Finding Solutions for a Greener Tomorrow," *Reviews in Environmental Science and Biotechnology*, Vol. 9, No. 1, 2010, pp. 33-37. doi:10.1007/s11157-010-9187-6

[45] F. E. Trainer, "Can Renewable Energy Sources Sustain Affluent Society?" *Energy Policy*, Vol. 23, No. 12, 1995, pp. 1009-1026. doi:10.1016/0301-4215(95)00085-2

[46] E. E. Thorhallsdottir, "Environment and Energy in Iceland: A Comparative Analysis of Values and Impacts," *Environmental Impact Assessment Review*, Vol. 27, No. 6, 2007, pp. 522-544. doi:10.1016/j.eiar.2006.12.004

[47] H. Lund, "Renewable Energy Strategies for Sustainable Development," *Energy*, Vol. 32, No. 6, 2007, pp. 912-919. doi:10.1016/j.energy.2006.10.017

[48] H. Lund and B. V. Mathiesen, "Energy System Analysis of 100% Renewable Energy Systems—the Case of Denmark in Years 2030 and 2050," *Energy*, Vol. 34, No. 5, 2009, pp. 524-531. doi:10.1016/j.energy.2008.04.003

[49] V. Smil, "Energy Transitions: History, Requirements, Prospects," Praeger, California, 2010.

[50] L. Rybach and M. Mongillo, "Geothermal Sustainability—A Review with Identified Research Needs," *GRC Transaction*, Vol. 30, 2006, pp. 1083-1090.

[51] B. A. Goldstein, G. Hiriart, J. Tester, B. Bertani, R. Bromley, L. Guierrez-Negrin, C. J. Huenges, H. Ragnarsson, A. Mongillo, M. A. Muraoka and V. I. Zui, "Great Expectations for Geothermal Energy to 2100," *Proceedings 36th Workshop of Geothermal Reservoir Engineering*, Stanford, 31 January-2 February 2011.

[52] D. Dicaire and F. H. Tezel, "Regeneration and Efficiency Characterization of Hybrid Adsorbent for Thermal Energy Storage of Excess and Solar Heat," *Renewable Energy*, Vol. 36, No. 3, 2011, pp. 986-992. doi:10.1016/j.renene.2010.08.031

[53] M. Greenberg and H. B. Truelove, "Energy Choices and Risk Beliefs: It Is Just Global Warming and Fear of a Nuclear Power Plant Accident?" *Risk Analysis*, Vol. 31, No. 5, 2011, pp. 819-831. doi:10.1111/j.1539-6924.2010.01535.x

[54] D. W. Keith, J. F. DeCarolis, D. C. Denkenberger, D. H. Lenschow, S. L. Malyshev, S. Pacala and P. J. Rasch, "The Influence of Large-Scale Wind Power on Global Climate," *Proceedings of the Natural Academy of Science of the United States of America*, Vol. 101, No. 46, 2004, pp. 16115-16120. doi:10.1073/pnas.0406930101

4

Adapting Business of Energy Corporations to Macro-Policies Aiming at a Sustainable Economy. The Case for New Powering of Automobiles

Jose M. "Chema" Martinez-Val Piera, Alfonso Maldonado-Zamora, Ramon Rodríguez Pons-Esparver

ETSI Minas, Universidad Politécnica de Madrid, Madrid, Spain

Email: chemaval@gmail.com

ABSTRACT

A portfolio of new energy technologies has emerged in the first decade of the 21st Century, and many of them could be used for restructuring the energy sector towards Sustainable Development. A key subject in this quest is the future of automobile, with possibilities on powering ranging from biofuels to Hydrogen Cars (HC), to Electric Vehicles (EV). In turn, the latter is closely connected with the need to deploy Renewable Energies (RE) for electricity generation. Within such new situation, countries and governments are aware that there are new tools for fighting Global Warming (GW), and new policies could be established for winning this battle against CO_2. All these initiatives will affect the future of energy corporations, notably hydrocarbon companies; and it should be noted that it will be difficult for the companies to define long-term strategies if energy policies convey upheavals, sudden changes in promoting alternatives and interruptions on activities. Hence, it is very important to adopt energy policies allowing a smooth evolution of the companies' activities to the new energy model. After analyzing the alternatives with a forecasting-backcasting methodology, an "eclectic approach" is proposed, with the Plug-in Hybrid car with Flexible Fuel (PiHFF) as the central paradigm in the coming promoting policies.

Keywords: Sustainable Energy Policies; Electric Vehicles; Corporation Adaptation

1. Sustainable Development as a Worldwide Quest

These initial years of the third millennium (A.D.) are evolving under a sort of paradox: the economic and financial perspective is rather obscure and without horizons, while the technology perspective is very positive and full of potential. The latter includes Bio-technologies, Communication and Information Technologies, Material Science and Nano-technologies, Space-related spin-offs and, of course, Energy technologies [1,2]. The list is not exhaustive, but Energy is a main block in it.

The technology perspective is so appealing that many people (scientists, journalists, politicians) speak about the upcoming Third Industrial Revolution [3], which would be based on Energy and Machines, as the previous ones. The First Industrial Revolution was enabled by Coal and Steam Engines, applied to industrial machines, navigation and trains. The Second one exploded from Internal Combustion Engines (ICE; see list of acronyms at the end of the article), petroleum products and cars, on the one hand, and electricity on the other hand. The Third one could convey EV and new energy sources, notably RE and Nuclear Fusion. Although mastering all these elements is still a challenge, it will take time, efforts and budget, and not all players will advance at the same rate. It is worth highlighting the potential merging of the two legs of the 2nd Industrial Revolution into a single stem in the 3rd one, if electric cars actually succeed in the new Revolution. Some people can consider this fact is a mere coincidence, but it seems there is a technical destiny in this merging.

This optimism about technology cannot run freely. A new and tight scenario has been set up by a strong paradigm inherited from the end of the 20th Century, namely, Sustainable Development [4,5]. It includes the need to stop human intervention in Climate Change (CC) [6-11], also known as the fight against Global Warming [12-14]. Some people (including politicians and journalists) consider that this problem is less acute and important than the Global Economy Crisis currently running, but there is a difference between both problems, which is similar to the difference between Physics and Economy. Laws of Economy can be changed or can evolve by human action. This is not the case for the laws of Physics. They cannot be changed.

A third type of perspective must be considered in order

to complete the premises of the study. It can be called the organizational perspective; *i.e.*, the capability to organize our forces at a corporate level, country level, continent level and finally, Worldwide, for modeling the future in a really good way, if not the best way. An important example in this domain is the Intergovernmental Panel on Climate Change (IPPC) [15] although it is not an executive body, but a purely scientific one. Even so, scientific evidence of GW was essential in the Kyoto Protocol, and it has been confirmed since then, what urges the IPCC to ask for a substantial reduction of Greenhouse Gases (GHG) emissions.

In the organizational perspective, a salient entity is the European Union (EU), which has likely been and continues to be the most committed body in the fight against GW, and in the promotion of sustainable development (Directives 2003/87/CE on CO_2 emission; and 2001/80/CE on Large Combustion Plants). As most of the political decisions, those of the EU concerning these principles can be considered very rhetoric, but the commitment has been established in terms of numeric objectives, particularly in the EU Directive "20/20/20" 2009/28 CE establishing binding goals for year 2020 in reduction of GHG emissions, use of RE, and improvements in efficiency and energy savings, all of them in 20% variations. Of course, this is easier saying than doing, and some people consider this policy is just wishful thinking. It is important to underline that the actual rate of change in the productive sectors seems to be too slow, particularly in Energy. Many of the new technologies are not competetive yet, and some of them (notably, Nuclear Fusion) seem they will need about half a century for getting a place in the applicable portfolio. In other cases, as RE, the deployment is only possible thanks to strong subsidies and feed-in tariffs (as Spain's decree RD 661/2007), which is a distorting factor for current markets.

This paper addresses the difficult adjustment between micro-economics of the corporations and the macro objectives of energy and economy policies, particularly in relation to hydrocarbon corporations. Many papers [16] are devoted to state-level analysis, but they do not contemplate the very practical and fundamental problem of how to implement selected policies by the action of the corporations working in this field. Moreover, in a short-term future, energy corporations of some countries will have to internalize the cost of GHG emissions, and a parallel market of emission rights will be established. This is the case of the EU, as regulated by Directive 2003/87, which includes a transient phase ending in January 1st 2013. That Directive has been transposed to domestic legislation in the Member Countries, as Spain's Law 1/2005 and the RD 1370/2006. Although not all emission sources are contemplated in the same way, and external "clean mechanisms" are included in that market as flexible actions, the actual fact is that CO_2 will have to be incorporated as another cost in the corporations of the energy industries. Additionally, a better scientific insight is needed for clarifying the impact of natural catastrophes (volcanoes mainly) in the evolution of CC.

It is difficult to anticipate how strong and long lasting the enforcement of that legislation will be, because many specialists point out that GHG are a worldwide problem that cannot be addressed by individual decisions, even if they are taken by the EU. A key point is that coal has undergone the highest increase in consumption in the current Century [17] and one of the causes is that electricity development in China in the last decade has relied mainly on coal.

The next section presents a very brief summary of relevant facts of the Energy sector at large. Section 3 will be devoted to analyze some features and principles of corporations as living bodies that must survive in a competitive environment where the rules can be changed by laws, and laws can be changed by goals. The problem addressed is specifically focused on Energy, where two branches notably separated until this Century, the oil and electricity industries, can somehow merge in a complex new sector. The paper is then focused, in Section 4, on the global challenge created by the rising contents of equivalent CO_2 in the atmosphere. This section also includes some considerations about the role of technology evolution as a global answer to that challenge. Section 5 presents a finer analysis of the technology evolution that seems more likely to succeed. Finally, Section 6 is devoted to how a corporation can deal with the problem of reshaping itself in accordance to the anticipated future. If some decisions are taken too early, they can take the wrong path. If they are taken too late, they can be useless for getting (or keeping) market share. We conclude with some recommendations on addressing the global problem, proposing a stronger link between transportation and energy sectors.

It is particularly important to speed-up technology development and to identify in due time legal and regulatory changes to define the new rules of the game, which must include the internalization of environment costs, notably CO_2 emissions. However, such changes must follow a smooth track, so as not to induce a crisis is the hydrocarbon industry, which is a backbone of social and economical activities in most of the world. The smooth track proposed in this article can be called the eclectic approach.

2. The World of Energy

There is a broad consensus nowadays about the principles of Energy Policy:
- Security of supply (and the related objective of energy independence, in contrast to the existence of global

markets).

- Reasonable (if not minimum) costs (which conveys economic competitiveness).
- Environmental quality (at local, regional and global levels; the latter being in close connection with the problem of GHG Effect).

In spite of this consensus, it can be seen [17,18] that the actual situation of energy in the different countries presents a wide variety of cases, both in quantity (consumption per person) and quality (types of energy used). Few common points can be marked, notably the dependence of transport on oil products.

It is worth pointing out that the role technology could play in coping with the energy problems, as was pointed out in the EU "Green paper: Towards a European strategy for the security of energy supply" [19] where a complete analysis was attempted in order to meet the objective of the title without forgetting the restrictions in CO_2 emissions. At the end of its Executive Summary was written:

"Every form of technological progress will help to reinforce the impact of this outline energy strategy". However, the technological analysis was rather conventional, with two controversial poles (coal and nuclear) and a clear quest for RE and Natural Gas (NG), but RE were not foreseen in the way they have evolved. On the side of the demand, emphasis was put in energy savings in Housing and Transportation, without including any reference to HC or EV. In 2000, those inventions were considered out of the mainstream in terrestrial mobility.

It must be noted that in many countries, notably Spain, the strategy outlined in the Green Paper was clearly followed. In 2000, electricity generation by NG and RE was very limited, and it underwent an impressive development in one decade, reaching very high values at the end of 2010 [20] after an interesting deployment [21-28] including unanticipated effects [29]. In 2010, 23% was generated by Gas-fired combined cycles (GCC), and 22% by RE (wind power being the biggest contributor with 16%).

In the Technical Annex of the "Green Paper" it was again underlined the importance of technology, with the following strong statement: "Energy technology will be critical in meeting the needs of current and future generations" but at the same time there was a caveat about the price to pay for such development: "In the energy field, technological change does not come cheap", and some advice was given to promote the development of RE by state funding or feed-in tariffs.

In the last ten years, the technology scenario has changed. Looking ahead to the next decade, it is seen that the portfolio of emerging technologies has widened quite a lot, and some of them are being proposed as new tools for satisfying human demands in very traditional sectors,

as on-ground mobility, where HC [30-34] and EV [35-40] have appeared as contenders versus conventional ICE. Additionally, new electric grid devices have opened the way for the so called Smart Grids [41], which will use internet and telecommunication technologies, so helping to optimize the total electric system in a country or a macro-region [42,43].

Most of these technologies were not in the picture at the turn of the century. A new driving force was needed to refresh the interest for new energy technologies, and this force was the Kyoto Protocol. Many scientists and some governments (notably, the Bush's Administration in the USA) considered that restrictions per se (the Kyoto recipe) as a blunt answer to the problem, and pointed out that there were technologies to be analyzed or revisited to find a better solution to the GW problem. To some extent, the IPHE (30), and Nuclear Generation 4 (and the Generation 4 Forum [44]) were also a product of this situation after Kyoto. Above all, any low-carbon energy source or energy technology was considered a priority for restructuring the Energy sector, mainly RE [45] and Carbon (CO_2, properly speaking) Capture and Sequestration (CCS). Its most visible initiative is the Carbon Sequestration Leadership Forum [46].

Although most of these technologies are evolving very slowly, which is a main part of the problem that must be solved, it has become clear that a big change in the Energy sector is possible. However, the stiffness of the industry and the markets, and the lack of maturity and economic viability of the new technologies are keeping the status quo. In fact, the most relevant reaction to energy needs in the USA nowadays is unconventional gas (mainly shale gas). It is true that it is not conventional in a strict sense, but it is NG and can be used as such.

NG is going to be at the very center of the ongoing evolution in the Energy sector [47,48], and this fact includes both short-term initiatives (as the shale gas [49, 50], already running) and long term ones, as methane clathrates [51-55].

Interplay between Electricity and Hydrocarbons

Although a few countries have started earlier, the global move towards GCC exploded by the end of the last century. Reasons for such extended deployment can be found in some facts that must be kept in mind, because they are rather unique:

- NG global availability (although not cheap, indeed; but cheaper than oil in energy terms).
- Small specific cost investment (the lowest within power plants fully manageable and reliable).
- Short construction times (taking advantage of prefabrication of the main components, particularly the gas turbine).

- Operation flexibility (within some margins depending on the project). It is worth citing that in Spain, where a GCC capacity of 24,000 MW has been built in less than 12 years, these units act as back-up power for Wind parks and Photovoltaics.

The combined factor of time and money has been decisive for such fast deployment of these power plants, and it is not easy for any other thermal generation technology (nuclear, coal, oil) to compete with gas in an open market. **Table 1** presents a concept that can help explain the current evolution of generation capacity in liberalized markets. It is the "investment burden", expressed as the specific investment times the construction length. The smaller the "burden", the simpler the decision to invest. This concept has been coined in this work, and gathers a lot of qualitative information on the subject of time value of money.

A different situation is found in centrally planned economies, where coal or nuclear can receive a special deal as long-term programs, in order to guarantee better price stability in the long term. This is the case of China, particularly for coal [17,18]. In fact, coal has undergone the fastest growing rate of all energy products in these years of the new millennium, and China was the explanation. However, such long-term construction programs are not suitable for a free market, where gas power plants will very likely preserve its position of dominance [48].

3. Challenges for Energy Corporations

Commercial corporations are driven by profit. This does not mean short-term profit only, although any corporation will try and avoid short-term losses so as not to start a phase with so many negative features that could seem of going broke. In fact, a second fundamental objective of corporations is long-term viability, which is the guarantee to make profits to satisfy the corporate plan. Many corporations have specialized departments in Business Development and others rely on external consultants, but the final decisions on the future are obviously taken in

Table 1. "Investment burden" evaluating the adaptability of different power plants to a free market on the basis of construction time and specific investment cost.

Plant type	Const. time (yr)	Spec. invest. ($/W)	Investm. burden
GCC	1.5	0.75	1.13
Wind power	1.0	1.2	1.2
Photovoltaic	0.5	4.0	2.0
Coal	3.0	1.5	4.5
Solar thermal	2.5	5.0	12.5
Nuclear	8.0	4.0	32.0

the very core of the corporation, including considerations on environmental factors, such as industrial ecology [56-69].

Those strategic lines must take into account the potential evolution of all aspects of life and technology that can have an influence on the market or sector where the corporation works. They can be considered as the boundary conditions of the problem, and they can be different in nature: social, financial, legal, technical and so on. And this is the point where a substantial difference is found between corporations in general and corporations working in the energy industry, mainly in the domain of primary energy and electricity generation: the difference is the strong influence of energy policy objectives in the definition of the sector, including the case of liberalized markets.

No other economic field is so dependent on current international or global decisions as energy industry, which is (or will be) totally affected by decisions as the Kyoto Protocol, or the EU directive 2009/28 usually known as 20/20/20, already cited. A new situation appears now, with liberalized but regulated sectors, which must comply with global objectives. One by one, corporations are not obliged by those decisions, but governments must do something to meet the objectives, and the first reaction (very simplistic, in some cases) is the so-called "indicative planning", where some types of investments receive subsidies and/or feed-in-tariffs, to stimulate some actions reducing CO_2 emissions and/or increasing RE consumption.

Such a new framework will represent a quantitative change, but this is only part of the problem. It can be anticipated that a qualitative change, associated with emerging technologies, will still have a major impact in the energy corporations. The technology change will not only affect the Electric Sector, as has happened in the previous decade with GCC and Renewable Energy Sources (RES), but the entire energy system, because it will affect automobile transportation, connecting it with the rest of the energy system, particularly electricity generation and distribution. It could be such a deep change, that corporations would have to transform themselves to adapt to a new situation, coming from different technology cultures, and moving to a new one. Technology evolution alone would likely not be so powerful as to motivate a qualitative transformation of the system, but it will be pushed ahead by Sustainable Development decisions.

The impact of this transformation on a given energy corporation cannot be addressed as an academic exercise or a research subject, and it falls into the domain of consultant strategies and the like, which must be run in a very confidential way.

The general impact on the corporations of the energy

sector is indeed a general concern that can and must be studied as an academic subject to try and find recommendations for obtaining the best outcome from the changes ahead. The empathy between the macro level of international and decisions and the micro level of corporation daily life is a fundamental point. Paving the way in a well-understood framework is much less risky than reacting against unexpected facts.

4. The Future of Energy: Forecasting-Backcasting

CC is considered as the first global challenge in human history, needing a global answer based on soundly established science, which points out the need of limiting the atmospheric contents of GHG [15]. A first reaction (the Kyoto spirit) was to establish emissions restrictions, but in a few years a deeper change has appeared, based on technology. The global challenge was clearly described by the IPCC. Although some critics have disagreed with these very long-term weather predictions, there is a rather general consensus on the need not to trespass a threshold of CO_2-equivalent contents in the atmosphere, be it 450 ppm, 500 ppm, or a similar figure. Enforcing this limitation by consumption restriction was something difficult to accept, in spite of said Kyoto spirit. Nevertheless, the situation is becoming totally different because of technology evolution, with two main factors, which can give a global answer to that global challenge:

- Low Carbon technologies, with three main routes (for the moment?):
  ◦ Renewable Energy Sources, notably for electricity generationp;
  ◦ CO_2 capture and sequestration;
  ◦ Nuclear technologies (Fission and Fusion, which present very different problems, different energy potential and very different time span to maturity).
- Technology evolution in automobiles, mainly on Green cars, which include a set of possibilities as:
  ◦ Biofuels [70];
  ◦ Gas to liquid processes;
  ◦ Hydrogen cars;
  ◦ Electric vehicles.

Most of the topics are quite connected with the hydrocarbon industry. RES need a back-up, which is mainly provided by gas power plants. CCS could play a role in shaping the future around CO_2, which will directly or indirectly affect also hydrocarbon consumption. For instance, a sizeable sequestration of CO_2 produced in coal-fired power plants would alleviate the pressure on CO_2 produced in transportation.

Cars are currently powered by petroleum products, and this is the field where a deeper change could take place. As fuel cells and new batteries emerge as industrial components, it seems that conventional fuels will have to compete with kWh or with H_2 production process.

Is it by chance that this technology evolution has gained momentum when it seemed it was needed? Of course, not. Some of the technologies were known for years, but they could not find any niche to compete. Technology evolution has been boosted by concerns on CC, and a sort of feedback loop has been closed.

Methodology to Study the Challenge. Forecasting

Forecasting can give us a picture (or a movie) of the future, starting from the current situation and projecting the inherent capabilities of each line towards the future. This technique has been widely used in many fields and with many purposes, particularly for anticipating the demand of energy, but it can not be considered just a statistical exercise. It is true that a good statistical model helps obtain better results, but the main requirement is to make a good estimate of the parameters featuring the statistical model. Adjustments to previous phases are a way to calculate these parameters, but in this case it does not seem applicable, because of the quantitative and qualitative changes in the behavior of the system.

Some lines of research present a threshold, or a breakthrough, and results are almost negligible if the threshold is not exceeded. This is what happens with Nuclear Fusion, which needs to reach ignition to become an actual promise. This threshold seems higher than anticipated, and the outcome is that Fusion seems to be always 40 years ahead. Forecasting in this case is extremely difficult, and Fusion does not fit in the picture for the moment. The rest of Low Carbon technologies and New Transportation technologies can fit into the forecasting without major difficulties, with the exception of social opposition to some of them, usually for environmental and safety reasons. This point mainly applies to Nuclear Fission and CCS, but it can also affect some RES as off-shore windmills and above all, Biofuels (or Biomass in general). A clear example of failed forecasting was Spain's Plan of RES Fostering (1999-2008), which put a lot of emphasis on electricity generation with biomass, and there was a severe failure in that field.

Backcasting has complemented Forecasting in recent years [63,64,71-73], to improve the tools for shaping the future in order to achieve a sought situation. This method has been activated to some extent by the CO_2 problem, because the goal was foreseen, $i.e.$, the level of CO_2 concentration in the atmosphere that would be acceptable, which could be taken as a goal. Backcasting could identify potential roads to reach that goal, for instance, by sharing the total emissions among countries, or among technologies, or both.

Forecasting is nonetheless more familiar to everyone,

and seems simpler to apply, although many predictions have gone totally wrong. Even so, it seems mandatory to carry out an exercise of forecasting (or prognosis) in any paper probing the future. The summary of that exercise is presented in **Table 2**. The four technologies previously considered in the field of automobile powering are assessed versus some fundamental criteria, chosen according to an extensive literature review [70,74-82], plus some expert judgment from the authors. The last column presents the average value of the appraisal of said technology against those criteria. The value is expressed in a black-grey-white scale. Darker color in a cell means poorer appraisal.

Criteria applied to **Table 2** correspond to the following concepts, considered for mid and long terms (up to mid-century):

1) Primary sources;
2) Conversion technologies;
3) Suitability to be stored in a car;
4) Technical marketability (including investments, both specific and global ones);
5) Public acceptance;
6) Actual external limitations (as materials for batteries);
7) Positive effect in CO_2 emission reduction.

Last column is a qualitative average.

The judgment underlying **Table 2** and the way to present it, only corresponds to the authors, but it has been elaborated after considering very many pieces of technical information. Appraisals of **Table 2** are arguable, but they indicate in a qualitative manner the inherent value of each technology.

This value should be the basis to mobilize a change in the automobile sector towards Sustainable Development, and identify the places that could convey stronger difficulties. In the table, an average evaluation has been given in last column, but there are alternative methodologies to define the final appraisal. For instance, multiplication instead of addition puts more emphasis on the weakness of a criterion evaluation, and gives clearer classification. Similar result can be obtained using, as global evaluation, the worst appraisal in any criteria, for a given technology. This is the principle of the weakness of a chain, which corresponds to the weakest link. If the difficulties associated to a criterion are not overcome, the full deployment cannot take place. In any case, for both ways of considering the appraisal, the global result of the exercise is the same, and the priority order given below:

1) Electric Vehicles;
2) Biofuels;
3) Gas to Liquid;
4) Hydrogen Cars;

This appraisal is an integrated view, but forecasting also needs to take into account the time-dependent evolution, and the interferences that can appear among different elements of the scenario, including interferences among technologies themselves. It is obvious that all technologies will not be developed at the same speed, and only one or two will be deployed up to commercial level, and this must also be part of this analysis.

Not all forecasting studies on the different technologies identify the EV as the most promising technology to reduce oil dependence and to reshape the transportation sector to a cleaner activity. Some studies done or supported by the USA Department of Energy [83] do not give too much credit to the deployment of EV and assume that Biofuels will be the natural substitute for petroleum products.

Biofuels [65,70] will mainly stem from sugar crane for making ethanol in a first stage, followed by corn for the same objective, and soy for making diesel, with a long term quest for cellulosic biofuels. One of the reasons for adopting that choice is that it requires the minimum transformation of the infrastructure of the whole system (related to Light Duty Vehicles, which will represent an inventory of 300 million cars in the USA by 2035). Indeed, gasoline engines have already some flexibility to accept small percentages of bio-ethanol, and full flex-fuel engines will run on E85 (85% of ethanol).

It is obvious that the "2011 Annual Energy Outlook" [83] does not make a bid for technology transformation, and relies completely on hydrocarbons and similar chemicals. It was already so in the previous reports, where the central issues were dominated by NG either from Alaska or from shales. The last report puts more emphasis on Biofuels, in spite of the problems they are already causing in Brazil and elsewhere [84-87].

The overall picture of energy values related to this problem seems to indicate that a "Biofuel industry" can be developed worldwide without too much distortion of the current agriculture activities; but a deeper analysis points to deeper problems. It is estimated that the total solar energy captured by living beings of all kind amounts to 3000 EJ/year. It is about 0.08% of the total solar radiation impinging on Earth. It is six times as large as the human demand of primary energy, which is close to 10 billion tons of oil equivalent per year, which is about 420

Table 2. Forecasting appraisal of technologies with potential of change in the automobile sector.

Tech & Criteria	1	2	3	4	5	6	7	avg
Biofuels	3	3	1	1	2	4	2	2
Gas to Liquid	2	5	1	1	2	3	4	3
Hydrogen Car	2	3	5	4	4	3	2	4
Electric Vehicles	1	2	3	3	1	4	1	1

EJ/year (without accounting for primary biomass consumed as an energy good, which is about 40 EJ/year additional). Oil products amount to a little less than 150 EJ/year, which is about 5% of the energy capture of biomass. So, the disturbance seems acceptable, but the full picture contains other elements which must be accounted for.

The energy contents of the biomass for our food are one twentieth (1/20) of the energy we consume for other uses, *i.e.*, 23 EJ/year. If harvesting and commercialization efficiencies are taken into account, the raw materials of our food chain amount to 50 EJ/year approximately. This value must be taken as the real reference for any man-made intrusion in the biomass world with the goal of producing new types of commercial products (some genetically modified cellulosic plants, for instance) or already known products (sugar cane, soy beans, palm oil) at a scale much larger than the total current effort for our food. In fact, if one third of the oil products should be replaced by biomass products, it would be necessary to double, at least, the total activity devoted today to agriculture. Of course, this is not an impossible quest, but some warnings should be expressed on the potential distortion caused in agriculture, and the potential impact caused in the environment (particularly, in some privileged ecosystems of high biological vitality).

Precautions on biofuels are mainly rooted in the natural roots. From the complementary technology point of view, some important efforts have been made, and high quality processes and products (www.nesteoil.com) are already available but not at a global scale. Technology evolution in this field has been boosted by suitable policies implemented in several well-developed countries. The EU directive on RE, for example, requires that they should account for at least 10% of the energy used in traffic and transport by 2020, National legislation in Finland has targeted 20% content by 2020, and legislation in the US will require 20% content by 2022. All these figures originally meant biofuels, although other alternatives such as electric cars charged with electricity from RES could also be accepted now to meet that goal.

Nevertheless, it should be kept in mind that total ethanol production is about 700 thousand barrels a day (673.5 thousand/day in 2004 [70]) and biodiesel production is close to 50 thousand (39.6 thousand in 2004, but it is increasing). These figures are much lower than the final consumption of gasoline, which is about 20 million barrels a day, and diesel, which is slightly higher, 21 million barrels/day. Although it is claimed by biomass proponents that they would require the smallest modification in automotive infrastructure as compared with other alternatives, Biomass could produce a huge perturbation in the downstream oil industry, although the main upheaval will likely happen in agriculture, forestry and the cycles

of natural nutrients.

5. Bridging the Gap between Now and the Sustainable Future. The Role of Technology

It has been pointed out by many authors that we have in the history of humanity, a first global challenge, which is the fight against GW. Indeed, other challenges affecting billions of people are very important, as the fight against Hunger and Poverty (Objectives of Millennium, United Nations) but they are localized in space, and should be considered specific challenges, strongly related to the political arena. It has also been said that several lines of research have been proposed for developing new systems and processes that could contribute to meet the goals of that challenge.

A backcasting exercise can shed some additional light on this problem, including in this case the electricity part of the picture, which will become much more intermixed with hydrocarbons than ever before.

Figure 1 shows the final goal in the left-hand-side of the picture. The goal is the origin of the plot, which develops towards the right-hand-side, representing going back-wards in time, until the current day.

Three ways are identified as potential causes of GHG emission reduction, namely;

Technology evolution & revolution;

Limiting quotas per country;

Changes in habits & working conditions.

Last line, "changes in habits & working conditions" is outside the scope of this paper. It sounds like science fiction, but some of those things are likely to happen. The second line is also out of our domain, although it was the pathway taken in the Kyoto Protocol. It has had many problems for confirmation by some important countries such as the USA and China. It is not a simple story, because an equal quota for all persons will not be fair, if weather conditions, geography, population densities and other factors are not taken into account. The Kyoto spirit of fighting the problem by establishing restrictions was a simple and prompt response, and not much more was possible in the time span available to prepare proposals. Much more time was needed to set up alternatives to that rough reaction, but in a few years the scenario changed a lot, and many technological proposals represented very sound options to really cope with the challenge.

To a large extent, technology had created the problem, and to a large extent, technology is ready to solve it (by the way, neither creating confrontation among countries, nor anxiety on people. Individuals just can change habits by their own will, not by force).

The technology domain points out that there are a set

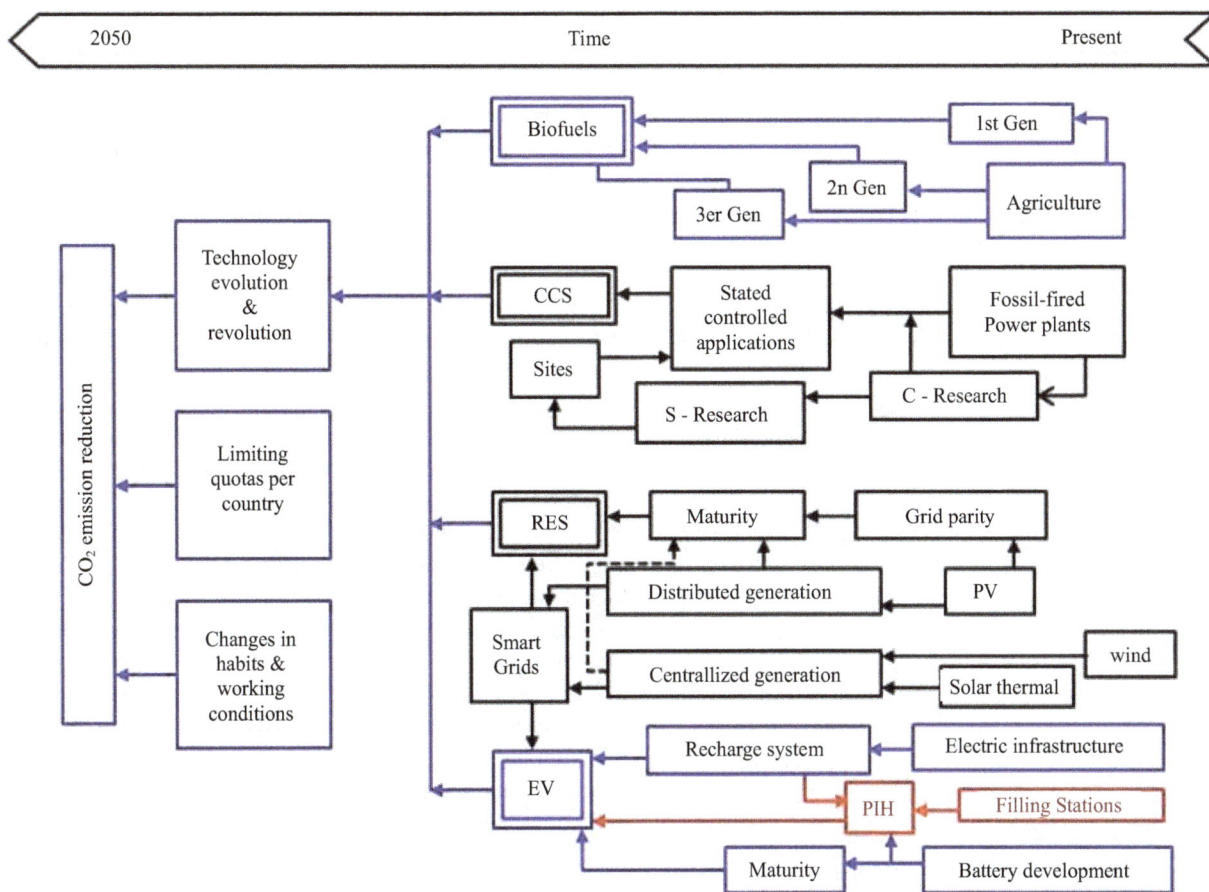

Figure 1. Backcasting exercise starting from the goal of reducing significantly CO_2 emissions by 2050, going backwards in time until present.

of technologies which can produce a significant reduction of CO_2 emissions, but they have to overcome some problems (schematically indicated in the figure). A detailed exercise on that figure should include the expected reductions of emissions under different hypotheses of technology development and primary energy sources availability, but in our general analysis the figure is limited to qualitative descriptions.

It is worth commenting on the special case of CCS. Sites for deep underground CO_2 storage are the critical point in this field. Early experiences in The Netherlands, Norway and Germany, where public opposition promptly sprung, are highly indicative. Some of the opponents seem to act under the "lake Nyox" syndrome [88,89], remembering an eruption of gigantic CO_2 bubbles from the Nyox volcanic lake (Cameroon), in 1986, with 1746 casualties. A much smaller catastrophe had happened six years before in lake Manoun, also in Cameroon. Nowadays, those lakes are under surveillance for early detection and emergency declaration, but that risk was poorly known before those unexpected emissions. Of course, the small but existing volcanic activity of those lakes is in the very root of the emissions. The situation would be absolutely different in a depleted reservoir, replenished with CO_2, or in a saline dome that has not suffered any geologic disturbance for several hundred million years. Enhanced Oil Recovery projects that considered CSS goals could be added in to this list.

In any case, the special features of CCS led us to advise its application in a totally transparent way, under regulation and control of public authorities, following the example of the Nuclear Regulatory Commission [90]. It is not clear at this moment to which extent CCS will be needed to meet the requirements of the fight against GW, but it is a technique that can work with similar guarantees to any other modern technique. From this viewpoint, CCS must be developed and the sites must be characterized, for the sake of having a back-up solution to the rest of low-carbon technologies under consideration. Hence, it seems an authority-controlled or at least strongly regulated activity is required, much more than a field where private initiative could be carried out by corporations.

On the contrary, RES is a field fully open to private initiative, although public-funded programs of R&D seem absolutely necessary for advancing in the learning curve of each technology. A similar advancement is needed in EV, which can be considered complementary to RE.

The whole electric system will be deeply affected by the deployment of EV. The complete recharge system will still be denser than the current net of filling stations, including privately owned individual recharge posts. Electricity infrastructure will expand quite a lot, and electricity consumption will undergo a dramatic increase, whilst gasoline and diesel will undergo a significant decline for ICE vehicles. Of course, this tremendous upheaval, which is expected by many as a blessing, can evolve into a crisis with many negative consequences. This change, the 3rd Industrial Revolution for some prophets, has to be properly managed by government and international authorities so as not to transform a great opportunity into a disaster.

Energy corporations have to prepare themselves for an efficient adaptation to such a "revolution". Of course, technology evolution cannot be so fast as to hamper said adaptation, but lazy and badly prepared corporations will face stronger problems for survival than well equipped corporations. The equipment should include activities of technology surveillance and, above all, internal analysis to identify weaknesses, strengths, risks and opportunities. Those analyses should be done for any conceivable scenario, as the ones exposed in many publications (notably WEO [18]) of forecasting and backcasting. Governments and international bodies will establish policies according to the political and environmental pressure, (as the Directive "20/20/20" 2009/28/CE of the EU), and the reality is that technology is already here to face that challenge, although the specific technologies dominating the market in a given phase cannot be anticipated in advance. Nevertheless, some additional investigation can be launched into the future.

6. A Choice of Strategic Lines

Corporations will have to use the conventional tools for corporate planning in order to prepare for adaptation established in this field, including consideration of market inertia and producers' reactions against threats to the established industry.

The current car industry and the ground transportation structure fully relies on the conventional petroleum industry, and the very many appealing factors of this industry have been the main pillar for large scale social and economical development for more than one century, at least in the industrial and post-industrial economies. A change in this sector (petroleum + ground mobility) could entail deep perturbations in social and economic welfare, if the system does not evolve smoothly from the current situation to a carbon-free economy with a much lower direct dependence on oil (as we approach the oil-peak) and more efficient use of natural resources (particularly methane, including clathrates, permafrost and other unconventional gas).

As already stated in previous sections, biofuels is an important field of interest. Bio-ethanol and bio-diesel are already embodied in many commercial products for transportation. A controversy appeared from the so-called First Generation Biofuels, very closely connected with standard agriculture products. The dream was to develop a Bio-refinery, but the critical problem for this lies in its roots, i.e., the vegetable world [70,85,87]. The current level of hydrocarbon consumption is about 8 times as large as consumption of traditional biomass for modest energy application, notably in developing countries. It is very hard to identify how to evolve from standard agriculture and forestry to bio-refineries. In most of the cases (as Finland's wood industry) the added value of other applications (furniture, paper industry) is even higher than the energy value.

Global (well to wheel) analysis of different technology scenarios point out that EV [75-77] with a relevant role of plug-in hybrids as a long-lasting intermediate step, will offer the highest efficiency, and will minimize contamination in populated areas. The plug-in hybrid approach seems to have the better characteristic for a smooth, although fast evolution in automobile for reducing CO_2. As an illustrative case for this quest, we can consider that a new vehicle of medium class will contribute an average emission of 125 g/km. An electric car would consume around 0.14 kWh/km. This value will rise to 0.17 if the efficiency in charging the battery is accounted for. In Spain, an average value of CO_2 emission in the electric generation system is slightly less than 400 g/kWh; which means less than 70 g/km. The situation will improve if coal-fired power plants decline in activity (they generate 1 kg of CO_2 per kWh) and RES continue to increase their generation. As a reference, GCC produce between 350 and 400 grams of CO_2 per kWh. With a gasoline engine, a car can have CO_2 emissions between 110 and 140 g/km. Should gas be burnt in a Combined Cycle, an electric car fed with that electricity would have (indirect) CO_2 emissons between 55 and 70 g/km.

The outcome of this change will be a cleaner environment in cities and much lower CO_2 emissions in the planet, but the economic side cannot be ignored, because RES need for subsidies or feed-in tariffs (or both) and such a situation is not "sustainable" at all. On the contrary, despite some side effect generated by OPEC, hydrocarbons constitute a perfectly-running market, supporting a fiscal burden as no other product, at least in Europe.

The Hydrocarbon industry will face a change in this "energy and technology revolution", because there will be a decrease in direct consumption in vehicles, and an increase in electricity generation in high-efficiency plants, as combined "Brayton + Rankine" cycles and new free-pinning gas turbines. Battery recharge will be as popular as gasoline filling, and both services could be done in the same place (for instance, current filling stations abiding by some safety rules [91-94], will include battery recharge and replacement [42,95]. This scenario would convey a profound transformation in many industries, and it will likely be dramatic in energy infrastructures, where a true interpenetration will likely appear connecting gasoline and diesel distribution to electricity, with "dual filling stations" providing liquid hydrocarbons and battery replacing. Customers will have to pay for the difference in charge between the replaced battery and the substitute, and for the difference in quality, because of performance deterioration with time (which will have to be evaluated by a rapid measurement of relevant variables in each battery).

Hydrocarbon corporations will have to follow an "adaptation mechanism" to evolve at the required speed. Of course, very long-distance and autonomous transportation units, as ships and airplanes, will remain oil-dependent, directly; but a sizeable fraction of the hydrocarbons will be consumed through electricity. Moreover, the formerly cited scenario of "dual filling stations" distributed for all the geography of an advanced country will have to be deployed before the actual deployment of EV.

Of course, hydrocarbon corporations will have to define their "adaptation mechanisms" which can be commercial (diversification of services) or technological (participation in the new energy conversion mechanisms). Merging with complementary corporations could also be a right tool. In any case, it seems advisable to start an exploratory phase in order to have a deeper insight into the future.

The uncertainty level is still very high about the impact and effects of those futuristic scenarios, but the warnings on CC, the problems with petroleum availability at a reasonable price, and the appearance of new technologies seem to aim at the same target of evolving towards sustainability. In that quest, actions and reactions of hydrocarbon corporations will be critical; and they are still to be identified.

Business as usual is out of question in the oil industry, but a last point must be taken into account; there are also new roads to be explored for increasing oil reserves, particularly shale and sand oil. Heavy oil and ultra-deep off-shore oil have been in the portfolio of corporations for a long time, but they must be accounted for in long-term policies with the perspective of special strategic assets. Some industry experts such as Dr. Kazemi [96] estimate that about one trillion barrels of oil is waiting to be discovered, in addition to the similar amount that is

"proven" reserves waiting to be produced. However, these sources would be more expensive than conventional ones, and would likely require a modified refinery processes. They must be considered, but they will likely remain as resources, for being exploited in a long distance future. This is also the case for gas-hydrates, that could have estimates of original gas-in-place exceeding 10,000 trillion m^3 [97-100]; with estimated resources over one hundred trillion barrels of oil equivalent [100]. Note that current oil consumption is 85 million barrels of oil per day (mmbpd) and a peak could be reached in 12 or 15 years at 100 mmbpd. The IPCC and other forecasters point out the necessity to reduce that value to less than 50 mmbpd by 2050 for limiting GW to acceptable levels.

We know now that technology can drive us to that goal, and a sort of competition has been outlined between Biofuels and EV. Pros and Cons of each option have been briefly commented (particularly in relation to DOE's "Energy Outlook" 2011), and have been treated in the bibliography [64-67,73]. It seems therefore that a choice has to be made between those options, and that situation is not desirable for any corporation, because the general framework will be decided by political opinions, presumable at a high international level, and this situation will not allow a corporation to adopt long lasting strategies, because of fears of policy changes.

At this foreseeable crossroads, the following "eclectic proposal" is introduced in this paper as a smother and more controlled entry into this period of competition and possible turmoil. The proposal can give time for better founded and cheaper technologies (for instance, for electricity generation from RES); can relax the anxiety of the "recharge syndrome" created by an autonomy range in the EV, which is much shorter than the standard autonomies in current cars. The "eclectic proposal" is to select the "plug-in hybrid with flexible fuel" as the dominant car in the coming decades. The term "flexible fuels" [101] means that they can run with a range of mixtures between oil products (either gasoline or diesel) and the corresponding biofuels (either alcohols or bio-diesels). Some relevant reviews on that topic are found in the literature [102-112]. In that way, the current effort on Biofuels [113-119] will not suffer a sudden halt, and the promising world of EV will have the possibility to mature without big expenses in the short term. Indeed, those technologies that are presented as confronting ones can be complementary.

7. An Eclectic Summary

At the end of the 20th Century, Sustainable Development [4] was proposed as a new paradigm to guide general policies, and within that context CC was recognized as a fundamental global problem, connected with the human emissions of CO_2 [6-9]. First reactions were based solely on establishing emission quotas; but early in the 21st Century some important technology proposals were presented as new ways for creating a Sustainable Energy Sector. This was a profound change of philosophy, and will have a profound impact on policy making in Energy, which in turn will produce a deep change in the Energy industry. Nevertheless, non-anthropogenic CO_2 emission, mainly those released from volcanoes, should be strongly considered as boundary conditions in the resolution of the CC problem. Theories on this subject [120-131] point out the importance of those natural catastrophes on weather variations. Those theories are supported by observations of modern eruptions [132-142] and by indirect evaluations of volcanic activity in the past [143-157].

Of course, such a big transformation will need a previous development of the required technologies, ranging from RES (including Biofuels) to EV, as well as huge investments, notably in the deployment of recharge infrastructure and, even more, in vehicle manufacturers and component makers. Even hydrocarbon consumption will also be partially oriented to electricity generation, because the global efficiency for car transportation will be better, if batteries finally achieve a level of maturity able to support such a substantial change [81-85]. In this scenario with so many questions marks, which will not reverse to past structures, the pathway of the PiHFF seems to be the most likely one, and the most convenient for all stakeholders in the energy industry.

A selection of keywords and corresponding challenges can be made from the previous sections, but the main conclusion is the technology capability to reshape the future of Energy, including an dramatic reduction of CO_2 even if hydrocarbons continue to play a very prominent role, because the general efficiency will increase by a factor of two in transportation, if the EV deployment succeeds. This approach between electricity and hydrocarbons opens a lot of possibilities, and corporations from both industries should take advantage of them, overcoming the traditional communication problems between both sectors.

The macro-economy of the new situation will be defined by general policies, but this leaves a main question to be solved: how corporations can adapt to the new situation? There must be an internal coherence between macro-economy and micro-economy, and some reflections and guidance are needed on that.

From previous sections it is obvious that a review has to be made on the different set of options available to a corporation, from technology surveillance to commercial trades, trying to reconcile the main objectives of energy policy with the success of corporations. However, if PiHFF is generally adopted, the situation will be easier, in the sense that everyone can continue with their own development, and the PiHFF will go accommodating advancements in a smooth way.

There is nonetheless a domain where oil corporations will have to develop a very active and pioneering role, because they have their own network of filling stations, which should be converted to "dual stations" including infrastructures for recharging batteries, and services for replacing them.

This is an opportunity not to miss and not to fail at, but the task is not easy. The effort ahead belongs to the field of electricity distribution and consumption management, and the typical expertise of oil corporations is very modest in those fields.

Many tasks involving standards will be needed to make such a big change possible, and some previous work can be anticipated, such as making some conceptual designs of the futuristic "dual filling stations", so that an early identification of problems, issues and needs could be done. It is obvious that this activity will be highly confidential, when applied to a given corporation and a given market, but general principles and analysis, and the figures of merits to qualify the options according to the selected criteria will be of general concern, and they will be treated in the open literature, to which this paper wishes to contribute.

8. Acknowledgements

The contents of this paper fully belongs to the Ph.D. Thesis of the first author (JMM-VP) who recognizes that former studies at The Colorado School of Mines and The French Institute of Petroleum offers unique opportunities to understand the world of Energy, notably the hydrocarbon sector.

REFERENCES

[1]   Lawrence Livermore National Laboratory, "Science and Technology Review," 2011. https://str.llnl.gov

[2]   Michigan Institute of Technology, "Technology Review," 2011. www.technologyreview.com

[3]   J. Rifkin, "The Third Industrial Revolution: How Lateral Power Is Transforming Energy, the Economy, and the World," Palgrave Macmillan, Hampshire, 2011.

[4]   Brundtland, "Our Common Future," Oxford University Press, New York, 1987.

[5]   W. Hafele, "Energy in a Finite World. Paths to a Sustainable Future. Energy in a Finite World. A Global System Analysis," Ballinger Pu. Co., Pensacola, 1981.

[6]   T. J. Crowley and R. A. Berner, "CO_2 and Climate Change," *Science*, Vol. 292, No. 5518, 2001, pp. 870-872. doi:10.1126/science.1061664

[7]   T. P. Barnett, D. W. Pierce and R. Schnur, "Detection of Anthropogenic Climate Change in the World's Oceans,"

Science, Vol. 292, No. 5515, 2001, pp. 270-274. doi:10.1126/science.1058304

[8] M. Allen, "Constraints on Future Changes in Climate and the Hydrologic Cycle," *Nature*, Vol. 419, 2002, pp. 224-227. doi:10.1038/nature01092

[9] L. Kump, "Reducing Uncertainty about Carbon Dioxide as a Climate Driver," *Nature Insight*, Vol. 419, No. 6903, 2002, pp. 188-190. doi:10.1038/nature01087

[10] S. F. B. Tett, J. F. B. Mitchell, D. E. Parker and M. R. Allen, "Human Influence on the Atmospheric Vertical Temperature Structure: Detection and Observations," *Science*, Vol. 247, 1996, pp. 1170-1173. doi:10.1126/science.274.5290.1170

[11] G. Marchuk, K. Kondratyev and V. Kozoderov, "Earth Radiation Budget," Nauka Pub, FSU, 1990.

[12] H. B. Dulal, G. Brodnig and C. G. Onoriose, "Climate Change Mitigation in the Transport Sector through Urban Planning: A Review," *Habitat International*, Vol. 35, No. 3, 2011, pp. 494-500. doi:10.1016/j.habitatint.2011.02.001

[13] F. Grazi and J. C. J. M. van den Bergh, "Spatial Organization, Transport, and Climate Change: Comparing Instruments of Spatial Planning and Policy," *Ecological Economics*, Vol. 67, No. 4, 2008, pp. 630-639. doi:10.1016/j.ecolecon.2008.01.014

[14] H. Huang, M. von Lampe and F. van Tongeren, "Climate Change and Trade in Agriculture," *Food Policy*, Vol. 36, 2011, pp. S9-S13. doi:10.1016/j.foodpol.2010.10.008

[15] Intergovernmental Panel on Climate Change, 2007. www.ipcc.ch Fourth Assessment Report

[16] D. Keles, D. Most and W. Fichtner, "The Development of the German Energy Market until 2030—A Critical Survey of Related Scenarios," *Energy Policy*, Vol. 39, No. 2, 2011, pp. 812-825. doi:10.1016/j.enpol.2010.10.055

[17] BP Statistical Review of World Energy, 2011. http://www.bp.com/sectionbodycopy.do?categoryId=7500&contentId=7068481

[18] World Energy Outlook, 2009. www.worldenergyoutlook.com

[19] European Union, "Green Paper: Towards a European Strategy for the Security of Energy Supply," 2000.

[20] Red Eléctrica de España. http://www.ree.es/sistema_electrico/pdf/infosis/sintesis_REE_2010.pdf

[21] European Commission, "EUROSTAT: Energy Yearly Statistics," Office for Official Publications of the EU, Luxembourg, 2008.

[22] P. del Río and G. Unruh, "Overcoming the Lock-Out of Renewable Energy Technologies in Spain: The Cases of Wind and Solar Electricity," *Renewable and Sustainable Energy Reviews*, Vol. 11, No. 7, 2007, pp. 1498-1513. doi:10.1016/j.rser.2005.12.003

[23] G. M. Montes, E. P. Martín and J. O. García, "The Current Situation of Wind Energy in Spain," *Renewable and Sustainable Energy Reviews*, Vol. 11, No. 3, 2007, pp. 467-481. doi:10.1016/j.rser.2005.03.002

[24] Y. Perez and F. J. Ramos-Real, "The Public Promotion of Wind Energy in Spain from the Transaction Costs Perspective 1986-2007," *Renewable and Sustainable Energy Reviews*, Vol. 13, No. 5, 2009, pp. 1058-1066. doi:10.1016/j.rser.2008.03.010

[25] G. M. Montes, M. M. S. López, M. C. R. Gámez and A. M. Ondina, "An Overview of Renewable Energy in Spain. The Small Hydro-Power Case," *Renewable and Sustainable Energy Reviews*, Vol. 9, No. 5, 2005, pp. 521-534. doi:10.1016/j.rser.2004.05.008

[26] F. Hernández, M. A. Gual, P. Del Río and A. Caparrós, "Energy Sustainability and Global Warming in Spain," *Energy Policy*, Vol. 32, No. 3, 2004, pp. 383-394. doi:10.1016/S0301-4215(02)00308-7

[27] F. Foidart, J. Oliver-Solá, C. M. Gasol, X. Gabarrell and J. Rieradevall, "How Important Are Current Energy Mix Choices on Future Sustainability? Case Study: Belgium and Spain—Projections towards 2020-2030," *Energy Policy*, Vol. 38, No. 9, 2010, pp. 5028-5037. doi:10.1016/j.enpol.2010.04.028

[28] C. Batle and P. Rodilla, "A Critical Assessment of the Different Approaches Aimed to Secure Electricity Generation Supply," *Energy Policy*, Vol. 38, No. 11, 2010, p. 7169. doi:10.1016/j.enpol.2010.07.039

[29] F. Moreno and J. M. Martinez-Val, "Collateral Effects of Renewable Energies Deployment in Spain: Impact on Thermal Power Plants Performance and Management," *Energy Policy*, Vol. 39, No. 10, 2011, pp. 6561-6574. doi:10.1016/j.enpol.2011.07.061

[30] International Partnership for Hydrogen and Fuel Cells in the Economy, 2011. www.iphe.net

[31] D. Keith and A. Farrell, "Rethinking Hydrogen Cars," *Science*, Vol. 301, No. 5631, 2003, pp. 315-316. doi:10.1126/science.1084294

[32] L. Barreto, A. Makihira and K. Riahi, "The Hydrogen Economy in the 21st Century: A Sustainable Development Scenario," *International Journal of Hydrogen Energy*, Vol. 28, No. 3, 2003, pp. 267-284. doi:10.1016/S0360-3199(02)00074-5

[33] H. S. Lee, K. S. Jeong and B. S. Oh, "An Experimental Study of Controlling Strategies and Drive Forces for Hydrogen Fuel Cell Hybrid Vehicles," *International Journal of Hydrogen Energy*, Vol. 28, No. 2, 2003, pp. 215-222. doi:10.1016/S0360-3199(02)00038-1

[34] S. G. Chalk and J. E. Miller, "Key Challenges and Recent Progress in Batteries, Fuel Cells, and Hydrogen Storage for Clean Energy Systems," *Journal of Power Sources*, Vol. 159, No. 1, 2006, pp. 73-80. doi:10.1016/j.jpowsour.2006.04.058

[35] H. Turton, "Sustainable Global Automobile Transport in the 21st Century: An Integrated Scenario Analysis," *Technological Forecasting & Social Change* Vol. 73, No. 6, 2006, pp. 607-629. doi:10.1016/j.techfore.2005.10.001

[36] A. Ford, "Electric Vehicle and the Electric Utility Company," *Energy Policy*, Vol. 22, No. 7, 1994, pp. 555-570. doi:10.1016/0301-4215(94)90075-2

[37] R. Cowan and S. Hulten, "Escaping Lock-In: The Case of the Electric Vehicle," *Technology Forecasting and Social Change*, Vol. 53, No. 1, 1996, pp. 61-79. doi:10.1016/0040-1625(96)00059-5

[38] G. G. Harding, "Electrical Vehicles in the Next Millennium," *Journal of Power Sources*, Vol. 78, No. 1-2, 1999, pp. 193-198. doi:10.1016/S0378-7753(99)00037-3

[39] M. Wada, "Research and Development of Electrical Vehicles for Clean Transportation," *Journal of Environmental Science*, Vol. 21, No. 6, 2009, pp. 745-749. doi:10.1016/S1001-0742(08)62335-9

[40] S. Brown, D. Pyke and P. Steenhof, "Electric Vehicles: The Role and Importance of Standards in an Emerging Market," *Energy Policy*, Vol. 38, No. 7, 2010, pp. 3797-3806. doi:10.1016/j.enpol.2010.02.059

[41] R. Webster, "Can the Electricity Distribution Network Cope with an Influx of Electric Vehicles?" *Journal of Power Sources*, Vol. 80, No. 1-2, 1999, pp. 217-225. doi:10.1016/S0378-7753(98)00262-6

[42] A. K. Srivastava, B. Annabathina and S. Kamalasadan, "The Challenges and Policy Options for Integrating Plug-in Hybrid Electric Vehicle into the Electric Grid," *The Electricity Journal*, Vol. 23, No. 3, 2010, pp. 83-91. doi:10.1016/j.tej.2010.03.004

[43] W. Kempton and S. Letendre, "Electric Vehicles as a New Source of Power for Electric Utilities," *Transportation Research*, Vol. 2, No. 3, 1997, pp. 157-175.

[44] Generation 4 Forum Website, 2011. www.gen-4.org

[45] B. Sorensen, "Renewable Energies," 2nd Edition, Academic Press, Inc., New York, 2002.

[46] Carbon Sequestration Leadership Forum Website, 2011. www.cslf.org.

[47] H. Terrell, "US Gas Reserves Estimated at Record High," *World Oil*, Vol. 232, No. 5, 2011, p. 13.

[48] K. Costello, "Going 'Long' with Natural Gas?" *The Electricity Journal*, Vol. 24, No. 5, 2011, pp. 42-49. doi:10.1016/j.tej.2011.05.005

[49] T. C. Kinnaman, "The Economic Impact of Shale Gas Extraction: A Review of Existing Studies," *Ecological Economics*, Vol. 70, No. 7, 2011, pp. 1243-1249. doi:10.1016/j.ecolecon.2011.02.005

[50] R. McIlvaine and A. James, "The Potential of Shale Gas," *World Pumps*, Vol. 7, 2010, pp. 16-18. doi:10.1016/S0262-1762(10)70195-4

[51] B. Buffet and D. Archer, "Global Inventory of Methane Clathrate: Sensitivity to Changes in the Deep Ocean," *Earth and Planetary Science Letters*, Vol. 227, 2004, pp. 185-199. doi:10.1016/j.epsl.2004.09.005

[52] A. Demirbas, "Methane Hydrates as Potential Energy Resource: Part 1—Importance, Resource and Recovery Facilities," *Energy Conversion and Management*, Vol. 51, No. 7, 2010, pp. 1547-1561. doi:10.1016/j.enconman.2010.02.013

[53] A. Demirbas, "Methane Hydrates as Potential Energy Resource: Part 2—Methane Production Processes from Gas Hydrates," *Energy Conversion and Management*, Vol. 51, No. 7, 2010, pp. 1562-1571. doi:10.1016/j.enconman.2010.02.014

[54] K. A. Kvenvolden, "Methane Hydrate—A Major Reservoir of Carbon in the Shallow Geosphere?" *Chemical Geology*, Vol. 71, No. 1-3, 1988, pp. 41-51. doi:10.1016/0009-2541(88)90104-0

[55] S. Lee and G. D. Holder, "Methane Hydrates Potential as a Future Energy Source," *Fuel Processing Technology*, Vol. 71, No. 1-3, 2001, pp. 181-186. doi:10.1016/S0378-3820No. 01)00145-X

[56] K. H. Robert, B. Schmidt-Bleek, J. A. De Larderel, G. Basile, J. L. Jansen, R. Kuehr, P. P. Thomas, *et al.*, "Strategic Sustainable Development—Selection, Design and Synergies of Applied Tools" *Journal of Cleaner Production*, Vol. 10, No. 3, 2002, pp. 197-214. doi:10.1016/S0959-6526No. 01)00061-0

[57] J. Korhonen, "Industrial Ecology in the Strategic Sustainable Development Model: Strategic Applications of Industrial Ecology," *Journal of Cleaner Production*, Vol. 12, No. 8-10, 2004, pp. 809-823. doi:10.1016/j.jclepro.2004.02.026

[58] K. H. Robèrt, "Tools and Concepts for Sustainable Development, How Do They Relate to a General Framework for Sustainable Development, and to Each Other?" *Journal of Cleaner Production*, Vol. 8, No. 3, 2000, pp. 243-254. doi:10.1016/S0959-6526No. 00)00011-1

[59] J. R. Ehrenfeld, "Industrial Ecology: Paradigm Shift or Normal Science?" *American Behavioral Scientist*, Vol. 44, No. 2, 2000, pp. 229-244.

[60] T. E. Graedel and B. R. Allenby, "Industrial Ecology," *Academy of Management Review*, Vol. 20, No. 1, 1995, pp. 1968-1975.

[61] S. Erkman, "Industrial Ecology: Historical View," *Journal of Cleaner Production*, Vol. 5, No. 1-2, 1997, pp. 1-10. doi:10.1016/S0959-6526No. 97)00003-6

[62] H. E. Daly, "Beyond Growth: The Economics of Sustainable Development," Beacon Press, Boston, 1996.

[63] K. L. Anderson, "Reconciling the Electricity Industry with Sustainable Development: Backcasting—A Strategic Alternative," *Futures*, Vol. 33, No. 7, 2001, pp. 607-623. doi:10.1016/S0016-3287No. 01)00004-0

[64] J. Kuisma, "Backcasting for Sustainable Strategies in the Energy Sector: A Case Study in FORTUM Power and Heat," The International Institute for Industrial Environmental Economics, Sweden, 2000.

[65] R. Williams, "Roles for Biomass Energy in Sustainable Development," *Industrial Ecology and Global Change*, 1994, pp. 199-228.

[66] F. Figgea and T. Hahn, "Sustainable Value Added—Measuring Corporate Contributions to Sustainability beyond Eco-Efficiency," *Ecological Economics*, Vol. 48, 2004, pp. 173-187. doi:10.1016/j.ecolecon.2003.08.005

[67] E. Heiskanen, "The Institutional Logic of Life Cycle Thinking," *Journal of Cleaner Production*, Vol. 10, No. 5, 2002, pp. 427-437. doi:10.1016/S0959-6526No. 02)00014-8

[68] N. Darnall, I. Henriques and P. Sadorsky, "Do Environmental Management Systems Improve Business Performance in an International Setting?" *Journal of International Management*, Vol. 14, No. 4, 2008, pp. 364-376. doi:10.1016/j.intman.2007.09.006

[69] R. D. Klassen and C. P. McLaughlin, "The Impact of Environmental Management on Firm Performance," *Management Science*, Vol. 42, No. 8, 1996, pp. 1199-1214. doi:10.1287/mnsc.42.8.1199

[70] L. Matzny, "Biofuels for Transport: Global Potential and Implications for Energy and Agriculture," Worldwatch Institute, Earthscan Ltd., 2007.

[71] T. Mattila and R. Antikainen, "Backcasting Sustainable Freight Transport Systems for Europe in 2050," *Energy Policy*, Vol. 39, No. 3, 2011, pp. 1241-1248. doi:10.1016/j.enpol.2010.11.051

[72] P. Moriarty and D. Honnery, "Low-Mobility: The Future of Transport," *Futures*, Vol. 40, No. 10, 2008, pp. 865-872. doi:10.1016/j.futures.2008.07.021

[73] D. Giurco, B. Cohen, E. Langham and M. Warnken, "Backcasting Energy Futures Using Industrial Ecology," *Technological Forecasting and Social Change*, Vol. 78, No. 5, 2011, pp. 797-818. doi:10.1016/j.techfore.2010.09.004

[74] P. Moriarty and D. Honnery, "The Prospects for Global Green Car Mobility," *Journal of Cleaner Production*, Vol. 16 No. 16, 2008, pp. 1717-1726. doi:10.1016/j.jclepro.2007.10.025

[75] C. E. Thomas, "Fuel Cell and Battery Electric Vehicles Compared, International," *Journal of Hydrogen Energy*, Vol. 34, No. 15, 2009, pp. 6005-6020. doi:10.1016/j.ijhydene.2009.06.003

[76] K. Ç. Bayindir, M. A. Gözüküçük and A. Teke, "A Comprehensive Overview of Hybrid Electric Vehicle: Powertrain Configurations, Powertrain Control Techniques and Electronic Control Units," *Energy Conversion and Management*, Vol. 52, No. 2, 2011, pp. 1305-1313. doi:10.1016/j.enconman.2010.09.028

[77] G. J. Offer, M. Contestabile, D. A. Howey, R. Clague and N. P. Brandon, "Techno-Economic and Behavioural Analysis of Battery Electric, Hydrogen Fuel Cell and Hybrid Vehicles in a Future Sustainable Road Transport System in the UK," *Energy Policy*, Vol. 39, 2011, pp. 1939-1950. doi:10.1016/j.enpol.2011.01.006

[78] G. Gutmann, "Hybrid Electric Vehicles and Electrochemical Storage Systems—A Technology Push–Pull Couple," *Journal of Power Sources*, Vol. 84, No. 2, 1999, pp. 275-279. doi:10.1016/S0378-7753No. 99)00328-6

[79] T. Kojimaa, T. Ishizua, T. Horibaa and M. Yoshikawa, "Development of Lithiumion Battery for Fuel Cell Hybrid Electric Vehicle Application," *Journal of Power Sources*, Vol. 189, No. 1, 2009, pp. 859-863. doi:10.1016/j.jpowsour.2008.10.082

[80] J. Van Mierlo, G. Maggetto and Ph. Lataire, "Which Energy Source for Road Transport in the Future? A Comparison of Battery, Hybrid and Fuel Cell Vehicles," *Energy Conversion and Management*, Vol. 47, No. 17, 2006, pp. 2748-2760. doi:10.1016/j.enconman.2006.02.004

[81] T. H. Bradley and A. A. Frank, "Design, Demonstrations and Sustainability Impact Assessments for Plug-in Hybrid Electric Vehicles," *Renewable and Sustainable Energy Reviews*, Vol. 13, No. 1, 2009, pp. 115-128. doi:10.1016/j.rser.2007.05.003

[82] S. Amjad, S. Neelakrishnan and R. Rudramoorthy, "Review of Design Considerations and Technological Challenges for Successful Development and Deployment of Plug-in Hybrid Electric Vehicles," *Renewable and Sustainable Energy Reviews*, Vol. 14, No. 3, 2010, pp. 1104-1110. doi:10.1016/j.rser.2009.11.001

[83] Annual Energy Outlook 2011, with Projections to 2035 No. 2011, Report #:DOE/EIA-0383, 2011. www.eia.gov

[84] W. Liao, R. Heijungs and G. Huppes, "Is Bioethanol a Sustainable Energy Source? An Energy-, Exergy-, and Emergy-Based Thermodynamic System Analysis," *Renewable Energy*, Vol. 36, No. 12, 2011, pp. 3479-3487. doi:10.1016/j.renene.2011.05.030

[85] L. Luo, E. Van Der Voet and G. Huppes, "Life Cycle Assessment and Life Cycle Costing of Bioethanol from Sugarcane in Brazil," *Renewable and Sustainable Energy Reviews*, Vol. 13, No. 6-7, 2009, pp. 1613-1619. doi:10.1016/j.rser.2008.09.024

[86] E. Hanff, M.-H. Dabat and J. Blin, "Are Biofuels an Efficient Technology for Generating Sustainable Development in Oil-Dependent African Nations? A Macroeconomic Assessment of the Opportunities and Impacts in Burkina Faso," *Renewable and Sustainable Energy Reviews*, Vol. 15, No. 5, 2011, pp. 2199-2209. doi:10.1016/j.rser.2011.01.014

[87] R. Melamu and H. Von Blottnitz, "2nd Generation Biofuels a Sure Bet? A Life Cycle Assessment of How Things Could Go Wrong," *Journal of Cleaner Production*, Vol. 19, No. 2-3, 2010, pp. 138-144. doi:10.1016/j.jclepro.2010.08.021

[88] Lake Nyox Data. http://www.geo.arizona.edu/geo5xx/geos577/projects/kayzar/html/lake_nyos_disaster.html

[89] http://www.geology.sdsu.edu/how_volcanoes_work/Nyos.html

[90] Nuclear Regulatory Commission, 2011. www.nrc.gov

[91] A. Ritchie and W. Howard, "Recent Developments and Likely Advances in Lithium-Ion Batteries," *Journal of Power Sources*, Vol. 162 No. 2, 2006, pp. 809-812. doi:10.1016/j.jpowsour.2005.07.014

[92] T. M. Bandhauer, S. Garimella and T. F. Fuller, "A Critical Review of Thermal Issues in Lithium-Ion Batteries," *Journal of the Electrochemical Society*, Vol. 158, No. 3, 2011, pp. R1-R25. doi:10.1149/1.3515880

[93] M. Park, X. Zhang, M. Chung, G. B. Less and A. M. Sastry, "A Review of Conduction Phenomena in Li-Ion Batteries," *Journal of Power Sources*, Vol. 195, No. 24, 2010, pp. 7904-7929. doi:10.1016/j.jpowsour.2010.06.060

[94] L. Damen, M. Lazzari and M. Mastragostino, "Safe Lithium-Ion Battery with Ionic Liquid-Based Electrolyte for Hybrid Electric Vehicles," *Journal of Power Sources*, Vol. 196, No. 20, 2011, pp. 8692-8695. doi:10.1016/j.jpowsour.2011.06.005

[95] M. Amiri, M. Esfahanian, M. Reza Hairi-Yazdi and V. Esfahanian, "Minimization of Power Losses in Hybrid Electric Vehicles in View of the Prolonging of Battery Life," *Journal of Power Sources*, Vol. 190, No. 2, 2009, pp. 372-379. doi:10.1016/j.jpowsour.2009.01.072

[96] P. Roberts, "The Last Drops: How to Bridge the Gap between Oil and Green Energy," *Popular Science*, 2011.

[97] Y. F. Makogon, "Perspectives for the Development of Gas Hydrate Deposits," 1982. http://pubs.aina.ucalgary.ca/cpc/CPC4-299.pdf

[98] T. S. Collett, "Energy Resource Potential of Natural Gas

Hydrates," *AAPG Bulletin*, Vol. 86, No. 11, 2002, pp. 1971-1992.

[99] G. J. Moridis, *et al.*, "Toward Production from Gas Hydrates: Current Status, Assessment of Resources, and Model-Based Evaluation of Technology and Potential," 2008, SPE 114163.

[100] Y. F. Makogon, "Natural Gas Hydrates—A Promising Source of Energy," *Journal of Natural Gas Science and Engineering*, Vol. 2, No. 1, 2010, pp. 49-59. doi:10.1016/j.jngse.2009.12.004

[101] T. M. Odell, "High Efficiency Flexfuel Internal Combustion Engine," USA Patent Application US 2008/0041057 A1, 2009.

[102] C. Park, Y. Choi, C. Kim, S. Oh, G. Lim and Y. Moriyoshi, "Performance and Exhaust Emission Characteristics of a Spark Ignition Engine Using Ethanol and Ethanol-Reformed Gas," *Fuel*, Vol. 89, No. 8, 2010, pp. 2118-2125. doi:10.1016/j.fuel.2010.03.018

[103] H. Lee, C.-L. Myung and S. Park, "Time-Resolved Particle Emission and Size Distribution Characteristics during Dynamic Engine Operation Conditions with Ethanol-Blended Fuels," *Fuel*, Vol. 88, No. 9, 2009, pp. 1680-1686. doi:10.1016/j.fuel.2009.03.007

[104] C. P. Cooney, J. J. Worm and J. D. Naber, "Combustion Characterization in an Internal Combustion Engine with Ethanol-Gasoline Blended Fuels Varying Compression Ratios and Ignition Timing," *Energy and Fuels*, Vol. 23, No. 5, 2009, pp. 2319-2324. doi:10.1021/ef800899r

[105] G. Festel, "Bio Motor Fuels—A Comparative Analysis of Manufacturing Costs and Market Opportunities," *Chemie-Ingenieur-Technik*, Vol. 78, No. 9, 2006, pp. 1175. doi:10.1002/cite.200650010

[106] G. Valentino, F. E. Corcione, S. E. Iannuzzi and S. Serra, "Experimental Study on Performance and Emissions of a High Speed Diesel Engine Fuelled with N-Butanol Diesel Blends under Premixed Low Temperature Combustion," *Fuel*, Vol. 92, No. 1, 2012, pp. 295-307. doi:10.1016/j.fuel.2011.07.035

[107] Y.-H. Chen, T.-H. Chiang and J.-H. Chen, "An Optimum Biodiesel Combination: Jatropha and Soapnut Oil Biodiesel Blends," *Fuel*, Vol. 92, No. 1, 2012, pp. 377-380. doi:10.1016/j.fuel.2011.08.018

[108] S. K. Hoekman, A. Broch, C. Robbins, E. Ceniceros and M. Natarajan, "Review of Biodiesel Composition, Properties, and Specifications," *Renewable and Sustainable Energy Reviews*, Vol. 16, No. 1, 2012, pp. 143-169. doi:10.1016/j.rser.2011.07.143

[109] X. Fan and R. Burton, "Recent Development of Biodiesel Feedstocks and the Applications of Glycerol: A Review," *Open Fuels and Energy Science Journal*, Vol. 2, 2009, pp. 100-109. doi:10.2174/1876973X00902010100

[110] S. K. Hoekman, A. W. Gertler, A. Broch, C. Robbins and M. Natarajan, "Biodistillate Transportation Fuels 1. Production and Properties," *SAE International Journal of Fuels and Lubricants*, Vol. 2, No. 2, 2010, pp. 185-232.

[111] R. Sarin, R. Kumar, B. Srivastav, S. K. Puri, D. K. Tuli, R. K. Malhotra and A. Kumar, "Biodiesel Surrogates: Achieving Performance Demands," *Bioresource Technology*, Vol.

100, No. 12, 2009, pp. 3022-3028. doi:10.1016/j.biortech.2009.01.032

[112] S. Bezergianni, K. Kalogeras and P. A. Pilavachi, "On Maximizing Biodiesel Mixing Ratio Based on Final Product Specifications," *Computers and Chemical Engineering*, Vol. 35, No. 5, 2011, pp. 936-942. doi:10.1016/j.compchemeng.2011.01.034

[113] S. Lim and L. K. Teong, "Recent Trends, Opportunities and Challenges of Biodiesel in Malaysia: An Overview," *Renewable and Sustainable Energy Reviews*, Vol. 14, No. 3, 2010, pp. 938-954. doi:10.1016/j.rser.2009.10.027

[114] A. Sarin, R. Arora, N. P. Singh, R. Sarin, R. K. Malhotra and S. Sarin, "Blends of Biodiesels Synthesized from Non-Edible and Edible Oils: Effects on the Cold Filter Plugging Point," *Energy and Fuels*, Vol. 24, No. 3, 2010, pp. 1996-2001. doi:10.1021/ef901131m

[115] L. Azócar, G. Ciudad, H. J. Heipieper and R. Navia, "Biotechnological Processes for Biodiesel Production Using Alternative Oils," *Applied Microbiology and Biotechnology*, Vol. 88, No. 3, 2010, pp. 621-636. doi:10.1007/s00253-010-2804-z

[116] S. K. Hoekman, A. W. Gertler, A. Broch, C. Robbins and M. Natarajan, "Biodistillate Transportation Fuels 1. Production and Properties," *SAE International Journal of Fuels and Lubricants*, Vol. 2, No. 2, 2010, pp. 185-232.

[117] S. Jain and M. P. Sharma, "Stability of Biodiesel and Its Blends: A Review," *Renewable and Sustainable Energy Reviews*, Vol. 14, No. 2, 2010, pp. 667-678. doi:10.1016/j.rser.2009.10.011

[118] R. D. Misra and M. S. Murthy, "Straight Vegetable Oils Usage in a Compression Ignition Engine—A Review," *Renewable and Sustainable Energy Reviews*, Vol. 14, No. 9, 2010, pp. 3005-3013. doi:10.1016/j.rser.2010.06.010

[119] M. Y. Koh and T. I. Mohd. Ghazi, "A Review of Biodiesel Production from *Jatropha curcas* L. Oil," *Renewable and Sustainable Energy Reviews*, Vol. 15, No. 5, 2011, pp. 2240-2251. doi:10.1016/j.rser.2011.02.013

[120] A. Bostrom, R. E. O'Connor, G. Böhm, D. Hanss, O. Bodi, F. Ekström and P. Halder, "Causal Thinking and Support for Climate Change Policies: International Survey Findings," *Global Environmental Change*, Vol. 22, No. 1, 2011, pp. 210-222. doi:10.1016/j.gloenvcha.2011.09.012

[121] C. Williams, "Earth Shattering: How Global Warming Will Shake Up the Planet," *New Scientist*, Vol. 211, No. 2832, 2011, pp. 38-42. doi:10.1016/S0262-4079No. 11)62404-4

[122] S. Solomon, J. S. Daniel, R. R. Neely III, J.-P. Vernier, E. G. Dutton and L. W. Thomason, "The Persistently Variable 'Background' Stratospheric Aerosol Layer and Global Climate Change," *Science*, Vol. 333, No. 6044, 2011, pp. 866-870. doi:10.1126/science.1206027

[123] A. Robock, "Volcanic Eruptions and Climate," *Reviews of Geophysics*, Vol. 38, No. 2, 2000, pp. 191-219. doi:10.1029/1998RG000054

[124] B. M. Harris and E. J. Highwood, "A Simple Relationship between Volcanic Sulfate Aerosol Optical Depth and Surface Temperature Change Simulated in an Atmosphere-Ocean General Circulation Model," *Journal of Geophysical Research: Atmospheres*, Vol. 116, No. 5, 2011, Arti-

cle ID: D05109. doi:10.1029/2010JD014581

[125] G. M. Miles, R. G. Grainger and E. J. Highwood, "The Significance of Volcanic Eruption Strength and Frequency for Climate," *Quarterly Journal of the Royal Meteorological Society*, Vol. 130, No. 602, 2004, pp. 2361-2376. doi:10.1256/qj.03.60

[126] A. V. Eliseev and I. I. Mokhov, "Influence of Volcanic Activity on Climate Change in the Past Several Centuries: Assessments with a Climate Model of Intermediate Complexity," *Atmospheric and Ocean Physics*, Vol. 44, No. 6, 2008, pp. 671-683. doi:10.1134/S0001433808060017

[127] M. Free and A. Robock, "Global Warming in the Context of the Little Ice Age," *Journal of Geophysical Research D: Atmospheres*, Vol. 104, No. D16, 1999, pp. 19057-19070. doi:10.1029/1999JD900233

[128] G. A. Zielinski, "Use of Paleo-Records in Determining Variability within the Volcanism-Climate System," *Quaternary Science Reviews*, Vol. 19, No. 1-5, 2000, pp. 417-438.

[129] D. I. Axelrod, "Role of Volcanism in Climate and Evolution," Special Paper, 185, Geological Society of America, Boulder, CO., 1981.

[130] J. R. Bray, "Volcanic Triggering of Glaciation," *Nature*, Vol. 260, No. 5550, 1976, pp. 414-415. doi:10.1038/260414a0

[131] R. A. Bryson and B. M. Goodman, "Volcanic Activity and Climatic Changes," *Science*, Vol. 207, No. 4435, 1980, pp. 1041-1044. doi:10.1126/science.207.4435.1041

[132] B.-K. Moon, D. Youn, R. J. Park, S.-W. Yeh, W.-M. Kim, Y.-H. Kim, J. I. Jeong, *et al.*, "Meteorological Responses to Mt. Baekdu Volcanic Eruption over East Asia in an Offline Global Climate-Chemistry Model: A Pilot Study," *Asia-Pacific Journal of Atmospheric Sciences*, Vol. 47, No. 4, 2011, pp. 345-351. doi:10.1007/s13143-011-0021-z

[133] T. Yuan, L. A. Remer and H. Yu, "Microphysical, Macrophysical and Radiative Signatures of Volcanic Aerosols in Trade Wind Cumulus Observed by the A-Train," *Atmospheric Chemistry and Physics*, Vol. 11, No. 14, 2011, pp. 7119-7132. doi:10.5194/acp-11-7119-2011

[134] F. Yang and M. E. Schlesinger, "Identification and Separation of Mount Pinatubo and El Niño-Southern Oscillation Land Surface Temperature Anomalies," *Journal of Geophysical Research D: Atmospheres*, Vol. 106, No. D14, 2001, pp. 14757-14770. doi:10.1029/2001JD900146

[135] A. G. Marshall, A. A. Scaife and S. Ineson, "Enhanced Seasonal Prediction of European Winter Warming Following Volcanic Eruptions," *Journal of Climate*, Vol. 22, No. 23, 2009, pp. 6168-6180. doi:10.1175/2009JCLI3145.1

[136] D. E. Parker, H. Wilson, P. D. Jones, J. R. Christy and C. K. Folland, "The Impact of Mount Pinatubo on World-Wide Temperatures," *International Journal of Climatology*, Vol. 16, No. 5, 1996, pp. 487-497. doi:10.1002/No.SICI)1097-0088No.199605)16:5<487::A ID-JOC39>3.0.CO;2-J

[137] J. K. Angell, "Impact of El Nino on the Delineation of Tropospheric Cooling Due to Volcanic Eruptions," *Jour-nal of Geophysical Research*, Vol. 93, No. D4, 1988, pp. 3697-3704. doi:10.1029/JD093iD04p03697

[138] S. Bekki, J. A. Pyle, W. Zhong, R. Toumi, J. D. Haigh and D. M. Pyle, "The Role of Microphysical and Chemical Processes in Prolonging the Climate Forcing of the Toba Eruption," *Geophysical Research Letters*, Vol. 23 No. 19, 1996, pp. 2669-2672. doi:10.1029/96GL02088

[139] G. J. S. Bluth, S. D. Doiron, C. C. Schnetzler, A. J. Krueger and L. S. Walter, "Global Tracking of the SO$_2$ Clouds from the June, 1991 Mount Pinatubo Eruptions," *Geophysical Research Letters*, Vol. 19, No. 2, 1992, pp. 151-154. doi:10.1029/91GL02792

[140] R. D. Cadle, "Comparison of Volcanic with Other Fluxes of Atmospheric Trace Gas Constituents," *Reviews of Geophysics and Space Physics*, Vol. 18, No. 4, 1980, pp. 746-752. doi:10.1029/RG018i004p00746

[141] J. Cole-Dai, E. Mosley-Thompson and L. G. Thompson, "Quantifying the Pinatubo Volcanic Signal in South Polar Snow," *Geophysical Research Letters*, Vol. 24, No. 21, 1977, pp. 2679-2682. doi:10.1029/97GL02734

[142] M. P. McCormick, L. W. Thomason and C. R. Trepte, "Atmospheric Effects of the Mt Pinatubo Eruption," *Nature*, Vol. 373, No. 6513, 1995, pp. 399-404. doi:10.1038/373399a0

[143] G. Gu and R. F. Adler, "Large-Scale, Inter-Annual Relations among Surface Temperature, Water Vapour and Precipitation with and without ENSO and Volcano Forcings," *International Journal of Climatology*, Vol. 32, No. 12, 2012, pp. 1782-1791. doi:10.1002/joc.2393

[144] J. Boulon, K. Sellegri, M. Hervo and P. Laj, "Observations of Nucleation of New Particles in a Volcanic Plume," *Proceedings of the National Academy of Sciences of the United States of America*, Vol. 108, No. 30, 2011, pp. 12223-12226. doi:10.1073/pnas.1104923108

[145] S. Blake, "Correlations between Eruption Magnitude SO$_2$ Yield, and Surface Cooling," *Geological Society Special Publication*, Vol. 213, 2003, pp. 371-380.

[146] D. T. Shindell, G. A. Schmidt, M. E. Mann and G. Faluvegi, "Dynamic Winter Climate Response to Large Tropical Volcanic Eruptions since 1600," *Journal of Geophysical Research D: Atmospheres*, Vol. 109, No. 5, 2004, pp. D05104 1-12.

[147] J. K. Angell and J. Korshover, "Surface Temperature Changes Following the Six Major Volcanic Episodes between 1780 and 1980," *Journal of Climate & Applied Meteorology*, Vol. 24, No. 9, 1985, pp. 937-951. doi:10.1175/1520-0450No. 1985)0242.0.CO;2

[148] R. S. Bradley, "The Explosive Volcanic Eruption Signal in Northern Hemisphere Continental Temperature Records," *Climatic Change*, Vol. 12, No. 3, 1988, pp. 221-243. doi:10.1007/BF00139431

[149] R. S. Bradley and P. D. Jones, "Records of Explosive Volcanic Eruptions over the Last 500 Years," *Climate since A.D. 1500*, 1992, pp. 606-622.

[150] J. R. Bray, "Glacial Advance Relative to Volcanic Activity since 1500 AD," *Nature*, Vol. 248, No. 5443, 1974, pp. 42-43. doi:10.1038/248042a0

[151] K. R. Briffa, P. D. Jones, T. S. Bartholin, D. Eckstein, F.

H. Schweingruber, W. Karlén, P. Zetterberg and M. Eronen, "Fennoscandian Summers from AD 500: Temperature Changes on Short and Long Timescales," *Climate Dynamics*, Vol. 7, No. 3, 1992, pp. 111-119. doi:10.1007/BF00211153

[152] K. R. Briffa, P. D. Jones, F. H. Schweingruber and T. J. Osborn, "Influence of Volcanic Eruptions on Northern Hemisphere Summer Temperature over the Past 600 Years," *Nature*, Vol. 393, No. 6684, 1998, pp. 450-455. doi:10.1038/30943

[153] T. J. Crowley, T. M. Quinn, F. W. Taylor, C. Henin and P. Joannot, "Evidence for a Volcanic Cooling Signal in a 335-Year Coral Record from New Caledonia," *Paleoceanography*, Vol. 12, No. 5, 1997, pp. 633-639. doi:10.1029/97PA01348

[154] R. D. D'Arrigo and G. C. Jacoby, "Northern North American Tree-Ring Evidence for Regional Temperature Changes after Major Volcanic Event," *Climatic Change*, Vol. 41, No. 1, 1999, pp. 1-15. doi:10.1023/A:1005370210796

[155] R. J. Delmas, M. Legrand, A. J. Aristarain and F. Zanolini, "Volcanic Deposits in Antarctic Snow and Ice," *Journal of Geophysical Research*, Vol. 90, No. D7, 1985, pp. 12901-12920. doi:10.1029/JD090iD07p12901

[156] K. Grönvold, N. Óskarsson, S. J. Johnsen, H. B. Clausen, C. U. Hammer, G. Bond and E. Bard, "Ash Layers from Iceland in the Greenland GRIP Ice Core Correlated with Oceanic and Land Sediments," *Earth and Planetary Science Letters*, Vol. 135, No. 1-4, 1995, pp. 149-155. doi:10.1016/0012-821XNo. 95)00145-3

[157] H. H. Lamb, "Volcanic Dust in the Atmosphere, with a Chronology and Assessment of Its Meteorological Significance," *Philosophical Transactions of the Royal Society of London Series A: Mathematical and Physical Science*, Vol. 266, No. 1178, 1970, pp. 425-533. doi:10.1098/rsta.1970.0010

List of Acronyms

CCS	Carbon Capture and Sequestration
CC	Climate Change
EV	Electric Vehicles
EU	European Union
GCC	Gas-Fired Combined Cycles
GW	Global Warming
GHG	Greenhouse Gases
HC	Hydrogen Cars
IPP.C	Intergovernmental Panel on Climate Change
ICE	Internal Combustion Engines
MMBPD	Million Barrels of Oil per Day
NG	Natural Gas
PiHFF	Plug-In Hybrid Car with Flexible Fuel
RE	Renewable Energies
RES	Renewable Energy Sources

5

Energy Audit of a Brewery—A Case Study of Vitamalt Nig. Plc, Agbara

Olugbenga Olanrewaju Noah[1], Albert Imuentinyan Obanor[2], Mohammed Luqman Audu[3]

[1]Department of Mechanical Engineering,University of Lagos, Lagos, Nigeria
[2]Department of Mechanical Engineering, University of Benin, Benin, Nigeria
[3]Department of Mechanical Engineering, Auchi Polytechnic, Benin, Nigeria
Email: noaholugbenga@gmail.com, {aiobanor, auduluqman}@yahoo.com

ABSTRACT

The efficient use of energy is of prime importance in all sector of the economy. Energy cost is a significant factor in economic activity on par with factors of production like capital, land and labor [1]. The imperative of an energy shortage situation calls for energy conservation measure, which essentially means using less energy for the same level of activity. A comprehensive energy audit of Vitamalt Nigeria Plc, Agbara was carried out using portable thermal and electrical instruments with the objective of studying the present pattern of energy consumption and identifying the possibilities of saving energy in the plant. Collected, was a five year (2000-2004) data on energy consumption of Vitamalt Nig. Plc. The data were evaluated and analyzed to determine the present energy performance level of the firm. A complete energy balance of the factory was carried out to relate energy input, conversion efficiency with production output in order to identify areas of energy wastages/losses and savings that can be achieved. Energy performance parameters such as Energy intensity, Energy productivity and Normalized performance indicator (NPI) were used as a measure of assessing the energy performance of the plant. The NPI calculated over the span of five years gave an average of 1.2 GJ/m^2 indicating a FAIR range in energy performance level classification (1.0 - 1.2) while significant savings and improvement in energy usage is achievable. Maximizing efficiency of existing system, optimizing energy input requirement and significant capital investment in procuring new energy conserving equipment must be made for the energy performance level to fall into a good range classification (less than 0.8).

Keywords: Efficiency; Energy Intensity; Energy Productivity; Cost of Energy Input

1. Introduction

The advent of high crude oil prices resulted in a global energy crisis leading to huge cost in generating power, running of boilers and internal combustion engines, necessitating a need for energy management by industrial sector for efficient energy use, maximization of profit and enhanced competitive position [2].

Energy audit concept is a measure of the efficiency of energy utilization in a manufacturing process, thus leading to interest in energy performance of machines and plants directly associated with production process [3]. It is important to account for total consumption, cost and how energy is used for each commodity such as steam, water, air and natural gas. Attention is focused by Energy managers on how to reduce energy consumption per unit of production. To obtain best possible savings, good audit and survey must be carried out. An energy audit helps in energy cost optimization, pollution control, safety aspects and suggests the methods to improve the operating and maintenance practices of the system. Energy Audit

attempts to balance the total energy inputs with its use and serves to identify all the energy streams in the systems and quantifies energy usages according to its discrete function [4]. Proper maintenance helps conserve energy by keeping operational efficiencies at their best level.

Energy Surveys and audit are carried out to investigate ways employees can save energy and to identify areas that require high level of energy efficiency. Data were collected for a period of five years (2000-2004) on energy performance of Vitamalt Nig. Plc, Agbara and the analysis has been carried out.

2. Materials and Method

The factory has a total floor area 19,146 m^2 and a treated floor area of 12813.57 m^2. The Company's primary source of power supply is the Power Holding Company of Nigeria (PHCN) and 2 giant generating sets as back up. It has two fire-tube (shell) boilers that uses gas and black oil (low pour fuel oil) as its source of energy, some equipment/machines e.g. pumps, motors, and compressor that uses electricity as their source of energy. Portable test equipment like the flow meter, infrared thermocouple, manometer and multimeter were used in determining flow rates, temperature and electrical readings. The following data were collected:

- Electricity, diesel, and gas consumed per month over a 5 years period;
- Production rate of the Company per month over a 5 years period;

- Number of working hour per day;
- Number of occupancy (shift) per day;
- Floor area of the factory;
- Power rating of all machines/equipment powered by electricity.

All data were presented in tabular and graphical forms as seen in **Tables 1-5** and **Figures 1-5**.

Percentage Energy of Electricity and Fuel (diesel, black oil and gas) consumption were obtained for the 5 years period which can be seen in **Table 8**.

Table 1. Energy consumption and production output (2000).

MONTH	ELECTRICITY (NEPA)		GEN. SET FUEL (DIESEL)		BOILER (LPFO)		PRODCTION
	kwh	GJ	vol. (ltr)	GJ	vol. (ltr)	GJ	CARTONS
JAN	86.930	312.948	20.920	847	192.050	7854.845	157.693
FEB	83.690	301.284	37.520	1.520	270.321	11056.129	164.496
MAR	91.920	330.912	23.012	932	249.520	10205.368	175.983
APR	88.690	319.284	31.205	1.264	189.206	7738.5254	140.114
MAY	89.120	320.832	28.084	1.137	215.221	8802.5389	180.391
JUN	87.690	315.684	35.665	1.444	188.765	7720.4885	170.281
JUL	91.260	328.536	20.803	843	179.510	7341.959	152.228
AUG	71.520	257.472	53.337	2.160	271.020	11084.718	165.396
SEP	75.265	270.954	60.611	2.455	257.150	10517.435	160.754
OCT	76.030	273.708	51.123	2.070	290.210	11869.589	146.162
NOV	108.540	390.744	48.564	1.967	263.821	10790.279	180.379
DEC	74.540	268.344	32.760	1.327	261.590	10699.031	175.211
TOTAL	1025.195	3690.702	443.604	17.966	2828.384	115680.91	1969.088

Table 2. Energy consumption and production output (2001).

MONTH	ELECTRICITY (NEPA)		GEN. SET FUEL(DIESEL)		BOILER (LPFO)		PRODUCTION
	kwh	GJ	vol. (ltr)	GJ	vol. (ltr)	GJ	ctns
JAN	110.284	397.0224	40.845	1.654	311.250	12730.125	141.539
FEB	74.180	267.048	37.520	1.520	223.930	9158.737	138.472
MAR	108.300	389.88	63.273	2.563	232.240	9498.616	136.911
APR	81.220	292.392	60.730	2.460	251.135	10271.422	140.124
MAY	104.420	375.912	67.045	2.715	263.480	10776.332	141.672
JUN	104.260	375.336	55.300	2.240	281.891	11529.342	142.390
JUL	95.010	342.036	57.005	2.309	215.120	8798.408	139.643
AUG	90.920	327.312	29.765	1.205	172.510	7055.659	135.098
SEP	92.200	331.92	32.990	1.336	172.510	7055.659	139.641
OCT	118.220	425.592	32.940	1.334	188.867	7724.6603	142.421
NOV	116.140	418.104	11.224	455	193.201	7901.9209	145.007
DEC	118.731	427.4316	34.636	1.403	260.410	10650.769	155.550
TOTAL	1213.885	4369.986	523.273	21.193	2766.544	113151.65	1698.468

Table 3. Energy consumption and production output 2002.

MONTH	ELECTRICITY (NEPA)		GEN. SET FUEL(DIESEL)		BOILER (LPFO)		PRODUCION
	kwh	GJ	vol. (ltr)	GJ	vol. (ltr)	GJ	ctns
JAN	116.924	420.9264	4.250	172	30.873	1262.7057	100.75
FEB	119.600	430.56	30.130	1.220	207.038	8467.8542	136.764
MAR	118.630	427.068	26.580	1.076	175.425	7174.8825	134.631
APR	121.280	436.608	44.219	1.791	179.602	7345.7218	135.125
MAY	119.350	429.66	45.000	1.823	220.245	9008.0205	138.113
JUN	112.680	405.648	8.010	324	194.836	7968.7924	129.843
JUL	129.610	466.596	9.270	375	72.457	2963.4913	110.536
AUG	139.870	503.532	985	40	126.770	5184.893	132.590
SEP	116.100	417.96	8.600	348	109.629	4483.8261	128.319
OCT	262.922	946.5192	49.077	1.988	193.300	7905.97	133.354
NOV	332.671	1197.6156	49.197	1.992	125.271	5123.5839	141.109
DEC	277.220	997.992	43.128	1.747	191.936	7850.1824	132.764
TOTAL	1966.857	7080.6852	318.446	12.897	1827.382	74739.924	1319.927

Table 4. Energy consumption and production output (2003).

MONTH	ELECTRICITY (NEPA)		GEN. SET FUEL(DIESEL)		BOILER (LPFO/GAS)		PRODUCTION
	kwh	GJ	vol. (ltr)	GJ	vol. (ltr)	GJ	cartons
JAN	364,677	1312.8372	51,430	2083	145,822	5294.65	301,486
FEB	263,461	948.4596	45,777	1854	42,442	1681.76	295,723
MAR	277,188	997.8768	24,400	988	106,613	4125.9231	288,946
APR	310,311	1117.1196	6570	266	78,347	3032.0289	290,110
MAY	301,817	1086.5412	19,465	788	95,547	3697.6689	320,228
JUN	310,116	1116.4176	30,355	1229	74,373	2878.2351	319,520
JUL	261,449	941.2164	24,540	994	81,091	3138.2217	305,054
AUG	226,369	814.9284	40,900	1656	86,931	3364.2297	298,760
SEP	271,425	977.13	24,540	994	65,851	2548.4337	299,459
OCT	216,497	779.3892	60,825	2463	96,150	3721.005	302,428
NOV	317,885	1144.386	32,775	1327	96,456	3732.8472	318,230
DEC	251,872	906.7392	63,210	2560	110,264	4267.2168	310,734
TOTAL	3,373,067	12143.041	424,787	17,204		41482.22	3,650,678

Table 5. Energy consumption and production output (2004).

MONTH	ELECTRICITY (NEPA)		GEN. SET FUEL(DIESEL)		BOILER (GAS)		PRODUCTION
	kwh	GJ	vol. (ltr)	GJ	vol. (ltr)	GJ	ctns
JAN	310,987	1119.5532	83,455	3,380	122,935	4757.5845	284,741
FEB	193,810	697.716	36,850	1,492	50,332	1947.8484	283,670
MAR	184,436	663.9696	29,960	1,213	47,372	1833.2964	287,532
APR	218,332	785.9952	40,196	1,628	84,329	3263.5323	288,753
MAY	196,476	707.3136	58,162	2,356	144,988	5611.0356	286,470
JUN	190,764	686.7504	42,370	1,716	80,536	3116.7432	261,985
JUL	183,147	659.3292	44,300	1,794	38,676	1496.7612	279,553
AUG	189,241	681.2676	39,159	1,586	87,610	3390.507	284,043
SEP	254,016	914.4576	40,297	1,632	82,230	3182.301	288,411
OCT	240,683	866.4588	41,011	1,661	93,735	3627.5445	286,023
NOV	209,169	753.0084	40,985	1,660	98,858	3825.8046	289,642
DEC	302,039	1087.3404	51,279	2,077	145,560	5633.172	280,649
TOTAL	2,673,100	**9623.16**	548,024	**22,195**	1,077,161	**41686.131**	3,401,472

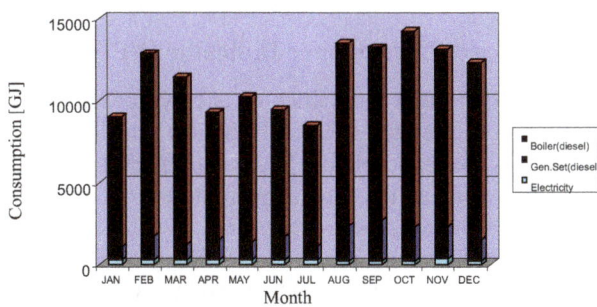

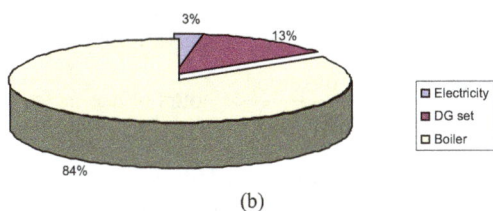

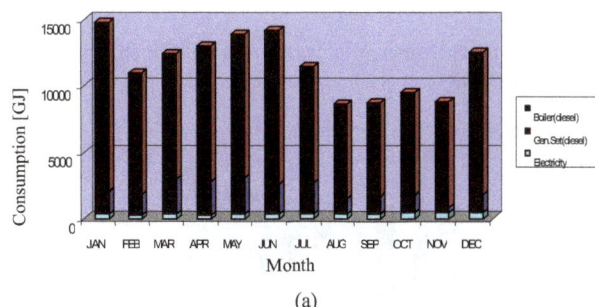

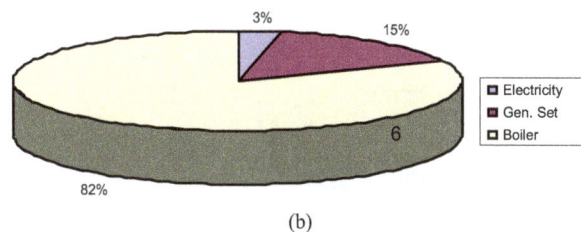

Figure 1. (a) Monthly energy consumption (2000); (b) Total energy inputs (2000).

Figure 2. (a) Monthly energy consumption (2001); (b) Total energy inputs (2001).

(a)

(b)

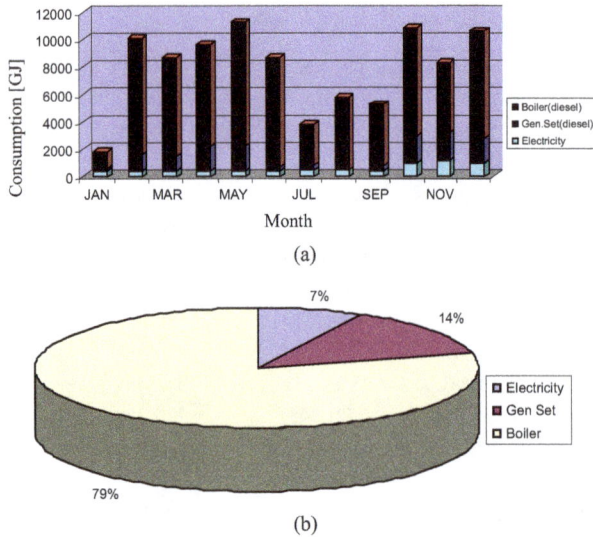

Figure 3. (a) Monthly energy consumption (2002); (b) Total energy inputs (2002).

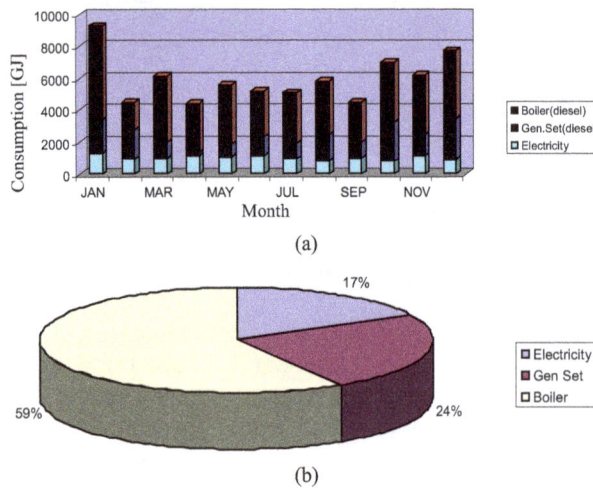

(a)

(b)

Figure 4. (a) Monthly energy consumption (2003); (b) Total energy inputs (2003).

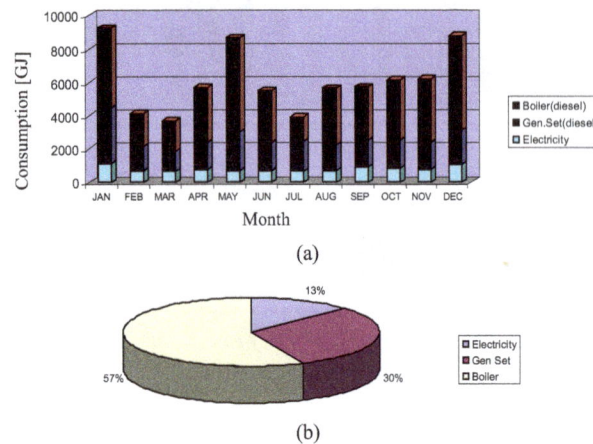

(a)

(b)

Figure 5. (a) Monthly energy consumption (2004); (b) Total energy inputs (2004).

Energy performance parameters such as Energy intensity, Energy productivity and Normalized performance indicator (NPI) are useful yardstick to assess the energy performance level of the company [5].

2.1. Energy Intensity

This is the ratio of the energy consumed per year in GJ to the floor area of the Factory in square meters.

$$\text{Intensity of Energy}\left(\text{GJ/m}^2\right) = \frac{\text{Total Energy Consumed}\left(\text{GJ}\right)}{\text{Treated Floor Area}\left(\text{m}^2\right)}$$

This was calculated for a 5 years period (2000-2004) and a summary presented in **Table 6**. Average Energy Intensity over the 5 years (2000-2004) was 7.852 GJ/m^2.

2.2. Energy Productivity

This is the total energy consumed per unit of production.

$$\text{Energy Productivity}\left(\text{MJ/ctn}\right) = \frac{\text{Total Energy Consumed}\left(\text{MJ}\right)}{\text{Output or unit of production}\left(\text{ctn}\right)}$$

Average Energy Productivity for the 5 years period (2000-2004) was 52.14 MJ/ctn (See **Table 6**).

2.3. Cost of Energy Input into Unit Production

This is the cost of energy to produce a unit product.

$$\text{Cost of Energy Input} = \frac{\text{Total Energy Cost}}{\text{Total Energy}} \times \text{Energy Productivity}$$

Cost of Energy for different energy sources utilized in the plant were summed together to obtain the Total Energy cost. Values of Cost of Energy input for the five years (2000-2004) are given in **Table 6**. Average Cost of Energy input/Product was N34.72/carton for the five year period.

2.4. Normalized Performance Indicator (NPI)

Performance Indicators are values of energy consumption which can be used to indicate whether the actual consumption is low or high relative to similar typical building. It is expressed as the total annual site energy consumption for a building per unit treated floor area and multiplied by the hour of use factor. Value obtained is compared with standard NPI value quoted by the Energy Efficiency Office [6,7] for such factory. If a building is rated as "good", then a further investigation may be required unless there are no obvious areas of improvement.

$$\text{NPI} = \frac{\text{Total Energy Consumed}}{\text{Total Floor Area}\left(\text{m}^2\right)} \times \text{hours of use factor}$$

Average Normalizes Performance Indicator over 5 years period (2000-2004) was 1.2 GJ/m^2 which is rated as fair for the factory size. NPI values calculated for the five years are summarized in **Table 6**.

A line graph was used to express the trend and also to compare the different energy efficiency performance Parameters for the 5 years period as can be seen in **Figure 6**.

3. Discussion of Result

Results obtained from analysis were based on data provided by the company and those taken using measuring equipment. Total energy consumed per annum was on a decline from 137.4 GJ in year 2000 to 71.4 GJ in 2003 and a slight increase in 2004. This decline is attributed to change in boiler fuel. Total energy cost per annum which was expected to drop due to decline in energy consumed was on the contrary as there was an increase from N60.1 million in year 2000 to N96.9 million in 2004. This energy cost increase was due to yearly increase in electricity unit charge by the main power supplier PHCN. A summary of total energy consumed, percentages and cost are presented in **Tables 7-10** and **Figure 7**.

Table 6. Energy efficiency performance result of the factory.

ENERGY PERAMETERS	2000	2001	2002	2003	2004	Average
Total Energy Consumed (GJ)	137337.61	125983.22	94717.61	71441.19	73504.29	100596.78
Production Output (Mctn)	1.969	2.351	2.196	3.651	3.401	2.71
Energy Intensity (GJ/m2)	10.72	9.83	7.39	5.58	5.74	7.852
Energy Productivity MJ/ctn	67.75	80	71.76	19.57	21.61	52.14
Cost of Energy Input/Product (ctn) (N)	29.64	44.12	49.47	21.99	28.48	34.74
Normalized Performance Indicator GJ/m2	1.36	1.37	1.29	1.04	1	1.2

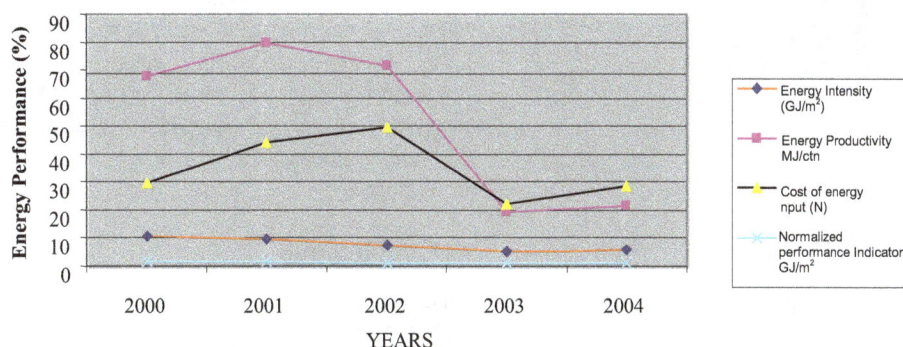

Figure 6. Comparison of different energy efficiency performance parameters.

Table 7. Summary of total energy consumed.

Year/Energy	2000 (GJ)	2001 (GJ)	2002 (GJ)	2003 (GJ)	2004 (GJ)
Electricity	3690.702	4369.986	7080.69	12143.041	9623.16
LPFO	115680.91	102880.23	74739.92	5294.65	-
AGO	17,966	18,733	12,897	17,204	22,195
Gas	-	-	-	36188.36	41686.131
Total	**137337.612**	**125983.22**	**94717.61**	**71441.191**	**73504.29**

Table 8. Summary of percentage energy consumption.

YEAR	ENERGY CONSUMPTION		
	Electricity (%)	LPFO/Natural Gas (%)	AGO (%)
2000	2.69	84.23	13.08
2001	3.47	81.66	14.87
2002	7.48	78.90	13.62
2003	17.00	8.27/50.65	24.08
2004	13.09	56.71	30.20
Avg. (2000-2004)	8.75	72.08	19.17

Table 9. Energy cost.

Year	Electricity (N)	LPFO (N)	AGO (N)	Natural Gas	Total
2000	7,135,805	39,597,376	13352480.40		60,085,662
2001	8111212.02	44,264,704	17111027.10		69486943.12
2002	19513241.38	32,892,876	12897063.00		65303180.38
2003	32581623.99	2,624,796	24212859.00	20848971.31	80268250.30
2004	25,394,450		33977488.00	37485544.31	96,857,482
Average	18,547,267	38,918,319	20310183.50	29167257.81	74,400,304

Table 10. Summary of percentage energy cost.

YEAR	ENERGY COST		
	Electricity (%)	LPFO/Natural Gas (%)	AGO (%)
2000	12.88	61.90	25.22
2001	12.67	59.70	27.63
2002	30.88	46.37	22.75
2003	41.59	25.24	33.17
2004	26.63	35.64	37.73
Avg. (2000-2004)	24.93	44.77	29.3

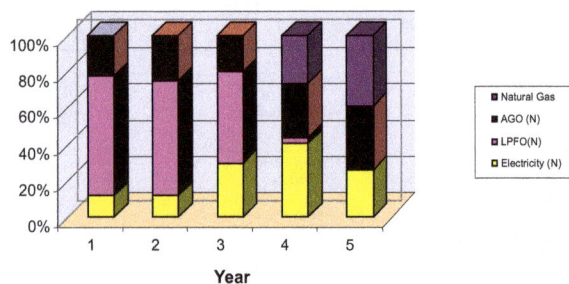

Figure 7. Percentage energy cost (2000-2004).

In the year 2004, the Company's boilers were completely fired by natural gas and the use of LPFO was phased out due high cost of the product. Consumption information for the year 2004 is presented in a Sankey diagram as illustrated in **Figure 8**.

The company looses N1.4 million worth of fuel annually to increase in blowdown rate. This was due to inappropriate feed water treatment resulting into high level of TDS present in the feedwater. A saving of N1.2 million will be achieved in fuel cost annually by raising boiler efficiency to 75.3% as calculated if all losses are reduced and an economizer is incorporated. Further saving of N1.6 m worth of fuel can also be made annually if the level of water treatment is improved upon and the proportion of returned condensate is increased thereby reducing the rate of blowdown. An average saving of N0.3 million can be realized if the firm reduces her compressed air discharge pressure by 10% as calculated to meet actual production requirement. Substantial cost in power consumption can be made if obsolete and non-efficient electric motors and pumps are replaced with modern, effective and less energy rated ones.

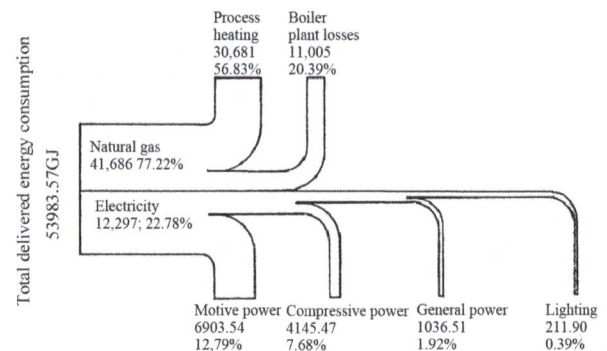

Figure 8. Sankey diagram for energy account in 2004.

The Normalized Performance Indicator (NPI) values calculated for the five years gave an average of 1.2 GJ/m^2, this indicated a "fair" range which implies an average performance [8] while significant savings and improvement in energy usage is achievable. There is room for improvement to a satisfactory and good classification range.

4. Conclusions

Based on the energy efficiency study carried out on Vitamalt Nigeria Plc, Agbara, the following conclusions were arrived at:

1) For a treated floor area of 12813.57 m^2 over a five-year period (2000-2004), average annual energy consumption was 100596.78 GJ was obtained.

2) This consumption was made up of Electricity-8.75%, LPFO/Natural Gas—72.08% and diesel fuel—19.17% while in terms of energy cost, electricity, LPFO/Gas and diesel accounted for 24.93%, 45.76% and 29.3% respectively.

3) Average annual production output (million cartons) for the five years studied was 2.407 while the NPI (normalized performance Indicator) for the factory in GJ/m^2 was 1.2 which is in the "fair" range indicating that there is still room for improvement [9] in terms of energy utilization and savings.

4) Average Intensity of energy (GJ/m2) was 7.852 while average energy productivity in (MJ/ctn) was 52.14 as seen in **Table 6**.

5) The Average cost of energy input/product was calculated to be N34.74/carton with a lowest value of

N21.99/ctn in year 2003.

6) The average cost of annual energy consumption was N74.40 million while an average yearly savings on fuel in boiler, improved feedwater quality, reduction in compressed air pressure and improved boiler efficiency to 75.3% would amount to N4.3 million.

7) Other factors that must be critically looked into are:

- Install electric meters in major production and administrative units tomonitor/curtail power wastages in each unit thereby reducing energy cost on power;
- Improve boiler feedwater quality to eliminate cost incurred in extra blowdown;
- Raise boiler efficiency by reducing flue gas losses and other losses;
- Reduce steam leakages along pipelines and improve lagging of steam pipes;
- Procurement of test equipment for energy monitoring in the factory;
- Good maintenance and control must be put in place in order to improve the energy performance of the factory and rating to "good";
- Motivation for energy conservation among workers;
- Significant capital investment should be made in replacement of inefficient energy consuming equipment to reduce the energy consumption.

5. Acknowledgements

Our acknowledgement goes to the Management and Staff of Vitamalt Nig. Plc for the opportunity given to us to conduct the energy survey research work in their factory.

REFERENCES

[1] Petroleum Conservation Research Association, "What Is Energy Audit," 2004. www.pcra.org/english/aboutus/default.htm

[2] Microsoft Corporation Encarta Premium Suite, 2004. http://encarta.msn.com/

[3] P. O. Aiyedun and O. B. Ologunye, "Energy Efficiency of a Private Sector with Cadbury Nigeria Plc., Ikeja, Lagos as a Case Study," *Nigeria Society of Engineers Technical Transaction*, Vol. 36, No. 2, 2001, pp. 59-63.

[4] C. O. Ojo, "Energy Audit of a Refinery—Case Study of Port-Harcourt Refining Company," Master's Thesis, University of Benin, Benin, 1995.

[5] Chartered Institution of Building Services, "CIBS Building Energy Code Part 4, Measurement of Energy Consumption and Comparison with Targets for Existing Buildings and Services," Chartered Institution of Building Services, London, 1982.

[6] The Construction Information Service, "Energy Audit for Industries," Building Research Energy Conservation Support Unit, Department of Energy, Watford, 1993.

[7] The Construction Information Service, "Economic Use of Gas-Fired Boiler Plant," Building Research Energy Conservation Support Unit, Department of Energy, Watford, 1979.

[8] Chartered Institution of Building Services, "CIBS Building Energy Code Part 2, Measurement of Energy Consumption and Comparison with Targets for Existing Buildings and Services," Chartered Institution of Building Services, London, 1984.

[9] D. K. Hale, D. Vincent and P. T. Clarke, "The Industrial Energy Thrift Scheme, Department of Industry, Energy Unit," National Physical Laboratory, Middlesex, 1979.

6

An Energy-Based Centrality for Electrical Networks

Ruiyuan Kong[1], Congying Han[2], Tiande Guo[1], Wei Pei[3]
[1]School of Mathematical Sciences, University of Chinese Academy of Sciences, Beijing, China
[2]College of Humanities and Social Sciences, University of Chinese Academy of Sciences, Beijing, China
[3]Institute of Electrical Engineering, Chinese Academy of Sciences, Beijing, China
Email: xyzkong@126.com

ABSTRACT

We present an energy-based method to estimate centrality in electrical networks. Here the energy between a pair of vertices denotes by the effective resistance between them. If there is only one generation and one load, then the centrality of an edge in our method is the difference between the energy of network after deleting the edge and that of the original network. Compared with the local current-flow betweenness on the IEEE 14-bus system, we have an interesting discovery that our proposed centrality is closely related to it in the sense of that the significance of edges under the two measures are very similar.

Keywords: Centrality; Energy; Effective Resistance; Current-flow Betweenness

1. Introduction

The electrical network is one of the most critical and complex infrastructure networks in modern society. There are some important issues which are keys to the performance of the network. Reliable electric power supply, for example, is crucial for many devices and its disturbances may disrupt the devices or even paralyze the network. This brings the concern about reliability and resilience to disturbances and failures of various types of infrastructure systems, and a corresponding demand for methods of analyzing the vulnerabilities of the electrical network [1]. Moreover, the blackouts of the North American and Italian electric power grids in 2003 exposed the weaknesses of the electrical network. The weakness and vulnerable analysis about the electrical network have been widely studied in the past years [1-4].

With recent advances in network and graph theory, many researchers have applied centrality measures to complex networks in order to study network properties. Various centrality measures have been defined. They draw links between the structure of networks and the vulnerability to certain types of failures, and are used to identify the most vulnerable elements of a network. Traditionally, there are four centrality measures within network analysis, i.e., degree centrality, betweenness, closeness, and eigenvector centrality. The degree metric utilizes the local information. Closeness and betweenness utilize the shortest path information. And the eigenvector metric rely on the Laplacian matrix of the group. All of them consider only the topological properties but not the

actual physical flow through the power system. Moreover, the betweenness and closeness centrality postulate that the information or flow transfer along the shortest path, but this is not true for the current in the electrical network. A series of centrality measures considering the physical flow are proposed. [5] proposed a so-called random-walk betweenness, counting how often a node is traversed by a random walk between two other nodes. This centrality is known to be useful for finding vertices of high centrality that do not lie on the shortest path. Actually, the random-walk betweenness is closely related to the current-flow betweenness proposed in [6]. The paper derives the metric straightforward from the electrical current and proves that the current-flow closeness is in fact identical with the information centrality [7]. Some papers proposed their measures which are actually of no difference with the current-flow betweenness though they didn't point that directly. For example, [8] proposed an electrical centrality measure based on the impedance matrix which is similar to the current-flow centrality. Besides, they pointed out the differences of the topology of power grids from that of Erdos-Renyi random graphs, the "small-world" networks or "scale-free" networks but the power networks appear to have a scale-free network structure under their proposed measure. However, as the indication of [9], the proposed electrical centrality measure in [8] was defined incorrectly. But a simple analysis shows that the revised measure was the right current-flow betweenness.

The betweenness above needs to take into account all pairs of nodes in the networks. [10] considered only the pairs of generations and load nodes. Besides, they considered some other features of power systems such as power transfer distribution and line flow limits, and got that according to the un-served energy after the network being attacked the nodes ranked highly is more vulnerable.

The centralities defined before are not so easy to understand. This paper will propose an easy-understanding method which is based on the effective resistance. The effective resistance between a pair of vertices s and t is the potential difference between them ensuring a current of size 1 from s to t, and can be seen as the total energy in the system. The effective resistance is local in some sense. Its global form is the Kirchhoff index, which is based on the resistance-distance matrix introduced in 1993 by Klein and Randic and defined as the effective resistance between pairs of vertices [11]. The Kirchhoff index is often used to quantify the structural attributes of the graph. See [12,13] for more information.

This paper is organized as follows. In Section 2, we introduce some preliminary concepts about centrality measures, the effective resistance and so on. In section 3, the definition of our measure based on energy and the variation of current-flow betweenness, that is, the local current-flow betweenness is given. Section 4 provides the comparisons with the current-flow betweenness centrality and other centrality measures. Conclusions are drawn in Section 5.

2. Preliminaries

2.1. Betweenness

Vertex betweenness, first introduced by Freeman in 1977 [14], is one of the most used centrality measure. It reflects the occurrence degree of a node on the shortest path between any pair of nodes. Given a undirected graph $G(V,E)$, where V is the set of vertices and E is the set of edges, the betweenness of a node v is defined by:

$$C_b(v) = \frac{\sum_{s \neq v \neq t \in V} \sigma_{st}(v)/\sigma_{st}}{(n-1)(n-2)/2},$$

where σ_{st} and $\sigma_{st}(v)$ are the number of shortest paths from s to t and the number of shortest paths from s to t through v. Girvan and Newman [15] generalized the vertex betweenness to edges and proposed edge betweenness which is defined as the number of shortest paths between pairs of vertices that run along it and used to find which edges are most important. If there is more than one shortest path between a pair of vertices, then take them as one path. The edge betweenness is given by:

$$C_b(v) = \frac{\sum_{s \neq t \in V} \sigma_{st}(e)/\sigma_{st}}{n(n-1)/2}.$$

It is found that the removal of the nodes or edges with

large betweenness will put the network at high risk to be disconnected.

2.2. Current-Flow Betweenness

The current-flow betweenness here is based on the definition of [6]. An electrical network is a graph $N = (V,E,c) = (G,c)$, together with a function $c :\to R^+$, where $c_e = c(e)$ is the reciprocal of the resistance of the edge e. Given a supply of size 1 from a source s to a sink t, the throughput of an edge e and an inner vertex v is defined by:

$$\tau_{st}(e) = |w(\vec{e})|,$$

$$\tau_{st}(v) = \frac{1}{2} \sum_{e:v \in e} |w(\vec{e})|.$$

Define the edge and vertex current-flow betweenness respectively by:

$$c_{CB}(e) = \frac{1}{n(n-1)} \sum_{s,t \in V} \tau_{st}(e),$$

$$c_{CB}(v) = \frac{1}{(n-1)(n-2)} \sum_{s,t \in V} \tau_{st}(v).$$

The current-flow betweenness is reasonable for that the current is unique by Lemma 1 of [6].

Besides, Brandes proposed also the current-flow closeness centrality which is a variation of closeness centrality. It is defined by:

$$c_{CC}(s) = \frac{n_C}{\sum_{t \neq s} p_{st}(s) - p_{st}(t)}$$

for all s in V, where $p(s)$ refers to the voltage in the vertex s. Moreover, Brandes proved that the current-flow closeness centrality equals information centrality.

2.3. Effective Resistances, Energy and Kirchhoff Index

Now we give the definition of effective resistances. The effective resistance is the potential difference between s and t ensuring a current of size 1 from s to t and denoted by $R_{st}(G)$.

The total energy in a network is defined as:

$$\sum_{xy \in E} w_{xy}^2 = \sum_{xy \in E} (V_x - V_y)^2 c_{xy} = \sum_{xy \in E} (V_x - V_y)w_{xy},$$

where V_x is absolute potential and $w_{xy} = (V_x - V_y)c_{xy}$ is the energy in the edge between x and y. By Lemma 2 of [6], there are unique potentials. So the definition of the total energy is reasonable.

Kirchhoff index is the sum of the effective resistances over all pairs of vertices:

$$Kf(G) = \sum_{i<j} r_{ij},$$

where r_{ij} is the effective resistance between i and j. It is proved that Kirchhoff index satisfies

$$Kf(G) = \sum_{i=1}^{n-1} \frac{1}{\lambda_i},$$

where $\{\lambda_i, i = 1, ..., n-1\}$ are the nonzero eigenvalues of the Laplacian matrix of G.

3. Centrality Metric Based on Energy

It is known that the total energy in an electric current with size 1 from s to t is the effective resistance [16]. If there is only one source s and one sink t, then the change in the energy is the change in the effective resistance between s and t. Therefore to measure the influence of removing one edge we can define a 'metric' based on the variant of energy by:

$$\Delta E(e) = R_{st}(G \setminus e) - R_{st}(G),$$

where $G \setminus e$ is the graph deleted by the edge e. Though $\Delta E(e)$ is not a strict definition of a metric, we regard it as a metric since we only focus on the results under the metric but not their values. The larger $\Delta E(e)$, the greater the risk for the network to be damaged when deleting the edge e. If there is no connection between s and t after deleting the edge e, then $\Delta E(e) = \infty$. In other words, the edge e with $\Delta E(e) = \infty$ is very important to the network. By the monotonicity principle, Corollary 7 of [16], $\Delta E(e)$ is nonnegative for all edges.

For the network with sources S and sinks T where $S \cap T = \varnothing$, the metric based on energy is defined by

$$\Delta E(e) = \frac{\sum_{s \in S, t \in T} R_{st}(G \setminus e) - R_{st}(G)}{|S||T|},$$

where $|S|$ denotes the cardinality of the set S. The energy-based centrality can be seen as a local metric to measure the importance of an edge. Analogous to the definition of betweenness, we consider the whole importance of an edge, that is,

$$\Delta E'(e) = \frac{\sum_{s \neq t} R_{st}(G \setminus e) - R_{st}(G)}{n(n-1)/2}. \tag{1}$$

If we define

$$\Delta Kf(e) = \frac{Kf(G \setminus e) - Kf(G)}{n(n-1)/2},$$

then it is easy to check that $\Delta Kf(e) = \Delta E'(e)$ by the definition of the Kirchhoff index. So we call the case defined by Equation (1) the edge Kirchhoff-based centrality. Analogous to the definition of vertex current-flow betweenness, we can also define the vertex energy-based centrality and vertex Kirchhoff-based centrality by

$$\Delta E_v(u) = \frac{\sum_{e \in \Gamma(u)} \Delta E(e)}{|\Gamma(u)|},$$

$$\Delta Kf_v(u) = \frac{\sum_{e \in \Gamma(u)} \Delta Kf(e)}{|\Gamma(u)|},$$

where $\Gamma(u)$ denotes by the adjacent edges of the vertex u.

Next we take Theorem 1 of [17] as a lemma which will be used to compute the energy-based centrality.

Lemma 1. *Let G be a connected graph on $n \geq 3$ vertices and $1 \leq i \neq j \leq n$. Let $L(i)$ be the submatrix obtained from the Laplacian matrix L of the graph G by deleting its ith row and ith column and $L(i, j)$ be the submatrix obtained from the Laplacian matrix L by deleting its ith and jth rows and the ith and jth columns. Then the effective resistance r_{ij} between i and j satisfies*

$$r_{ij} = \frac{\det(L(i, j))}{\det(L(i))}. \tag{2}$$

Note that the graph G in the lemma above can be seen as a graph with one unit resistance on each edge of the network. Following the steps of its proof, we can easily check that the result holds for the graph G with different resistances on the edges and the Laplacian matrix L being replaced by the admittance matrix of the network. Though in general the complexity of computing the effective resistance using equation (2) is $O(n^3)$, the remarkably simple expression is still very valuable.

Recall the argument of the reference [10] in the introduction. The authors utilized a local idea considering the electrical betweeness only between the pairs of generations and loads with some other restriction, but they didn't point out that clearly and didn't give the corresponding simulation results. Thus this paper gives the definition and simulation clearly, and calls it the local current-flow betweenness. Analogous to the current-flow betweenness, for the network with sources S and sinks T we define the edge local current-flow betweenness by

$$LC_b(e) = \frac{\sum_{s \in S, t \in T} \sigma_{st}(e)}{|S||T|}.$$

The vertex local current-flow betweenness is defined similarly and denoted by $LC_v(u)$.

4. Numerical Analysis

In this section, the edge (vertex) energy-based metric and the edge (vertex) Kirchhoff-based metric are compared with the edge (vertex) current-flow betweenness, the edge (vertex) local current-flow betweenness and the closeness centrality on the IEEE 14-bus. The IEEE 14-bus consists of 20 lines and 14 buses including 2 generators and 2 loads, as shown in **Figure 1**. And **Figure 2** is the graph representation of IEEE 14-bus transmission network. The circles with labeled G represent the generator nodes, the circles with labeled L represent the load

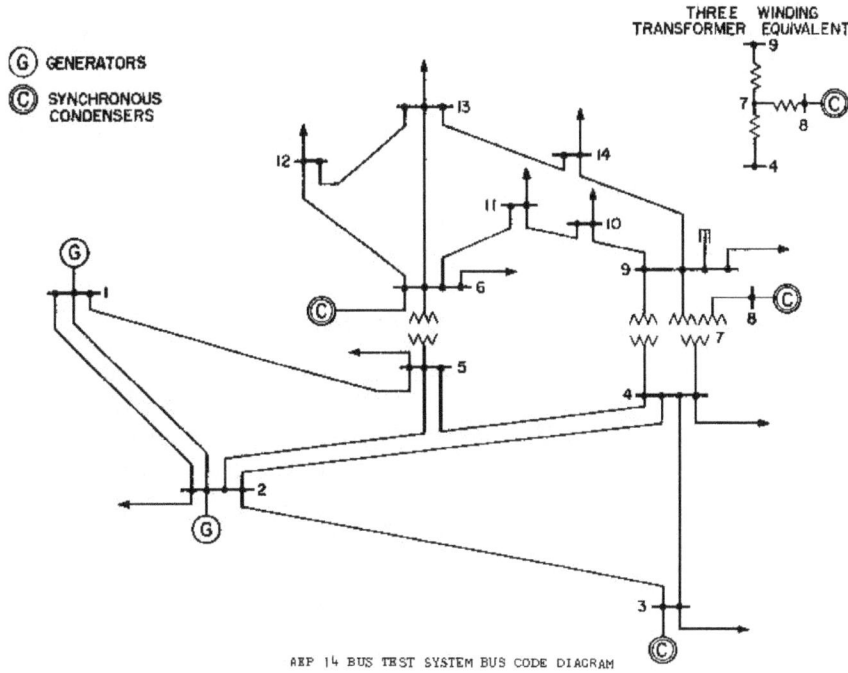

Figure 1. Transmission network IEEE 14-bus.

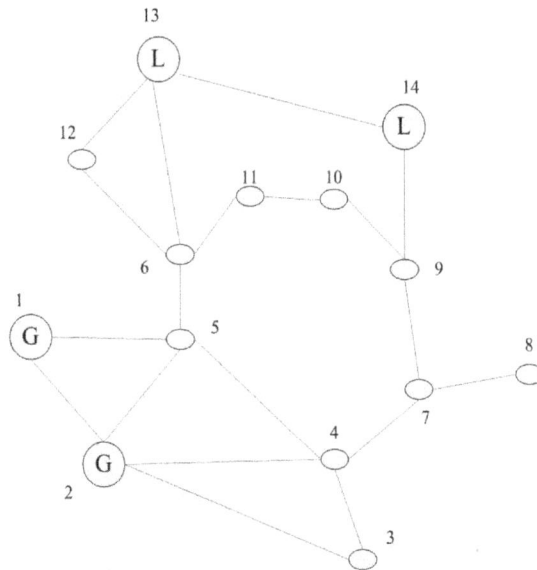

Figure 2. The graph representation of IEEE 14-bus transmission network.

nodes and the ellipses represent the transmission nodes. It is known that for the high-voltage transmission network in a power grid the reactance is usually the dominant component of a line impedance. Thus for the purpose of simplicity, we take the reactance as the edge weights. And the bus 8 will not affect the effective resistance between the generations and loads, thus we don't consider it. To keep in accord, we compute other centrality metrics based on the case above. Besides, the bus 8 is of low betweenness centrality by [18].

Table 1 ranks the edges according to the four edge centralities. The edge current-flow betweenness, edge local current-flow betweenness, edge energy-based measure and edge Kirchhoff-based measure are abbreviated as B, Local-B, E-based and Kf-based respectively. All the four methods rank the edge 5-6 first, which says that the edge 5-6 is very possible to be the most important branch in this network. It shows that the edge betweenness and the edge Kirchhoff-based centrality are quite different with each other and with the other two measures.

Table 1. The importance order of edges from high to low.

Order	B	Local-B	E-based	Kf-based
1	5-6	5-6	5-6	5-6
2	1-2	6-13	9-14	9-10
3	7-9	9-14	13-14	9-14
4	9-10	1-2	6-13	7-9
5	4-7	13-14	1-2	6-11
6	4-5	7-9	4-7	10-11
7	6-11	4-7	7-9	4-7
8	10-11	2-4	1-5	13-14
9	9-14	2-5	4-9	6-13
10	6-13	1-5	2-4	1-2
11	13-14	4-9	2-5	3-4
12	1-5	3-4	6-12	12-13
13	2-4	2-3	12-13	6-12
14	2-5	12-13	9-10	4-5
15	2-3	6-12	10-11	4-9
16	3-4	9-10	6-11	2-3
17	12-13	10-11	2-3	1-5
18	6-12	6-11	3-4	2-4
19	4-9	4-5	4-5	2-5

Table 2. The importance order of nodes from high to low.

Order	V-B	VLocal-B	C	VE-based	VKf-based
1	4	2	12	14	10
2	5	13	14	6	9
3	6	14	3	9	7
4	9	1	11	13	14
5	2	6	13	5	11
6	7	5	1	7	6
7	1	4	10	1	5
8	13	9	7	2	13
9	10	7	2	4	12
10	11	3	6	12	4
11	14	12	9	10	3
12	3	10	5	11	1
13	12	11	4	3	2

But there is a strong correlation between the edge local current-flow betweenness and our edge energy-based centrality. The first 5 important edges are the same and their orders are of little difference. Moreover, the edges ranking from 5 to 11 are also consistent with their orders being little different. In fact, for any edge e in the first 11 edges, the difference between the ranking of e in the two cases is at most 2. However, the complexity of computing the edge energy-based centrality is lower than that of computing the edge local current-flow betweenness. And the complexities for computing the two centrality are both $O(n^3)$. But the latter has a much more clear expression. In practice, we can use both to measure the importance of two edges from different perspectives. If we give the edge ranking first the score 19, the edge ranking second the score 18, ..., the edge ranking last the score 1, and denote $i-j$ by the edge between the ver-

tex i and j. Then given the ordered sequence {1-2, 1-5, 2-3, 2-4, 2-5, 3-4, 4-5, 4-7, 4-9, 5-6, 6-11, 6-12, 6-13, 7-9, 9-10, 9-14, 10-11, 12-13, 13-14}, denote by X_1 and X_2 the ranking lists for energy-based centrality and edge local current-flow betweenness respectively. Using linear regression to check whether X_1 and X_2 are relevant, we get that the adjusted R^2 equals 0.814 and the p-value is very close to 0. This shows that they are strongly correlated.

5. Conclusions

This paper considers measures of centrality that are used to rank the importance of the nodes or edges in an electrical network. New methods of centrality are defined from the perspective of the energy of a network. More specifically, we use the variant of the effective resistances between the generations and loads after deleting an edge or a node to measure its importance, similar to which we also define a Kirchhoff-based measure with the effective resistances being replaced the Kirchhoff index. Besides, we propose the local current-flow betweenness in the most simple way more clearly.

Based on defined measures, experiments are performed on IEEE 14-bus and some interesting results are discovered. It has been found that our proposed edge energy-based measure is very similar to the local current-flow betweenness, in the sense that the importance rankings of the edges in the two measures are of little difference. While the expression of computing the energy-based measure is very simple and clear. Besides, from the experiments we get that the current-flow betweenness is very different from the local current-flow. However, it is difficult to judge which are more accurate. Moreover, we verify that the current-flow betweenness is closely related to the closeness centrality in our experiments. However, more tests and analysis need to be done in order to validate the proposed measure, to find the most effective measure and to dig deep to see which nodes or edges are the real most important nodes or edges.

6. Acknowledgements

This work is supported by CAS Knowledge Innovation Program (grant number KGCX2-RW-329) and National Natural Science Foundation of P.R. China (grant number 10831006, 71271204).

REFERENCES

[1] R. Albert, I. Albert and G. L. Nakarado, "Structural Vulnerability of the North American Power Grid," *Physical Review E*, Vol. 69, No. 2, 2004, pp. 1-4. doi:10.1103/PhysRevE.69.025103

[2] B. A. Carreras, V. E. Lynch, I. Dobson and D. E. New-

man, "Critical Points and Transitions in an Electric Power Transmission Model for Cascading Failure Blackouts," *Chaos*, Vol. 12, No. 4, 2002, pp. 985-994. doi: 10.1063/1.1505810

[3] P. Crucittia, V. Latorab and M. Marchioric, "A Topological Analysis of the Italian Electric Power Grid," *Physica A: Statistical Mechanics and its Applications*, Vol. 338, No. 1-2, 2004, pp. 92-97. doi:10.1016/j.bbr.2011.03.031

[4] M. Rosas-Casals, S. Valverde and R. V. Solé, "Topological Vulnerability of the European Power Grid Under Errors and Attacks," *International Journal of Bifurcation and Chaos in Applied Sciences and Engineering*, Vol. 17, No. 7, 2007, pp. 2465-2475. doi: 10.1142/S0218127407018531

[5] M. E. J. Newman, "A Measure of Betweenness Centrality Based on Random Walks," *Social Networks*, Vol. 27, No. 1, 2005, pp. 39-54. doi:10.1016/j.socnet.2004.11.009

[6] U. Brandes and D. Fleischer, "Centrality Measures Based on Current Flow," *Proceedings of the 22nd Symposium on Theoretical Aspects of Computer Science*, Lecture Notes in Computer Science, Springer, Berlin, Vol. 3404, 2005, pp. 533-544. doi:10.1007/978-3-540-31856-9_44

[7] K. A. Stephenson and M. Zelen, "Rethinking Centrality: Methods and Examples," *Social Networks*, Vol. 11, No. 1, 1989, pp. 1-37. doi:10.1016/0378-8733(89)90016-6

[8] P. Hines and S. Blumsack, "A Centrality Measure for Electrical Networks," *Proceedings of the 41st Hawaii International Conference on System Sciences*, Hawaii, 7-10 January 2008, p.185. doi:10.1109/HICSS.2008.5

[9] Z. H. Wang, A. Scaglione and R. J. Thomas, "Electrical Centrality Measures for Power Grids," *Control and Optimization Methods for Electric Smart Grids*, Power Electronics and Power System, Springer, New York, Vol.3, 2012, pp. 239-255.

doi:10.1007/978-1-4614-1605-0_12

[10] E. Bompard, D. Wu and F. Xue, "The Concept of Betweenness in the Analysis of Power Grid Vulnerability," *Complexity in Engineering*, Rome, 22-24 February, 2010, pp. 52-54. doi:10.1109/COMPENG.2010.10

[11] D. J. Klein and M. Randic, "Resistance Distance," *Journal of Mathematical Chemistry*, Vol. 12, No. 1, 1993, pp.81-95. doi:10.1007/BF01164627

[12] K. C. Das, A. D. Gungor and A. S. Cevik, " On Kirchhoff Index and Resistance-Distance Energy of a Graph," *MATCH Communications in Mathematical and in Computer Chemistry*, Vol. 67, No. 2, 2012, pp. 541-556.

[13] B. Zhou and N. Trinajstic, "On Resistance-Distance and Kirchhoff Index," *Journal of Mathematical Chemistry*, Vol. 46, No. 1, 2009, pp. 283-289. doi:10.1007/s10910-008-9459-3

[14] L. C. Freeman , "A Set of Measures of Centrality Based On Betweenness," *Sociometry*, Vol. 40, No. 1, 1977, pp. 35-41.

[15] M. Girvan and M. E. J. Newman, "Community Structure in Social and Biological Networks," *Proceedings of the National Academy of Sciences*, Vol. 99, No. 12, 2002, pp. 7821-7826. doi:10.1073/pnas.122653799

[16] B. Bollobás, "Modern Graph Theory,"3rd Edition, Springer-Verlag, New York, 2003.

[17] R. Bapat, I. Gutman and W. Xiao, "A Simple Method for Computing Resistance Distance," *Zeitschrift für Naturforschung*, Vol. 58a, No. 9-10, 2003, pp. 494-498.

[18] E. Zio , R. Piccinelli and M. Delfanti , "Application of the Load Flow and Random Flow Models for the Analysis of Power Transmission Networks," *Reliability Engineering and System Safety*, Vol. 103, 2012, pp. 102-109. doi:10.1016/j.ress.2012.02.005

7

Energy Harvesting Based on Magnetic Dispersion for Three-Phase Power System

Tarcisio Oliveira de Moraes Júnior, Yuri Percy Molina Rodriguez, Ewerton Cleudson de Sousa Melo, Maraiza Prescila dos Santos, Cleonilson Protásio de Souza

Federal Institute of Paraíba - IFPB, Cajazeiras, Brazil

Department of Electrical Engineering, Federal University of Paraíba - UFPB, João Pessoa, Brazil

Email: tarcisiocz@gmail.com, {molina.rodriguez}{protasio}{maraiza.santos}{ewerton}@cear.ufpb.br

ABSTRACT

This paper presents a comparative study on Magnetic-Dispersion based Energy Harvesting Systems (MD-EHS) on electrical conductors that supply power for a three-phase AC motor. It introduces two MD-EHS which are based on magnetic cores of different material, named, nanocrystalline, ferrite and iron powder. The first one consists of harvesting energy from magnetic flux through three symmetrical magnetic cores installed on each power conductors of a three-phase AC motor. The second one consists of a single magnetic core for harvesting energy from magnetic flux of only one of these conductors. Both ones have an AC/DC converter and a variable resistor based load. Experimental results have agreed with the theoretical analysis and show that the first proposed MD-EHS is capable of supplying 14 times more energy than the second MD-EHS, considering nanocrystalline cores and phase current of 3 A, and 7.5 times more energy, considering ferrite cores and phase current of 9 A. Such energy can be applied to various low-power devices, especially in wireless sensor network.

Keywords: Energy Harvesting; Magnetic Dispersion; Magnetic Cores

1. Introduction

Energy Harvesting is the process of capturing small amounts of energy from energy sources available in environment, for instance: thermal, solar, mechanical, magnetic induction, and others, and it is specially applied in supplying energy to low-power devices particularly those from wireless sensor networks. Energy harvesting system based on magnetic induction is receiving considerable attention since it is also applicable in measuring variable in power lines.

An example of application of energy harvesting by magnetic induction can be seen in [1] in which a system composed basically of a magnetic flux device is capable of transmitting to a base station the values of temperature variations of the power line where itself is installed. In [2] is proposed a system to harvest energy from electrostatic field created between a power line and the ground. Experimental results have shown that the system can harvest 16 mW. In [3] was studied an energy harvester based on magnetic induction on power line using a simple circuit model to validate the obtained theoretical results. As a result, 1mW of power was achieved considering air-core and 6.32 mW considering an iron core from a magnetic field of 21.2 uT. Recently, in [4] it was studied a tube shaped energy harvester from power lines where the power conditioning circuit is constraint for constant voltage. As a result, the circuit efficiency does influence the level of its output voltage. For constant output power, the voltage level of the power conditioning circuit decreases while the voltage of the transmission line increases.

In this work, it is presented a comparative study on Magnetic-Dispersion based Energy Harvesting Systems (MD-EHS) on electrical conductors that supply power for a three-phase AC motor. It was developed two MD-EHS's which are based on magnetic cores of different material, named, nanocrystalline, ferrite and iron powder. The first one, called 3F-MD-EHS, consists of harvesting energy from magnetic flux through three symmetrical magnetic cores installed on each power conductors of a three-phase AC motor. The second one, called 1F-MD-EHS, consists of a single magnetic core for harvesting energy from magnetic flux of only one of these conductors.

This work is organized as follows: Section II describes the essential theory of magnetic field regarding the cores specifications, Section III shows the experimental analysis, Section IV, the experimental results, and Section V, the main conclusions.

2. Magnetic Field Theory

According to Ampere's law, the magnetic flux density at a given distance r from a infinitely long conductor carrying an alternating current with a peak amplitude I and frequency ω is given by:

$$B = \frac{\mu_m I \sin(\omega t)}{2\pi r} \quad (1)$$

where B is the magnetic flux density in a distance r of the conductor and μ_m is the magnetic permeability of the material between the conductor and the point r.

Figure 1 shows a laminated core on the power conductor in a transversal way which the proposed MD-EHS's are based on. This core provides the magnetic path for the magnetic flux and consists of about 50 thin strips that are electrically separated by a thin layer of insulating material. A coin, not shown in the figure, is on the laminated core and is where the inducted voltage takes place.

Based on **Figure 1**, the concatenated magnetic flux, φ, across the laminated core, with sectional area A, is given by:

$$\phi = \int B dA \quad (2)$$

where

$$dA = w dr \quad (3)$$

Substituting (1) and (3) in (2), it is obtained:

$$\phi_L = \frac{\mu_m I \sin(\omega t)}{2\pi} w \int_{r_L}^{r_f} \frac{dr}{r} \quad (4)$$

which can be reduced as follows:

$$\phi_L = \frac{\mu_m I sen(\omega t) w}{2\pi} \ln\left(\frac{r_L + h}{r_L}\right) \quad (5)$$

where ϕ_L: magnetic flux of the magnetic L-th strip of the core (where $0 < L < N_L$, N_L = number of core strips)

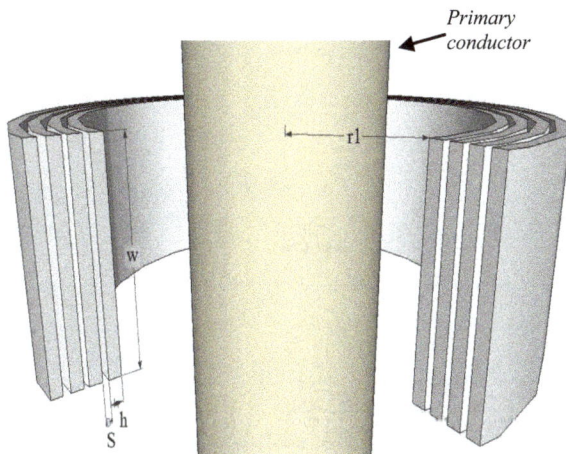

Figure 1. Laminated core on a power conductor.

and r_L is the radius of the L-th strip.

Concern the magnetic flux into the insulating material between the laminated core strips, it is obtained from a similar way and it is obtained:

$$\phi_P = \frac{\mu_0 I sen(\omega t) w}{2\pi} \ln\left(\frac{r_P + S}{r_P}\right) \quad (6)$$

where φ_P: magnetic flux of the magnetic P-th strip of the core (where $0 < P < N_{L-1}$) and r_P is the radius of the P-th strip.

Considering now that is a coin wounds around the core with N_2 turns. In this way, the output voltage of coin terminals, according to Faraday's Law, is expressed as:

$$V_S = -N_2 \frac{d\phi_T}{dt} \quad (7)$$

where ϕ_T is the total magnetic flux. ϕ_T can be obtained by the summation of the magnetic flux through the magnetic and insulating blades:

$$\frac{d\phi_T}{dt} = \sum_{L=1}^{L=n} \frac{d\phi_L}{dt} + \sum_{P=1}^{P=n-1} \frac{d\phi_P}{dt} \quad (8)$$

where:

$$\frac{d\phi_L}{dt} = \frac{\mu_m I \omega \cos(\omega t) w}{2\pi} \ln\left(\frac{r_L + h}{r_L}\right) \quad (9)$$

$$\frac{d\phi_P}{dt} = \frac{\mu_0 I \omega \cos(\omega t) w}{2\pi} \ln\left(\frac{r_P + S}{r_P}\right) \quad (10)$$

To obtain theoretical results, it was considered as parameters the core dimensions: w, h, S, and r_1 (internal radius), as shown in **Figure 1**.

3. Experimental Study

In order to design the proposed MD-EHS, it was used magnetic cores of three different materials: iron powder, ferrite and nanocrystalline. The first proposed MD-EHS, called 3F-MD-EHS, consists of harvesting energy from magnetic flux through three symmetrical magnetic cores installed on each power conductors of a direct start three-phase AC motor, as can be seen in **Figure 2**.

In **Figure 3**, it is shown the power conditioning circuits (PCC) for the 3F-MD-EHS and in **Figure 4** it is shown for the 1F-MD-EHS.

The second proposed MD-EHS, called 1F-MD-EHS, consists of a single magnetic core for harvesting energy from magnetic flux of only one of these conductors.

It was carried out several experiments considering cores of iron powder, ferrite and nanocrystalline and load values, R_v, ranging from 10 Ω up to 10 kΩ. In all experiments, the phase current of each conductor is 3 A for the nanocrystalline cores, 9 A for ferrite and 12 A for iron powder. The main parameters of the used cores are described in **Table 1**.

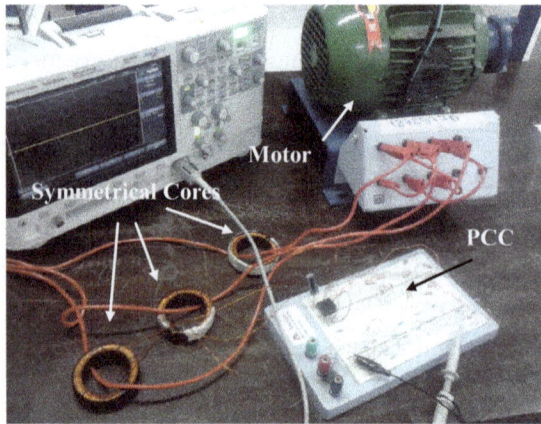

Figure 2. 3F-MD-EHS.

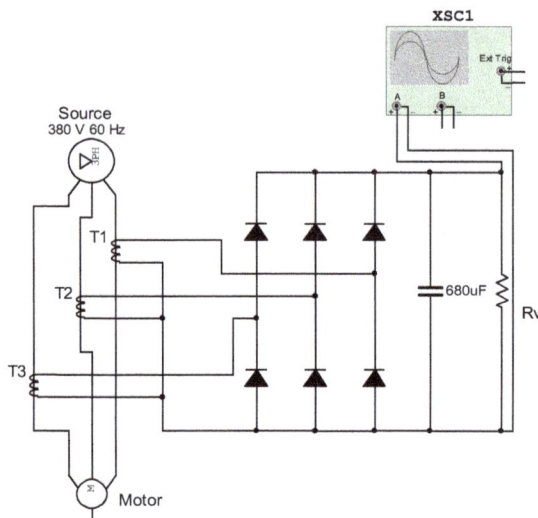

Figure 3. Power conditioning circuit of the 3F-MD-EHS.

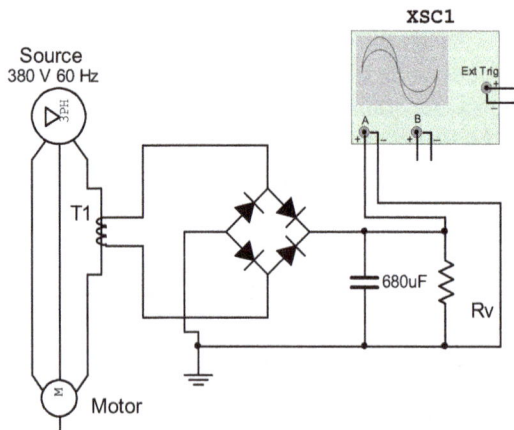

Figure 4. Power conditioning circuit of the 1F-MD-EHS.

4. Experimental Results

After performing experiments considering the different cores and the set of values of R_v, it was possible to obtain the maximum obtained power and the voltages values.

The best results are described in **Tables 2** and **3** and the results considering the variations of R_v can be seen in **Figure 5** for the 3F-MD-EHS and **Figure 6** for the 1F-MD-EHS showing the power obtained for each value of the load, Rv.

Table 1. Parameters of the used cores.

Material	Core's Parameters				
	w [mm]	h [mm]	r [mm]	S	μ_r
Ferrite	8	4.2	6.85	0.0006	2300
Nanocrystalline	10.35	4.4	22.5	0.0006	1000000
Iron Powder	18	11.3	12.05	0.0006	75

Table 2. Experimental results from 3F-MD-EHS.

Material	3F-MD-EHS				
	V [mV]	I [mA]	P [mW]	Rv [Ω]	Ip [Arms]
Ferrite	560	11.3	6.3	50	3.0
Nanocrystalline	374	37.4	14	10	9.0
Iron Powder	5.7	0.57	0.0032	10	12.0

Table 3. Experimental results from 1F-MD-EHS.

Material	1F-MD-EHS				
	V [mV]	I [mA]	P [mW]	Rv [Ω]	Ip [Arms]
Ferrite	310	4.6	1.5	70	3.0
Nanocrystalline	602	1.5	0.9	400	9.0
Iron Powder	18.5	1.85	0.034	10	12.0

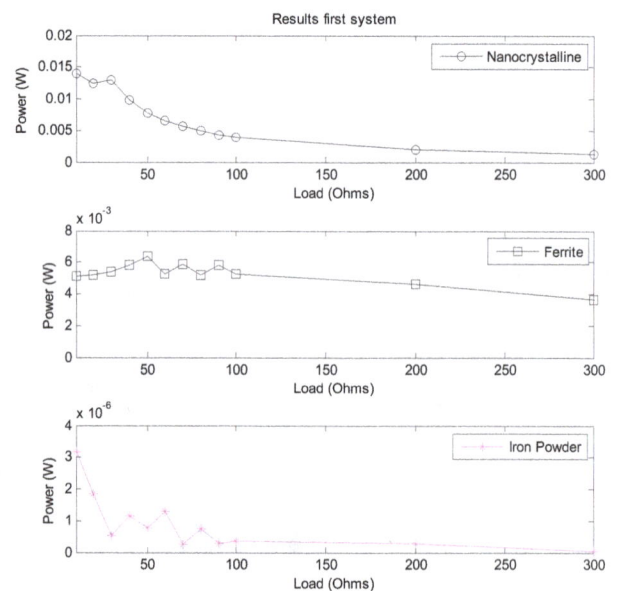

Figure 5. Power levels – first system.

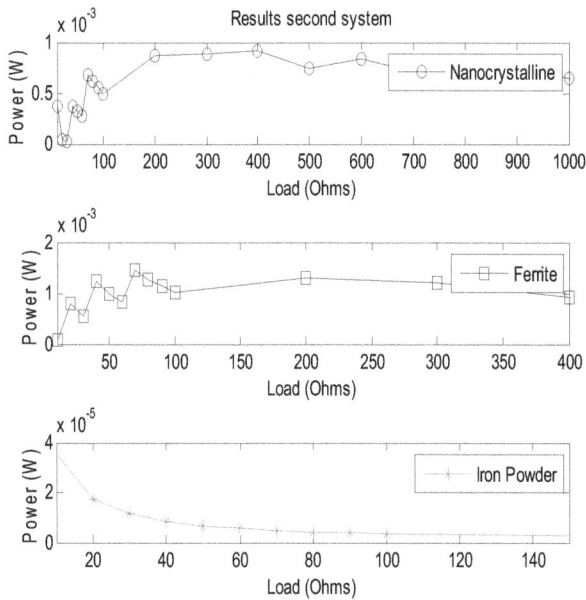

Figure 6. Power levels – second system.

5. Conclusions

In this work, it was presented two proposed Magnetic-Dispersion based Energy Harvesting Systems, called 3F-MD-EHS and 1F-MD-EHS, where the former works on three phase conductors and the latter on one phase conductor. Both are based on magnetic cores of different material, named, nanocrystalline, ferrite and iron powder. The obtained experimental results have shown that, considering the 3F-MD-EHS, it was capable of harvesting up to 14 mW at a load of 10 Ω for a nanocrystalline core with 3 A RMS current on the power line; up to 6.3 mW at a load of 50 Ω for a ferrite cores with 9 A RMS current on the power line; and up to 3.2 µW at a load of 10Ω for

a core of iron powder with 12 A RMS current. Considering the 1F-MD-EHS, it was capable of harvesting up to 0.9 mW at a load of 400 Ω for the nanocrystalline core with 3 A RMS current on the power line; up to 1.5 mW at a load of 70 Ω for ferrite cores with 9 A RMS current on the power line, and up to 34 µW for a load of 10 Ω for an iron powder core with 12 A current on the power line. In this way, it can be observed that the power harvested from the nanocrystalline core 3F-MD-EHS is 14 times higher than the nanocrystalline core 1F-MD-EHS and 7.5 times higher considering ferrite core. However, considering iron powder cores, the obtained power of the 3F-MD- EHS was 10 times less than in the 1F-MD-EHS.

REFERENCES

[1] D. M. Toma, J. Del Rio and A. M. Lázaro, "Self-Powered High-Rate Wireless Sensor Network for Underground High Voltage Power Lines," *Instrumentation and Measurement Technology Conference*, p. c1, 2012.

[2] X. Zhao, M. Ketuel, M. Baldauf and O. Kanaoun, "Energy harvesting for overhead power line monitoring," *Systems, Signals and Devices (SSD), International Multi-Conference on*, 2012, pp. 1-5.

[3] K. Tashiro, H. Wakiwaka, S. Inoue and Y. Uchiyama, "Energy Harvesting of Magnetic Power-Line Noise," *IEEE Transactions on Magnetics*, Vol. 47, No.10, 2011, pp. 4441-4444.
doi.org/10.1109/TMAG.2011.2158190

[4] F. Guo, H. Hayat and J. Wang, "Energy Harvesting Devices for High Voltage Transmission Line Monitoring," *Power and Energy Society General Meeting, IEEE*, 2011, pp. 1-8.

8

Strategies for Energy Efficiency Improvement in Zimbabwean Industries Using the Energy Audit

Wilson Mungwena[1], Cosmas Rashama[2]
[1]Department of Mechanical Engineering, University of Zimbabwe, Harare, Zimbabwe
[2]Department of Electrical Engineering, University of Zimbabwe, Harare, Zimbabwe
Email: wmungwena@gmail.com

ABSTRACT

Energy efficiency is a modern approach for using energy resources economically. Energy audit ensures that every unit of energy gives the maximum in terms of production. This paper brings out the advantages of using energy audit to save future installation of power generation capacity and load reduction of distribution systems. It also envisages the introduction of energy conservation techniques to eliminate sub-standard equipment.

Keywords: Energy Efficiency; Energy Audit; Energy Conservation

1. Introduction

As civilization grows, human beings consume more energy. **Table 1** gives the chronological growth of energy consumption by man [1].

The above detrimental effects can be mitigated by sustainable development. The objectives of sustainable development are: as man consumes more energy, the economic growth has caused rapid depletion of key natural resources like fossil fuels, forest, fresh water and air quality. There is also the possibility that large scale deaths from pollution related diseases may occur, because basic life-support systems of the planet are affected.

The Zimbabwean energy scenario is currently in a precarious state with available capacity almost 50% of national requirements. This is because there has been no new generation assets installed for the past thirty or so years. The critical shortage of energy has affected all sectors of the country with industry operations at 40% partly because of shortage of electricity and obsolete equipment.

- Abundant resources maybe a substitute for depleted natural resources, e.g., solar energy for oil.
- Economic growth leads to capital accumulation which can be used as a substitute for natural resources such as energy conservation requiring capital investment.
- Continuation of technical augmentation, like more efficient power systems, more efficient consumption systems

Table 2 summarizes resource options available for electric utilities [2].

Energy consumed in the plant is determined by whether we make or buy material. Materials used in industry have already an energy input built into it. It is required to estimate the energy content in products so that less energy intensive materials could be promoted to develop more energy effective products.

In industry, the reasons for high energy consumption though not exhaustive may be listed as:

- Inadequate modernization of plant.
- Continued use of obsolete technology.
- High specific energy consumption.
- Less efficient lighting systems.

Table 3 lists typical values of energies consumed in the manufacturing processes of materials in Zimbabwean industries [2].

Table 1. Energy consumption by man.

People	Time	Energy consumption per day
Primitive man	Before 1000 AD	4000 kilo calories
Man in agricultural community	Around 1000 AD	10,000 kilo calories
Man in industrial society	Around 1870 AD	60,000 kilo calories
Modern man	20th century	350,000 kilo calories

Table 2. Resource options for utilities.

Supply side		Demand side	
Option	Example	Option	Example
1. Conventional capacity	Gas turbine, combined cycle, hydro, pumped storage or upgrading of existing plants	1. Utility controls the load as needed	Interruptible consumer load, appliance control.
2. Advanced technology (high efficiency)	Fluidised bed combustion of coal, integrated gassifier combined cycle, mini-hydro photovoltaics etc	2. Consumer installations encouraged through utility incentives, captive generation	Captive generation must be mandatory for power intensive industries like cement, steel, aluminum, paper etc
3. Co-generation (high efficiency)	Gas fired combustion combined cycle, fluidized bed diesel combined cycle	1. New technologies	Computers, robotics, microwaves, magnets
		2. Elimination of energy theft	Effective metering and checking, vigilance on consumer connections

Table 3. Energy contents of materials.

Materials	Energy Consumed (Mj/Kg)	% of Cost of Product Attributable to Energy
Metals		
1. Steels	20 - 50	30
2. Aluminum alloys	60 - 270	40
3. Copper	25 - 30	5
4. Magnesium	80 - 100	10
Other products		
1. Glass	30 - 50	30
2. Plastic	10	4
3. Paper	25	30
4. Inorganic chemicals (average values)	12	20
5. Cement	9	50
6. Waste	4	10

Nowadays, a total review of weight/strength needs is undertaken by the automobile and aircraft industries using the Finite Element Analysis. As a result, more of plastic and aluminum is used in the industry. This gives better HP/weight ratio in the product and process of manufacture as well as in the lifecycle energy costs of the product.

Recycling of materials is energy economical. For example, recycling of old/broken glass pieces requires ¼ the energy required for manufacturing new glass; recycling of iron scrap requires ¼ the energy required to manufacture new iron metal; recycling of aluminum scrap requires ½ the energy required to manufacture new aluminum and recycling of used paper requires about ½ the energy required for new paper manufacture.

2. The Energy Audit

The energy audit is a survey done on an organization to ascertain the energy consumption and to examine energy conservation options. The typical objectives are:

- To review energy consumption patterns so as to evolve industry-wise norms and database.
- To classify consumers with respect to load demand who 75% of total large supply industrial consumption.
- To make energy audit mandatory once a year with HT consumers with maximum demand greater than 500 KVA.

The activities of the energy audit are: [3]

- Electrical energy consumption month-wise vis-à-vis the finished product.
- Power bill study for each month focusing on KWh, KVA, power factor and production throughput.
- Analysis of the load curve to curtail/shift some loads to off-peak periods.
- Monitoring energy consumption of various equipment separately to check efficiency, harmonics starting currents power factor and taking remedial measures to achieve higher efficiency.

American Case Study [4]

Evidence of the importance of monitoring the consumption of various equipment separately to check efficiency was observed at Pacific Gas and Electric Company (PG&E) that understood that businesses may need help with understanding their consumption patterns, and created the Pacific Energy Center (PEC) in San Francisco in 1991 to provide such assistance. The company initiated the Tool Lending Library as a service to customers to help customers understand and document their consumption patterns. Lending Library contained an array of measurement tools that were loaned out to California utility customers free of charge for load studies up to 30 days or more in length. In order to prove the importance of a detailed breakdown of energy audit consumption data, PG&E used its library equipment to measure the specific energy usage at the Pacific Energy Center building itself.

PEC staff monitored individual loads, and logged power levels, and cross-checked the results against overall energy usage to verify that the building's energy usage was indeed accounted for in the recordings. Once that was done, PEC staff compared the loads of the building to the common loads for other buildings of similar size and type, looking for consumption patterns that needed to be corrected. The energy audit found higher than expected baseline energy usage over the weekend. Analysis revealed that the high consumption was as a result of the amount of safety illumination during unoccupied hours, and the amount of refrigeration needed in their commercial kitchen. To address this issue, they reduced the wattage for safety illumination on weekends, and identified more efficient refrigeration equipment that would replace existing equipment when it reached the end of its useful life. They also found a boiler and an exhaust fan that were unnecessarily running constantly during the monitoring period.

Further, in 2011, the PG & E Tool Lending Library completed over 1250 test equipment loans to customers. Borrowers estimated that the monitoring projects supported by these loans helped reduce energy demand by 157 megawatts and saved 92.5 million kWh of electrical energy in the year 2011.

3. Achieving Energy Efficiency

3.1. Demand-Side Management

Here, the utilities seek to directly influence demand for electricity in predetermined ways. The programs are load management, strategic conservation, demand reduction and development of captive power with cogeneration. The main objective of demand-side management is to influence the consumption patterns and behavior of consumers towards efficient use of electrical energy. This is achieved in the following strategic steps.

3.1.1. Load Management
Under this strategy, we direct load control in which portions of the load are under the direct operational control of the utility with the agreement of the consumers; indirect load control where the consumers may control their loads voluntarily & alter the use of electricity in response to price signals and lastly power utilities implement tariffs with inbuilt mechanism to discourage wasteful energy consumption.

3.1.2. Time-of-Day Pricing
Under this strategy, the price elasticity of demand for electrical energy is assessed. This is quantified using an economic model for the electricity demand and then linear regression techniques are applied to estimate the price elasticity of demand. The tariff could indicate to the consumers when electricity is cheap/expensive. The aim is to produce tariffs which meet the utilities' financial targets. The price and demand can be coupled by the elasticity factor to achieve load shifting away from the peak time-of-day for different categories of consumers

Common forms of energy rate tariffs are given below:
a) Consumption limited tariff-consumer is only allowed to consume electricity to a limited extent.
b) Flat rate tariff-consumer is charged at a flat rate/unit no matter how much electricity they consume.
c) Block rate regressive/progressive tariff-cost/unit can decrease or increase per block of consumption.
d) Time-of day tariff-cost/unit depends on time of day or month of the year.
e) Bulk-rate tariff-special tariffs for large consumers.

3.1.3. Demand Reduction
The leveling of demands will decrease the maximum current flow. As loses vary with the square of the current, the lower current will result in reduction of the total energy requirements of the consumer and reduction of loses. Microprocessor base 'demand controllers' could be used to supervise the operations of the consumers' equipment. The first step is to obtain the consumers' demand profile. If the profile shows a few sharp peaks, then the equipment causing these peaks is identified and remedial measures are taken. In this connection, the consumers' major loads are classified into 4 categories like,
- Those which can be rescheduled.
- Those which can be deferred.
- Those which can be curtailed or eliminated.
- Those which are essential base loads.

Demand controllers increase the effectiveness by removing all the non-essential loads in addition to keeping the demand under a preset level. The control function can be by many types of systems namely,
- Instantaneous—controls all loads at any time during an interval if the rate of usage exceeds a preset value.
- Ideal rate—controls load when they exceed the set rate but allow a higher usage at the beginning of the interval.
- Converging rate—has a broad control bandwidth in the beginning of the interval, but tightens control at the end of the interval.
- Predictive rate—the controller is programmed to predict the usage at the end of the interval by the usage pattern along the interval and switches load to achieve the preset demand level.

Continuous interval—the controller looks into the past usage over a period equal to (or less than) the demand interval. Loads are switched in such a manner that no time period of an interval's duration will see an accumulation of KWh that exceeds the preset value.

Before any of the above controllers can be installed, a load survey should be made. This survey is an equipment/process audit. Each process and piece of equipment should be surveyed to find which loads can be switched off and to what extent they can be switched. Any loss of equipment life or mechanical problems associated with switching each load should be evaluated.

The simple fact is that no energy is used when equipment is shut off. Hence it is required to make sure that unused, redundant and idling equipment is shut off.

3.1.4. Cogeneration [5]

This is an important energy conservation strategy. Energy savings from cogeneration do not necessarily imply economic savings. Cogeneration will be an economical investment for a firm if the value of the electricity produced is greater than the incremental capital and operating costs incurred by the firm. Packaged cogeneration plants have potential applications in hospitals, hotels and industries. Cogeneration systems can be classified into two into 2 categories namely, topping systems and bottoming systems.

In topping systems, electricity is produced first in a turbine and some of the energy is exhausted and used in industrial processes.

In bottoming system, high temperature energy is produced first for applications like steel reheating process, cement kilns or aluminum furnaces; further heat is then extracted from the hot exhaust waste steam and transferred to a working fluid. The fluid is vaporized and used to drive a steam turbine to produce electrical energy. The figure below shows the fuel effectiveness.

a) Modern coal fired system
   Maximum efficiency = 35%;
   Losses in condenser = 48%;
   Boiler losses = 15%;
   Other losses = 2%.
b) Gas turbine cogeneration systems
   Maximum efficiency = 90%;
   Exhaust losses =10%.
c) Steam turbine cogeneration system
   Maximum efficiency = 84%;
   Boiler associated losses = 15%;
   Other losses = 1%.

3.2. Efficient Energy Use in Lighting

Lighting constitutes an appreciable load and consists of an inefficient system of lamps and luminaries. Use of energy effective products will lead to the ultimate possibility of halving this connected load, thereby avoiding waste in a cost effective manner. The energy saving measures could be:

- Compact fluorescent lamps as replacement for GLS light points in hotels, commercial, domestic and other applications where 20 W/40 W tube lights are too large.
- Electronic ballasts for fluorescent tube circuits as replacement in existing tubes and for incorporation in new lighting points. Theses operate at low voltages & have instant start with easy installation, high power factor and immediate saving in connected load.
- Upgrading of fluorescent tubes to high pressure sodium lamps as recommended for techno-commercial considerations. Intelligent lighting as a practice can be followed in terms of the following steps.
- In many security situations, lighting is simply left on throughout the high risk period which is generally at night. Passive infrared protection systems are available which automatically sense occupancy and switch on light to specific zones providing round the clock security lighting with a huge potential for energy saving. The infrascan device includes passive infrared sensors, photoelectric sensors and timers. The integration of all these control elements provides an intelligent solution which is capable of dealing with potential security risks as well as continually adjusting to daylight levels.
- Energy saving, fully automatic controllers are available. These are designed to be installed in place of existing wall switches and fir into standard wall boxes. Electricity is wasted when people neglect to switch-off when rooms are vacated or when daylight makes artificial lighting unnecessary. The controllers respond only to physical occupancy within the confines of the room or area controlled by the individual unit. Automatic switching is activated by means of a double-dual-passive-infrared sensor system giving a 180° beam coverage over a 170 m² area. The controllers are fully programmable with internal switches allowing adjustment at the time of installation to suit individual situations. The time delay between a room becoming unoccupied and the lights being switched off can be set to either 1, 4, or 16 minutes. Occupancy in where full daylight does not penetrate can be considered by setting the ambient-daylight-sensitivity function off.

3.3. Efficient Energy Use in Motors [6]

3.3.1. Efficiency

Motors are fairly efficient at rated loads. In general three-phase motors are more efficient than single-phase motors and larger motors are more efficient than the smaller ones. Motor voltage unbalance will increase motor losses due to the negative sequence voltage that the causes a rotating magnetic field in the opposite direction of the motor rotation. A 2% voltage unbalance will increase losses by 8%; a 3% unbalance will increase losses by 25% and a 5% unbalance will increase losses by 50%.

The power factor of three-phase motors is between 80% & 90% at full load & decreases as load is reduced. The installation of capacitors for power factor correction up to 0.95 or so) will decrease current requirements, thereby reducing I^2R losses in supply lines.

3.3.2. Oversized Motors
If a motor is operated at a reduced load; then its efficiency begins to fall, it has higher starting current, lower running power factor and higher capital costs. Oversized motors lead to energy wastage.

3.3.3. Soft Starters for Induction Motors
These regulate the voltage at the motor terminals so that the magnetizing forces just meet the load demand. This boosts the efficiency of the motors operating below their rated outputs. Energy savings are significant for motors operating at less than 50% load for about 50% of the time.

3.3.4. Efficient Motors Design
These motors consume 5% to 8% less electricity than standard motors but more material is used to reduce copper and iron losses. These motors have a higher efficiency because higher grade steel is used and have special low friction bearings, added copper windings, close tolerances & small air-gaps. They have a longer life because they run cooler than less efficient motors.

3.3.5. Delta to Star Connection
The winding of any under-loaded three-phase motor can be reconnected in star. This reduces the voltage across each winding to give 58% of its rated values. Motors constantly running at less 58% of full load will benefit.

3.3.6. Variable Speed Drives
They adjust the speed of the motor replacing constant speed motors. The variable-speed motors are energy effi-cient at reduced loads and reduced speeds to meet different load requirements. These drives are well established over a complete power range in all areas of Industry like basic industries, material handling plants, transport systems and utility companies for mechanical equipment such as machine tools, extruders, pumps fans compressors railways, elevators and conveyors.

4. Conclusion

Energy efficiency as a resource for saving future installation of power generation capacity & unloading of distribution systems is a modern approach for using resources efficiently, especially in Zimbabwe where there are an acute power shortage and no investment in new generating assets. The energy audit aides this process by identifying the deficiencies in the existing systems. Energy conservation management, load management, time-of day metering, electricity pricing & cogeneration efficient technologies are the various methods to reduce system demand and save system capacity. The benefit is an efficient system, increasing plant capacity and a big save on financial resources.

REFERENCES

[1] M. Gown and L. Baine, "Energy Saving Lighting Controllers," *Electrical Installation International*, Vol. 3, No. 9, 1999, p. 14.

[2] Department of Energy and Power Development, "Zimbabwe Energy Policy," 2013, p. 49

[3] Norweigian Institute of Technology, "Economic & Financial Analysis of Energy Systems," 1998, p. 122.

[4] PG & E Tool Lending Library, "California Companies in Measuring Energy Consumption," 2013. http://www.dentinstruments.com/case-study-library-energy-cost-savings.html

[5] W. Mungwena "Cogeneration in Zimbabwe Sugar Industries," *JESA*, Vol. 23, No. 1, 2012, pp. 67-71.

[6] A. S. Pabla, "Efficient Energy Use in Motors," 1996, pp. 11-17.

9

A Conceptual Framework Evaluating Ecological Footprints and Monitoring Renewable Energy: Wind, Solar, Hydro, and Geothermal

Joanna Burger[1,2], Michael Gochfeld[3,4]
[1]Division of Life Sciences, Rutgers University, Piscataway, USA
[2]Environmental and Occupational Health Sciences Institute, UMDNJ and Rutgers University, Piscataway, USA
[3]Consortium for Risk Evaluation with Stakeholder Participation,
Vanderbilt University and Rutgers University, Piscataway, USA
[4]Environmental and Occupational Medicine, UMDNJ-Robert Wood Johnson Medical School, Piscataway, USA
Email: gochfeld@eohsi.rutgers.edu, burger@biology.rutgers.edu

ABSTRACT

With worldwide increases in energy consumption, and the need to increase reliance on renewable energy, we must examine ecological footprints of each energy source, as well as its carbon footprint. Renewable energy sources (wind, solar, hydro, geothermal) are given as the best examples of "green" energy sources with low carbon emissions. We provide a conceptual model for examining the ecological footprint of energy sources, and suggest that each resource needs continued monitoring to protect the environment, and ultimately human health. The effects and consequences of ecological footprint need to be considered in terms of four-compartments: underground (here defined as geoshed), surface, airshed, and atmosphere. We propose a set of measurement endpoints (metrics may vary), in addition to CO_2 footprint, that are essential to evaluate the ecological and human health consequences of different energy types. These include traditional media monitoring (air, water, soil), as well as ecological impacts. Monitoring human perceptions of energy sources is also important for energy policy, which evolves with changes in population density, technologies, and economic consequences. While some assessment endpoints are specific to some energy sectors, others can provide cross-cutting information allowing the public, communities and governments to make decisions about energy policy and sustainability.

Keywords: Airshed; Atmosphere; Energy; Ecological footprint; Geoshed

1. Introduction

The United States and the World are moving toward complex and diversified means of producing energy for growing demands, particularly in rapidly advancing economies, such as China and India [1]. These needs relate to increasing populations and growing per capita demand for energy [2]. For decades, oil (and to a lesser extent coal-fire burning power plants) provided the major share of electricity for developed countries, with hydro and nuclear being important in some [3]. The public, scientists, managers, and public policy makers are interested in energy efficiency and conservation, and in diversifying energy sources, including renewable sources. The potential interruption of energy supply provides a threat to stable economies, national security, and global stability [4-6]. Dependence on foreign oil fuels these threats, and the increased use of nuclear energy is hampered by public perceptions of accidents and other risks

[7]. It will be many years before the full human and ecological health consequences of the Chernobyl accident will be known [8-10], and even more for the recent Fukushima disaster caused by the earthquake and tsunami [11,12].

In addition to the growing need for energy, the risk to the environment and human health from climate change caused by CO_2 emissions is an international scientific and policy challenge. CO_2 concentrations have risen from 280 ppm in the 19th century to 00 ppm now, with increases to 560 ppm expected in the next 50 - 60 years [13]. The European Community countries have experienced an 11% rise in energy consumption from 1995 to the mid-2000s [1]. One of the difficulties is tracking energy consumption per country over the same time period, and there is even less information on energy conservation effects.

Renewable energy is viewed as a potential future mitigation option for climate change through reduction of CO_2 emissions [14,15]. Renewable energy technologies are those that rely on primary energy resources not subject to depletion [16], and often include wind, solar, geothermal, biomass, hydropower, and tidal. In general, renewable energies are secure and environmentally benign compared to fossil fuels, and using them does not prevent future use. One difficulty is that governments have enacted laws and regulations that promote renewable energy and encourage sustainability, and both definitions, and what qualifies as renewable, are often inconsistent both within and among countries [16].

Future development of renewable energy partly depends upon assessment of potential global and regional resources. Such assessment is necessary for nearly all forms of renewable energy, particularly for geothermal [15], solar, wind, and hydro. In 2000, renewable energy sources supplied between 15% and 20% of the total world energy demand [17]. To be effective, renewable energy must have: 1) a track record of sustainable production using existing technology; 2) be applicable to developed and developing countries; 3) be useful for both industrial and residential or urban environments; 4) be reliable; 5) be effective for both heating and cooling; 6) allow for life cycle assessment; and 7) be cost effective [18-20]. Life cycle assessment should be as broad as possible of both front-end and back-end costs [18,20]. And we would add, it must be protective of human health and the environment, protective of landscape and Earth systems, and be acceptable to the public [21,22].

Considerable attention has been devoted to energy and carbon footprints, whereby the amount of land (and energy) required for keeping up the current lifestyle of a community or country is calculated [13,23,24]. Ecological footprint was originally defined as a measure of how much biocapacity a population, organization, or process requires to produce products and absorb its wastes using prevailing technology [24]. However, within an energy context, it usually refers to the CO_2 dynamics of energy production (as often measured by megawatts generated/ facility or structure). We suggest that the ecological or spatial physical footprint of the energy generation facility itself, with its associated facilities and transmission infrastructure, needs to be considered. Questions include, for example, how much physical space is required to generate the electricity, how can these spatial needs be compared among non-renewable and renewable energy sources, what are the ecological costs of those space needs, and what needs to be monitored to understand the effect of energy-generating facilities on ecosystems, human health, and social systems?

In this paper we develop a model for evaluating the ecological spatial footprint of four renewable energy types (wind, solar, geothermal, hydro) that can be used in conjunction with CO_2 footprint models to understand the real costs (and benefits) of renewable energy for societies,

and provide some metrics for monitoring among these facilities. The conceptual model can be applied to other energy sources, whether renewable or non-renewable, and the metrics can similarly be used for different energy sources. Further, we compare the importance of the metrics among the renewable energy sources considered, with those necessary for nuclear energy. The metrics are assessment endpoints, and metrics for a given assessment endpoint may differ among energy types. This paper does not address cost/benefit analysis in terms of economics, but rather in terms of the ecological costs, which ultimately can be put in terms of megawatt energy production for each ecological cost (e.g. CO_2 emissions, ecosystem conversion to energy production, space used).

2. Background

2.1. Environmental Evaluation and Assessments

Ecologists, conservationists, health professionals, and managers have been evaluating ecological health for decades, and farmers and fishing communities have done so for centuries. Healthy ecosystems are essential to provide the necessary goods and services for human communities, whether they are hunters or gathers, or live in dense cities. Ecological evaluations range from qualitative statements about the state of a habitat to quite formal processes, such as ecological risk assessment [25-28]. Less formal approaches are often used where sufficient data are not available for each of the required steps of ecological risk assessment (ERA), the problem being examined does not require a formal process, or a more complex series of problems need to be integrated (e.g. chemical contamination in areas with habitat loss, avian and bat mortality from wind mills, local versus migrant populations of birds and human disturbance).

The lack of consistency among evaluation or assessment methods led to confusion on the part of managers, regulators, decision-makers, and the public, which created a need for a formal risk assessment paradigm that could be applied uniformly. The National Research Council [29] formalized the human health risk assessment paradigm (HRA) to include four parts: hazard identification, dose-response assessment, exposure assessment, and risk characterization. Hazard identification is defining the agent (or condition) that has the potential to cause harm [25]. Dose-response usually involves laboratory tests with animals that indicate how the response varies with the exposed dose. Exposure assessment is determining the pathways (source, fate and transport) and routes (uptake) of exposure, both to humans themselves, and to target organs. It is identifying the pathway from source to receptor. Risk characterization is integrating the hazards, dose-response curves, and exposure data to describe or characterize the risk to given receptors (for

HRA = humans). The NRC formal risk assessment paradigm for humans was modified and adapted for ecological risk assessment [25,30-32], and modified to fit the needs of individual agencies, such as the US Environmental Protection Agency (EPA [33]). While the process varies among agencies, the overall steps are similar: problem definition or formulation, hazard identification, assessment of potential effects (and dose-response curves where possible), exposure assessment, and risk characterization (melding exposure with assessment of effects).

However, there are many situations in which sufficient data are not available to conduct a full, formal ecological risk assessment, or in which the system is sufficiently complex that a series of interlocking models are required to examine the risks to humans, the environment, and the earth's systems. In this situation, often assessment endpoints (metrics) are developed to evaluate ecological and human health risk from system changes. Using CO_2 as a measure of the ecological footprint of different energy sources has become the current currency.

In this paper, we suggest that in addition to CO_2 measures, other metrics (such as the spatial footprint and associated consequences) should be examined and considered by managers, scientists, health professionals, public policy makers, and the public. We will argue that the ecological footprint should include four compartments (subsurface or geoshed, surface, airshed, atmospheric) that have local, regional and global impacts. In this paper we define and describe geoshed, particularly as it applies to energy resources. Ecological footprint should involve metrics that can be used to compare the ecological consequences across energy types, and in this paper we use four renewable energy types (solar, wind, hydro and geothermal) to examine the conceptual model.

2.2. Monitoring

Environmental and human health monitoring are important tools to assess exposure to environmental hazards [34], and to judge whether energy types are sustainable, have low ecological footprints, and have low CO_2 emissions. Most monitoring schemes examine a limited group of pollutants or chemicals, and most biomonitoring studies concentrate on one or two bioindicators or biomarkers of exposure, but more complicated biomonitoring programs that include both exposure and effects are most effective [35,36]. This leads to a more holistic approach to assessing the potential risk to humans, ecosystems, and the Earth system. Further, the problem of exposure (human, ecological, and Earth system) to complex mixtures and interacting stressors has not been adequately examined, either in terms of exposures or effects [37]. Interacting stressors include CO_2 emissions, landscape scale changes, and conversion of natural habitats to anthropogenic ones or from farms to cities and industrial complexes.

There are many kinds and parameters for monitoring ecological health, including media (e.g. water, air, soil, sediment, biota), biological level (cellular to landscape and Earth systems), spatial scale (point source to landscape), temporal scale (variation in key aspects of biological and physical systems), and frequency (daily, weekly, monthly, yearly), as well as other logistical and techniques for monitoring. These in turn affect human health and societal systems.

Energy facilities, whether they are renewable or nonrenewable, are increasingly going to be held to monitoring schemes that are complex and inclusive of a full range of ecological, human, societal and Earth system variables. Climate change, CO_2 emissions, and ecological footprints are only one of many that will be required. For example, nuclear facilities will be required not only to monitor a complex series of pollutants (e.g. radionuclides, mercury and other heavy metals, organics, particulate matter), CO_2 emissions, media (e.g. groundwater, sediment, soil, air), and biota (eco-receptors and humans), but other ecological and human Earth system parameters (e.g. land occupied, landscape changes in habitats and ecosystems, global contribution to CO_2, SO_x and NO_x budgets and effects, local to regional deposition of wastes).

At present, Department of Energy facilities and commercial nuclear energy producers are dealing with the global issue of maintenance and safe storage of nuclear waste materials, as well as accidents and disasters caused by natural geological events (e.g. Fukushima [11,12]). The issue of land conversion (natural ecosystems to industrial, brownfields to industrial, brown-fields or already-contaminated systems to energy facilities) is a public policy issue as well as a practical one for communities and regions. The Department of Energy, for example, is considering establishing "energy parks" on the industrial, but re-mediated portions of their nuclear facilities [38]. This would achieve development of energy resources and facilities, without converting natural ecosystems (or farmlands) to industrial sites. Thus current contaminated footprints could be converted to productive energy facilities.

A weight-of-evidence approach to environmental assessment for energy facilities can be used within types of information, as well as among types of information for different energy sources. Renewable energy will be held to similar standards as nuclear, oil and coal sources, and it is essential to be able to compare energy sources and strategies to achieve sustainable development [39]. To achieve environmentally sustainable energy resources, with minimal or mitigatable effects on societal and Earth systems, it is essential to: 1) develop conceptual models for the four compartmental ecological footprints of each type; 2) develop monitoring schemes that assess current conditions, track changes, and provide early warning of potential problems; and 3) develop and institute monitoring schemes that allow comparisons within and among energy types. While extensive attention is being devoted to CO_2 emissions and monitoring, more needs to be devoted to other aspects of ecological footprints. Monitoring in the four compartments is essential to determining whether a given energy source is sustainable, what is needed to maintain it, and what the costs are to humans, ecosystems, and the Earth system.

3. Ecological Footprints

3.1. Definitions

Footprint usually refers to a spatial occupancy or physical space required or used by a given society, group, activity, plant, or energy source. For example, a chemical plant or nuclear plant may occupy 1000 hectares of land for their facilities and buffer lands [40]. Similarly, the wind turbines and electrical generation facilities of a wind facility may occupy a given amount of land. The ecological footprint concept, first popularized in the 1990's [41], expands the concept of footprint to include all the affected or impacted land area, such as the watershed involved in a hydroelectric plant [16], as well as the land affected to provide raw materials, transportation corridors, and waste disposal [42,43]. Ecological footprint can be calculated for a single family, facility, city, nation [42], or globally [44].

Basically, ecological footprint is a resource accounting tool that can have several different metrics [42]. As such, ecological footprints are often expressed in units of space (global hectares). Thus, global footprint demands can include land devoted to crops, grazing, fishing, forests, carbon, and built-up areas [42,44]. This method of accounting examines the state of current global lands. However, it does not examine energy use or production, or the costs/benefits of those demands and production.

3.2. A Conceptual Model for Examining Footprints

Often ecological footprints are examined for a given facility or type of facility, which addresses local environmental concerns, but does not provide a method of comparison among and between energy options and different kinds of facilities, nor does it include the ecological space or requirements of facilities for raw materials, transportation corridors, and waste disposal sites. There are many other methods for calculating footprints, particularly for energy sources. For example, solar footprints are often calculated in terms of a solar electric footprint (electric demand/solar energy density per area), or as a self-sufficiency footprint whereby all electricity used is derived from solar (electric demand [local + industrial]/ solar energy density per area (watts/m^2 [43]). Other indicators focus on sustainability (total inputs and outputs for a system) and on indicators that can be measured (productivity per inputs; energy produced per acre of land or fuel source), and thus compared across energy types [45].

There are clearly documented advantages and disadvantages to different renewable energy types depending both on geography and technology. The advantages of a renewable energy resource are low carbon emissions and few environmental effects, including low effluent and air pollution. However, there are well-known environmental effects of some renewable energy sources. These include injury and mortality to wildlife from wind facilities [46-48], waste heat (and sometimes sulfur gases) from geothermal energy [49], habitat loss to flooding, altered water flow, and obstruction to spawning fish and other wildlife from hydroelectric facilities [50,51], and conversion of wildlife habitat or farmland into biofuel production, impacting global food supply and prices [52].

Other effects may be less obvious, but require examination and monitoring. Examples that might illustrate the importance of sufficient monitoring (because effects have been found) include: the effect of wind facilities on local meterology [53] and humidity [54], the potential for local geothermal facilities to allow development on land not previously usable because of the surface springs and thermal activity [49], the importance of appropriate mixing of low- and high-temperature heating systems for geothermal [55], changes in soil microorganisms in the aquifers of geothermal well fields [56], and full life-cycle cost calculations for solar energy because of the use of fossil fuel based energy to produce materials for solar cells, modules and systems, and from smelting, production and manufacturing facilities [57]. The traditional two dimensional space, hectares or square kilometers of the Earth's surface required to support a particular activity or lifestyle, are no longer sufficient to evaluate effects on Earth systems.

We propose that ecological footprints should be expanded to four compartments that include the subsurface or underground, surface, airshed, and atmosphere because different energy sectors use the spatial environment differently (**Figure 1**). The subsurface, which henceforth we call geoshed, includes groundwater. The surface includes lands and waters, as well as traditional watersheds. The term airshed has gained recent attention, and recognizes that local or regional parts of the atmosphere share common features with respect to dispersion of emissions. Introduced to the air pollution literature in the early 1970 s [58,59], it became a popular approach around 2000 for analyzing and managing air pollution impacts or for setting geographic boundaries for air quality standards. Just

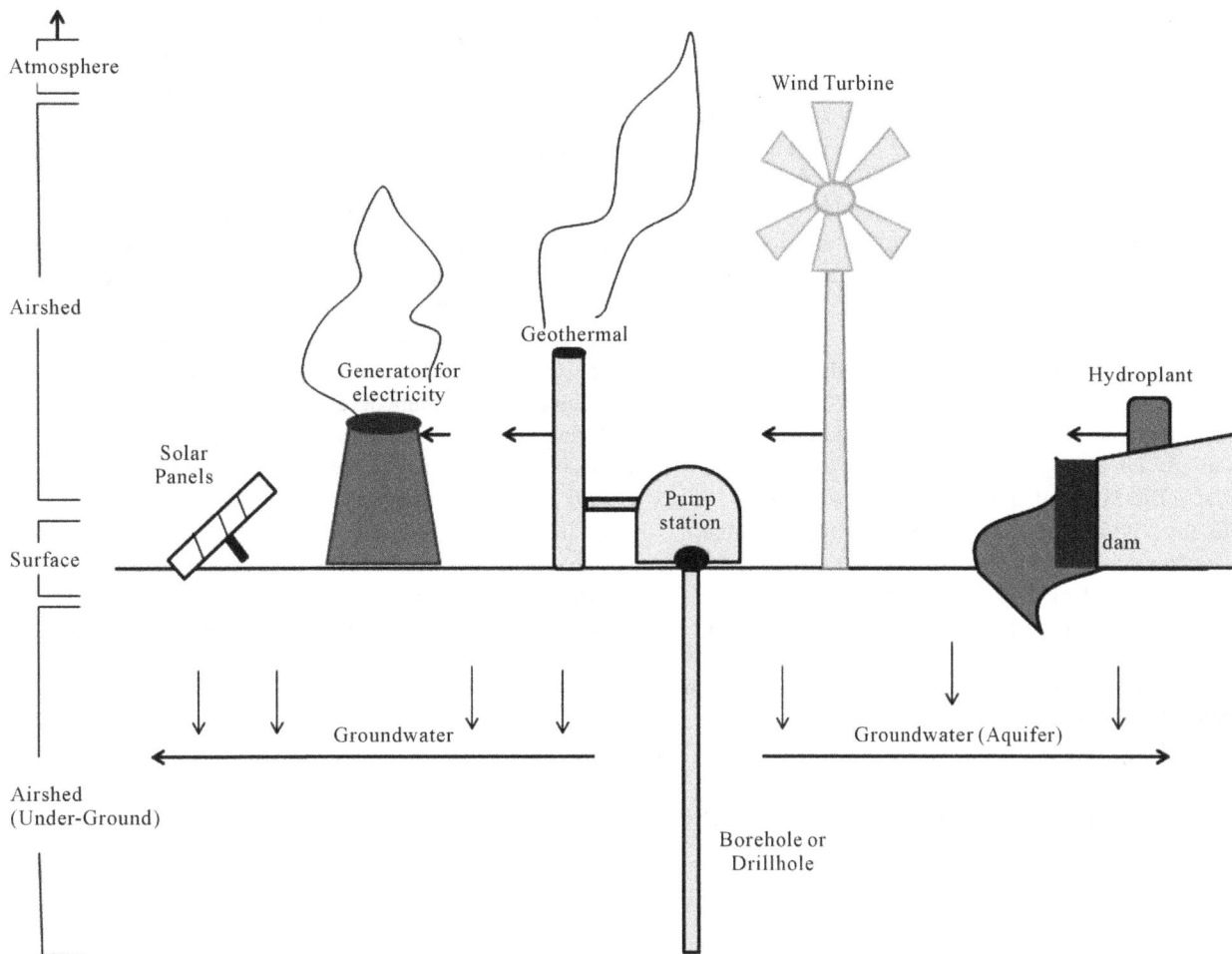

Figure 1. Illustration of the four compartments that require evaluation for solar, geothermal, wind and hydro production of energy. The arrow from the energy types indicates that they feed into a generator to produce electricity.

as watersheds typically encompass a valley or group of valleys draining into a water body, airsheds are bounded by geographical features that provide some limitation to pollution dispersion close to the ground surface. Watershed, however, is a drainage collection concept, while airshed has been an output expansion concept. By contrast, the larger atmosphere is unbounded as air mass movements occur over areas on the continental or global scale.

We propose geoshed for the subsurface (or underground) three dimensional space, particularly the aquifers drawn on or impacted by energy sources. It includes extraction of thermal energy, extraction of natural gas or coal, and disposal of waste, including nuclear and other wastes. The geoshed concept recognizes the importance of the underground ecology.

While each of the compartments can be divided further, the impacts and consequences of energy development on each compartment are critical to ecological (and human) health. For example, the geoshed could be divided into different geological formations and aquifers, varying

greatly in temperature and chemical composition both vertically and horizontally. Further, we suggest these compartments must include additional considerations for onshore and offshore development [60]. While each of the four compartments exists for offshore development, the marine environment creates additional challenges, with unique consequences and impacts for marine ecosystems (including estuarine, coastal, and pelagic zones). Ecological effects depend not only on amount of space used in each compartment by each energy sector, but the type of use (resource extraction, waste dilution), and conversion factors (land or water type before use).

Such an analysis is essential because the ecological effects and consequences of energy resource development in each compartment differ, and some effects may be overlooked if careful consideration is not given to each compartment. Recent attention has focused on Earth global systems, such as the effect of CO_2 emissions [1,61, 62], yet effects on surface and groundwater will become more important with increasing populations, and increasing concentrations of people that put higher demands on extracting potable water and water for agriculture.

The types of questions the public, scientists and governmental agencies will want answered, as well as the consequences, vary with the type of energy and the compartment involved (**Table 1**). Separate considerations for each compartment will focus attention on the unique ecological aspects of each. For example, if either soil structure or soil invertebrate communities are disrupted in the geoshed compartment, the land may be less usable for agriculture. If groundwater levels are lowered, or groundwater is contaminated, sufficient irrigation or drinking water may not be available. **Table 1** provides examples of some of the types of question that apply to each compartment, with possible consequences. It is meant to provide examples, and is not exhaustive.

The specific kinds of questions asked will relate not only to the compartment, but to the energy resource. Thus, considerations of wind resources will mainly focus on the surface and airshed compartments and possible effects on biota, ecological community structure, and compatible land uses, while hydro will focus mainly on atmosphere, surface (and subsurface) because there are few underground facilities, and few emissions into the airshed, although there may be spray and release of volatiles at falls. Questions for solar capture will focus on the surface and subsurface, as well potential effects on albedo (earth surface reflectance) and global temperature changes. Geothermal energy development potentially affects all compartments except for atmospheric because the energy source is underground and near the surface in some instances, and sulfur and other gases, including some CO_2, can be released to the airshed and may affect both ecological and human receptors.

During construction, operation and maintenance, these renewable energy sources have the potential for injuries to workers and the public, such as those involving vehicles, falls and crushing, electrocutions, and being struck by objects, as well as accidents during transportation of materials to and from energy facilities [63]. Such accidents can occur at any energy facility, whether renewable

Table 1. Questions (*) and implications (**) of a four-compartmental evaluation of ecological footprints (see Figure 1). These are meant only as examples.

Compartment	Types of questions
Geoshed[a]	*How much cubic space is actually used by pipes, boreholes, or other apparatus needed *How much is groundwater level affected? *How much bedrock is affected? *Can geological processes be interrupted? *Are pollutants released, and at what levels? *How will soil invertebrates be affected? *How will soil structure be affected? **Groundwater and drinking water quality or quantity can be adversely affected. **Operations can affect housing or other anthropomorphic surface activities. **Smell can be enhanced, rendering land unusable or unattractive
Surface	*How much cubic space is occupied by buildings and facilities? *How much surface water (including streams, rivers and lakes) is affected? *What contaminants are released, at what levels, when and where? *What natural ecosystems are destroyed or compromised? *How will surface activities affect human communities? **Will surface water pose a health risk? **Operations can affect health, housing, or other anthropogenic activities. **Operations can render water resources (lakes, streams) unusable for human recreation or extraction (hunting, fishing).
Airshed[b]	*How much cubic space is affected by surface activities? *What chemicals are released, at what levels, where and when? *Is visibility affected? *What chemicals filter to ground locally and regionally? **Pollution from facilities can adversely affect local and regional activities. **Sensory stimuli (visual, noise, smell) can adversely affect attractiveness of neighborhoods, as well as human and ecological health
Atmosphere[c]	*What are the CO_2 and other emissions that enter the atmosphere? *What chemicals enter the atmosphere, at what levels, when and where? *What particulates, smog or other materials enter the atmosphere, when and where? *How do chemicals or particulates enter global circulation? **Pollutants, particulates, and other emissions can enter the atmosphere, affecting other regions and global geochemical cycles. **CO^2 and other gases can affect regional and global climate

[a]Geoshed includes the subsurface and underground. Geoshed could be further divided into near-surface, vadose and below vadose; [b]For specific energy sources could be limited to 300 - 500 m above the ground surface; [c]Here loosely defined as the space where effects are no longer local, but affect regions or global environments.

or not. Most renewable energy facilities, however, do not have the potential for large-scale disasters that can impact human health, societies, and ecosystems. An exception is hydro because of large reservoirs located behind large dams. Dam failure, though rare, causing massive flooding, could result in loss of life and property, loss of towns and communities, and destruction of ecosystems. While devastating locally or to a region, they do not have the potential for global effects of high CO_2 or toxic chemical emissions (that affect Earth systems).

In contrast, non-renewable energy sources have the potential for high CO_2 emissions, mercury emissions, or catastrophic events that can cause massive effects. Coal-fire burning power plants have high CO_2 emissions. They also routinely release mercury into the atmosphere [64], which enters the food chain, resulting in high mercury levels in fish that can pose a threat to consumers, including humans [65-67]. Ash disposal catastrophes, such as the 2008 containment failure and massive spill at the Tennessee Valley Authority's, Kingston, Tennessee power plant [68] are uncommon, but have local consequences comparable to refinery fires or some nuclear releases.

While catastrophic events at nuclear plants such as Chernobyl (USSR in 1986) and Fukushima (Japan, 2011) are rare, they have the potential for high impact human and ecological effects, both at the regional and global level. The public generally has considerable "fear" or concern about the potential for low probability, high consequence events [7]. On first look, the ecological footprint of nuclear facilities in terms of absolute space is small compared to other energy sources, unless there are radionuclide emissions. There are few underground facilities (although this may change with new requirements for safety and cooling), and normally, little intrusion into either the airshed or atmosphere, although accidental or intentional venting may occur at some facilities during operations or maintenance. However, the full life-cycle cost of operation of a nuclear facility may be high. Life cycle costs include mining, milling, enrichment, and fabrication (with attendant ecological and human health effects), safe transportation corridors requiring new infrastructure, and the unsolved problem of waste disposal, which may entail significant subterranean (or geoshed) land use.

3.3. Application

The application of the model proposed requires an in-depth analysis for a given energy source, such as solar. Ideally, such an application is most useful from a given facility. As an example, we suggest steps for facilities to follow, which include: 1) Designation of a Project leader; 2) Appointment of a team for each compartment (geoshed, surface, airshed, atmosphere); 3) Selection of a member of each team as its Chair, and assignment of the chair to an Evaluation Board; 4) Assessment of each Compartment by the relevant team; and 5) Coordination and final evaluation by the Evaluation Board.

Each team should be composed of a range of scientific disciplines (as well as relevant stakeholders). The initial evaluation by each team would include the parameters held in common (e.g. cubic space involved, chemicals released, ecosystems and organisms at risk), followed by examination of the characteristics unique to that compartment (**Table 1**).

The initial assessment results in provision of status information. Status information can then be used to design the monitoring schemes necessary for each separate energy type (see **Table 2**). The actual monitoring necessary will depend not only on the energy source, but on local environmental and social conditions.

4. The Role of Monitoring

The monitoring required of energy facilities, at least in the United States, is governed by a series of local, state, and federal laws and regulations, and varies among energy types. Air quality, chemical and radiological releases, and effects on local fish, and wildlife are regularly monitored, and data reported to the appropriate agencies. However, we suggest that other assessments and monitoring of all energy facilities, whether renewable or non-renewable, would provide additional information to allow the public, ecologists and health professionals, and governmental agencies to compare and contrast among energy sources for the purposes of developing sound and sustainable energy policies.

We propose that monitoring should include ecological footprint monitoring of the four spatial compartments, as well as the traditional environmental quality monitoring of media (water, air, sediment, soil) and biota (**Table 2**). Not all monitoring will be appropriate for all energy resource sectors, but others will be applicable to all, both renewable and non-renewable energy resources. While the expected levels of CO_2, mercury, and other pollutants will vary greatly among energy sectors, all use fossil fuels in the transportation of materials to the facilities, which results in some release of CO_2. The four-compartmental ecological footprint measures are also applicable to all energy sources, although less so for the geoshed compartment. Similarly, biota monitoring parameters are applicable to all energy sectors, particularly metrics for human perceptions, mortality or injuries to wildlife, and other sublethal effects for humans and eco-receptors (**Table 2**).

The parameters described in **Table 2** are assessment endpoints, that is, aspects of energy resource use that should be monitored to assess potential ecosystem and

Table 2. Types of monitoring suggested for four types of renewable energy. Traditional involves environmental quality, including pollutant levels and particulate matter. X before the solar column indicate what might be necessary for nuclear as a comparison.

TYPE	N[a]	SOLAR	WIND	HYDRO	GEO-THERMAL
Traditional Media Assessing Environmental Quality[b]					
Water	X			X	X
Air	X			X	X
Sediment	X			X	X
Soil	X			X	X
Ecological Footprint (Space)					
Atmospheric	X	X	X	X	X
Airshed	X	X	X		X
Surface space	X	X	X	X	X
Geoshed (Underground)				X	X
Additional Evironmental Media					
Carbon emissions (from operations, transportation and raw materials = life cycle)	X	X	X	X	X
Wind speed, direction, and levels			X	X	X
Surface water flow, and levels	X			X	X
Groundwater flow, and levels	X			X	X
Solar radiation levels and directionality		X			
Changes in albedo		X			
Additional Parameters of Media Affecting Energy Source					
Temperature		X		X	X
Diurnal changes		X	X	X	X
Seasonal changes		X	X	X	X
Geological activities	X	X	X	X	X
Biota[c]					
Pollutant monitoring (absolute levels and health standards)	X			X	X
Pollutant monitoring (perception levels, e.g. visual, olfactory, noise)	X	X	X	X	X
Mortality or injuries to wildlife	X	X	X	X	X
Adverse effects to eco-receptors or humans	X	X	X	X	X
Ecosystem/Landscape					
Regional albedo		X	X		
Regional groundwater levels				X	X
Regional water levels and flow	X			X	X
Ecological life cycle costs	X	X	X	X	X
Human Perceptions of:					
Operations	X	X	X	X	X
Potential effects to humans and environment	X	X	X	X	X
Smells, noise and visual appearance		X	X	X	X
Personal risk or threat	X	X	X	X	X

[a]N = nuclear for comparison with a non-renewable energy source; [b]Monitoring of levels of contaminants or other chemicals; [c]Depending upon the receptor being considered, biota can be either media or endpoints.

human health effects, and to provide early warning of potential harm to humans, ecosystems, and the Earth system. Individual metrics for each assessment endpoint may differ among energy types. For example, effects to wildlife differ among energy sources. Wind facilities can have mortality of birds and bats [46], and geothermal can have effects on soil invertebrate diversity [56]. However, we propose that the assessment endpoint (effects to wildlife) needs to be examined across energy types.

5. Discussion and Conclusions

5.1. Need for Holistic Approach to Ecological Footprint and Monitoring

The imperative to develop a sustainable energy strategy for individual nations, as well as globally, must include conceptual models for ecological footprints that include the effects and implications of four compartmental space, and monitoring to allow for current assessment, trends analysis, and early warning of potential local, regional, and global effects. The recent emphasis on one or another aspect of a given energy sector (e.g. CO_2 emissions, other noxious gas emissions, use of fossil fuel) to the exclusion of a more holistic approach to assessment results in failure to consider other ecological implications. For example, conversion of natural ecosystems or farmland to massive solar facilities has long range implications for global temperature changes (through changes in reflectance and albedo), loss of food production areas, and loss of species diversity and ecosystem types (if massive solar facilities are placed in deserts). In contrast, considerations of the spatial footprint required in the surface compartment for solar facilities may encourage further research and development to reduce the physical space need for these facilities.

5.2. Integrating Different Ecological Consequences

Too often the focus of discussions about energy resources is one-compartmental, or at best, considers only a few factors, such as use, source or costs of fossil fuels or CO_2 emissions. We proposed a multi-compartmental approach that considers not only traditional environmental quality monitoring, but monitoring of ecological footprints in terms of four-compartmental space and ecosystem/landscape parameters. To some extent this will require a weight-of-evidence approach since comparing among parameters is difficult, although the common currency of megawatt production can be used. A weight-of-evidence approach to environmental assessment for energy facilities will not only be useful within types of information, but among types of information for different energy sources [68,69].

The ultimate measure of sustainability may well be our ability to provide energy without global change [13]. However, global change not only refers to global climate change, but to other measures as well. Amount of land used to keep up society's lifestyle, changes in species diversity, and what types of ecosystems are sacrificed to human development, may have equal implications for the societies and the earth [6,22]. Making societal decisions about energy strategies will require integrating among sustainability attributes (as measured by the monitoring indicators provided in **Table 2**), which in turn requires balancing among attributes. For example, how will managers and society balance individual versus population effects, ecological versus human health effects, local versus regional or global effects or consequences, and one type of global effect (climate change) against another (land use changes, ecosystem type changes). While the decisions about how to balance these are societal and within the realm of public policy makers, providing assessments and monitoring data on the necessary compartments is the role of ecologists, economists, and health professionals.

5.3. Conclusions

We suggest that there are a number of assessment endpoints or metrics that should be used when evaluating energy resource sectors, in addition to CO_2 emissions. One key metric is ecological footprint examined in terms of four-compartmental space: geoshed (underground or subsurface), surface, airshed, and atmosphere. Conversion of the earth's natural ecosystems into farming, residential, industrial, or energy production has conesquences for Earth system changes. The ecological issues, as well as the effects and consequences, will vary among the four compartments. Monitoring a range of metrics will provide a more balanced basis for making decisions about energy sources and long-term sustainability.

Balancing will require Earth system global change models (well examined and monitored by ICPP), conceptual models of ecological footprints (four-compartmental space), and monitoring metrics that can be applied within and among energy types, within and among countries, within and among different media (air, water, soil, sediment, biota), and within and among components (*i.e.* terrestrial, aquatic). While not all metrics suggested in this paper will be useful for all energy resource sectors, many will allow comparisons among energy types.

6. Acknowledgements

We are grateful to many people for valuable discussions about these topics over the years, including C. W. Powers, D. Kosson, J. Clarke, M. Greenberg, B. Goldstein, P. Lioy, E. Gunnalaugsson, C. Safina, S. Smallwood, J. Yee,

J. Estep, and S. Orloff and to T. Pittfield who prepared the figure. This research was partly funded by the Consortium for Risk Evaluation with Stakeholder Participation (CRESP) through the Department of Energy (AI # DE-FG 26-00NT 40938 and DE-FC01-06EW07053), NIEHS (P30ES005022), and EOHSI. The conclusions and interpretations reported herein are the sole responsibility of the author, and should not in any way be interpreted as representing the views of the funding agencies.

REFERENCES

[1] S. Perry, J. Klemes and I. Bulatov, "Integrating Waste and Renewable Energy to Reduce the Carbon Footprint of Locally Integrated Energy Sectors," *Energy*, Vol. 33, No. 10, 2008, pp. 1489-1497. doi:10.1016/j.energy.2008.03.008

[2] J. Sheffield, "World Population Growth and the Role of Annual Energy Use Per Capita," *Technological Forecasting and Social Change*, Vol. 59, No. 55, 1998, pp. 55-87. doi:10.1016/S0040-1625(97)00071-1

[3] A. McDonald, "Nuclear Power Global Status," International Atomic Energy Agency, Vienna, 2009. http://www.iaea.org/Publications/Magazines/Bulletin/Bul l492/49204734548

[4] A. L. Alm, "Energy Supply Interruptions and National Security," *Science*, Vol. 211, No. 4489, 1981, pp. 1379-1385. doi:10.1126/science.211.4489.1379

[5] A. E. Waltar, "Feeding the Nuclear Pipeline: Enabling a Global Nuclear Future," *IAEA Science Forum*, Vienna, 2002.

[6] M. Gochfeld, "Energy Diversity: Options and Stakeholders," In: J. Burger, Ed., *Stakeholders and Scientists: Achieving Implantable Solutions to Energy and Environmental Issues*, Springer, New York, 2011, pp. 207-208.

[7] M. Greenberg and H. B. Truelove, "Energy Choices and Risk Beliefs: It Is Just Global Warming and Fear of a Nuclear Power Plant Accident?" *Risk Analysis*, Vol. 31, No. 5, 2011, pp. 819-831. doi:10.1111/j.1539-6924.2010.01535.x

[8] T. G. Hinton, R. Alexakhin, M. Balonov, N. Gentner, J. Hendry, V. Prister, P. Strand and D. Woodhead, "Radiation Induced Effects on Plants and Animals: Findings of the United Nations Chernobyl Forum," *Health Physics*, Vol. 93, 2007, pp. 427-440. doi:10.1097/01.HP.0000281179.03443.2e

[9] A. Rantavaara, "Ingestion Doses in Finland Due to 90Sr, 134Cs, and 137Cs from Nuclear Weapons Testing and the Chernobyl Accident," Applied *Radiation and Isotopes*, Vol. 66, No. 11, 2008, pp. 1768-1774. doi:10.1016/j.apradiso.2007.12.018

[10] E. Ron and A. Brenner, "Non-Malignant Thyroid Diseases after a Wide Range of Radiation Exposures," *Radiations Research*, Vol. 174, No. 6, 2010, pp. 877-888. doi:10.1667/RR1953.1

[11] NPJ (Nuclear Plant Journal), "Information on Status of Nuclear Power Plants in Fukushima," *Nuclear Plant Journal*, E-News, 2011.

[12] J. P. Christodouleas, R. D. Forrest, C. G. Ainsley, Z. Tochner, S. M. Hahn and E. Glatstein, "Short-Term and Long-Term Health Risks of Nuclear Power Plant Accidents," *The New England Journal of Medicine*, Vol. 16, No. 24, 2011, pp. 2334-2341. doi:10.1056/NEJMra1103676

[13] G. Stoglehner, "Ecological Footprint—A Tool for Assessing Sustainable Energy Supplies," *Journal of Cleaner Production*, Vol. 11, No. 3, 2003, pp. 267-277. doi:10.1016/S0959-6526(02)00046-X

[14] J. A. Turner, "A Realizable Renewable Energy Future," *Science*, Vol. 285, No. 5428, 1999, pp. 687-688. doi:10.1126/science.285.5428.687

[15] C. J. Bromley, M. Mongillo, G. Hiriart, B. Goldstein, R. Bertani, E. Huenges, A. Ragnarsson, J. Tester, H. Muraoka and V. Zui, "Contribution of Geothermal Energy to Climate Change Mitigation: The IPCC Renewable Energy Report," *Proceedings of the World Geothermal Congress*, Bali, 25-30 April 2010, pp. 25-29.

[16] G. W. Frey and D. M. Linke, "Hydropower as a Renewable and Sustainable Energy Recource Meeting Global Energy Challenges in a Reasonable Way," *Energy Policy*, Vol. 30, No. 14, 2002, pp. 1261-1265. doi:10.1016/S0301-4215(02)00086-1

[17] J. P. Painuly, "Barriers to Renewable Energy Penetration; A Framework for Analysis," *Renewable Energy*, Vol. 24, No. 1, 2001, pp. 73-89. doi:10.1016/S0960-1481(00)00186-5

[18] M. Pehnt, "Dynamic Life Cycle Assessment (LCA) of Renewable Energy Technologies," *Renewable Energy*, Vol. 31, No. 1, 2006, pp. 55-71. doi:10.1016/j.renene.2005.03.002

[19] S. M. Benson and F. M. Orr, "Sustainability and Energy Conversions," *MRS Bulletin*, Vol. 33, 2008, pp. 297-304. doi:10.1557/mrs2008.257

[20] E. Martinez, F. Sanz, S. Pellegrini, E. Jimenez and J. Blanco, "Life Cycle Assessment of a Multi-Megawatt Wind Turbine," *Renewable Energy*, Vol. 34, No. 3, 2009, pp. 660-673. doi:10.1016/j.renene.2008.05.020

[21] J. Burger, "Environmental Management: Integrating Ecological Evaluation, Remediation, Restoration, Natural Resource Damage Assessment, and Long-Term Stewardship on Contaminated Lands," *Science of the Total Environment*, Vol. 400, 2008, pp. 6-19. doi:10.1016/j.scitotenv.2008.06.041

[22] J. Burger, "Stakeholders and Scientists: Achieving Implantable Solutions to Energy and Environmental Issues," Springer, New York, 2011.

[23] J. Holmberg, U. Lundqvist, K. H. Robert and M. Wackernagel, "The Ecological Footprint from a Systems Perspective of Sustainability," *International Journal of Sustainable Development and World Ecology*, Vol. 6, No. 1, 1999, pp. 17-33. doi:10.1080/13504509.1999.9728469

[24] M. Wackernagel and C. Monfreda, "Ecological Footprints and Energy," *Encyclopedia of Energy*, Vol. 2, 2004, pp. 1-11.

[25] National Research Council (NRC), "Issues in Risk Assessment," National Academy Press, Washington, 1993.

[26] W. N. Beyer, G. H. Heinz and A. W. Redmon-Norwood, "Environmental Contaminants in Wildlife," Lewis Publications, Winter Haven, 1996.

[27] G. W. Suter, R. A. Efroymson, B. E. Sample and D. S. Jones, "Ecological Risk Assessment for Contaminated Sites," Lewis Publications, Winter Haven, 2001.

[28] G. W. Suter, T. Vermeire, W. R. Munns and J. Sekizawa, "Framework for an Integration of Health and Ecological Risk Assessment," *Human and Ecology Risk Assessment*, Vol. 9, No. 1, 2003, pp. 281-301. doi:10.1080/713609865

[29] National Research Council (NRC), "Risk Assessment in the Federal Government," National Academy Press, Washington, 1983.

[30] National Research Council (NRC), "Ecological Knowledge and Environmental Problem-Solving," National Academy Press, Washington, 1986.

[31] J. Burger, "The Historical Basis for Ecological Risk Assessment," *Preventive Strategies for Living in a Chemical World*, New York Academy of Sciences, New York, 1997, pp. 360-371.

[32] M. Sorensen, W. Gala and J. Margolin, "Approaches to Ecological Risk Characterization and Management: Selecting the Right Tools for the Job," *Human and Ecology Risk Assessment*, Vol. 10, No. 2, 2004, pp. 245-269. doi:10.1080/10807030490438193

[33] S. B. Norton, D. J. Rodier, J. H. Gentile, W. H. van der Schalie, W. P. Wood and M. W. Slimak, "A Framework for Ecological Risk Assessment for EPA," *Environmental Toxicology Chemistry*, Vol. 11, 1992, pp. 1663-1672. doi:10.1002/etc.5620111202

[34] National Research Countil (NRC), "Human Biomonitoring of Environmental Chemicals," National Academy Press, Washington, 2006.

[35] J. Burger, "Bioindicators: Types, Development, and Use in Ecological Assessment and Research," *Environmental Bioindicators*, Vol. 1, No. 1, 2006, pp. 22-39. doi:10.1080/15555270590966483

[36] J. Burger, "Bioindicators: A Review of Their Use in the Environmental Literature 1970-2005," *Environmental Bioindicators*, Vol. 1, No. 2, 2006, pp. 136-144. doi:10.1080/15555270600701540

[37] K. Sexton and D. Hattis, "Assessing Cumulative Health Risks from Exposure to Environmental Mixtures: Three Fundamental Questions," *Environmental Health Perspective*, Vol. 115, No. 5, 2007, pp. 825-832. doi:10.1289/ehp.9333

[38] M. R. Greenberg, "Energy Parks for Former Nuclear Weapons Sites? Public Preferences at Six Regional Locations and the United States as a Whole," *Energy Policy*, Vol. 3, 2010, pp. 5098-5107. doi:10.1016/j.enpol.2010.04.040

[39] H. Lund, "Renewable Energy Strategies for Sustainable Development," *Energy*, Vol. 32, No. 6, 2007, pp. 912-919. doi:10.1016/j.energy.2006.10.017

[40] J. Burger, "A Framework for Analysis of Contamination on Human and Ecological Receptors at DOE Hazardous Waste Site Buffer Lands," *Remediation*, Vol. 17, No. 2, 2007, pp. 71-96. doi:10.1002/rem.20125

[41] W. Rees, "Ecological Footprints of the Future. Overview," *People Planet*, Vol. 5, No. 2, 1996, pp. 6-9.

[42] J. Kitzes, A. Peller, S. Goldfinger and M. Wackernagel, "Current Methods for Calculating National Ecological Footprint Accounts," *Science for Environment and Sustainable Society*, Vol. 4, No. 1, 2007, pp. 1-9.

[43] P. Denholm and R. M. Margolis, "Land-Use Requirements and the Per-Capita Solar Footprint for Photovoltaic Generation in the United States," *Energy Policy*, Vol. 36, No. 9, 2009, pp. 3531-3543. doi:10.1016/j.enpol.2008.05.035

[44] J. Kitzes, M. Wackernagel, J. Loh, A. Peller, S. Goldfinger, D. Cheng and K. Tea, "Shrink and Share: Humanity's Present and Future Ecological Footprint," *Philosophical. Transactions of the Royal Society of London Biological Sciences*, Vol. 363, 2008, pp. 467-475. doi:10.1098/rstb.2007.2164

[45] N. H. Afgan, M. G. Carvalho and N. V. Hovanov, "Energy System Assessment with Sustainability Indicators," *Energy Policy*, Vol. 28, No. 9, 2000, pp. 603-612. doi:10.1016/S0301-4215(00)00045-8

[46] K. S. Smallwood and C. G. Thelander, "Bird Mortality in the Altamont Pass Wind Resource Area," *Journal of Wildlife Management*, Vol. 72, No. 1, 2008, pp. 215-223. doi:10.2193/2007-032

[47] K. S. Smallwood and L. Neher, "Map-Based Repowering of the Altamont Pass Wind Resource Area Based on Burrowing Owl Burrows, Raptor Flights, and Collisions with Turbines," Sacramento, California Energy Commission, PIER Energy-Related Environmental Research Program, 2009.

[48] K. S. Smallwood, L. Ruggeb and M. L. Morrison, "Influence of Behavior on Bird Mortality in Wind Energy Developments," *Journal of Wildlife Management*, Vol. 73, No. 7, 2009, pp. 1082-1098. doi:10.2193/2008-555

[49] B. Snyder and M. J. Kaiser, "Ecological and Economic Cost-Benefit Analysis of Offshore Wind Energy," *Renewable Energy*, Vol. 34, No. 6, 2009, pp. 1567-1578. doi:10.1016/j.renene.2008.11.015

[50] J. E. Mock, J. W. Tester and P. M. Wright, "Geothermal Energy from the Earth: Its Potential Impact as an Environmentally Sustainable Resource," *Annual Review of Energy and the Environment*, Vol. 22, 1997, pp. 305-356. doi:10.1146/annurev.energy.22.1.305

[51] R. Waples, R. W. Zabel, M. D. Scheuerell and B. L. Sanderson, "Evolutionary Responses by Native Species to Major Anthropogenic Changes to Their Ecosystems: Pacific Salmon in the Columbia River Hydropower System," *Molecular Ecology*, Vol. 17, No. 1, 2007, pp. 84-96. doi:10.1111/j.1365-294X.2007.03510.x

[52] H. I. Jager and B. T. Smith, "Sustainable Reservoir Operation: Can We Generate Hydropower and Preserve Ecosystem Values," *River Restoration Applications*, Vol. 24, 2008, pp. 340-352. doi:10.1002/rra.1069

[53] P. W. Gerbens-Leenes, A. Y. Hoekstra and T. van der Meer, "The Water Footprint of Energy from Biomass: A Quantitative Assessment and Consequences of an In-

creasing Share of Bio-Energy in Energy Supply," *Ecology Economics*, Vol. 68, 2009, pp. 1052-1060. doi:10.1016/j.ecolecon.2008.07.013

[54] S. B. Roy, S. W. Pacala and R. L. Walko, "Can Large Wind Farms Affect Local Meteorology," *Journal of Geophysical Research*, Vol. 110, 2004, pp. 191-201.

[55] S. B. Roy, "Simulating Impacts of Wind Farms on Local Hydrometeorology," *Journal of Wind Engineering and Industrial Aerodynamics*, Vol. 99, No. 9, 2010, pp. 491-499.

[56] W. Nowak and A. Stachel, "Systems of Simultaneous Operation of Low- and High-Temperature Heating Installations and the Effect on the Degree of Geothermal Energy Utilization in a Geothermal Heating Plant," *International Geothermal Conference*, Reykjavik, September 2003, 61 p.

[57] K. P. York, Z. M. G. S. Jahangi, T. Solomon and L. Stafford, "Effects of a Large Scale Geothermal Heat Pump Installation on Aquifer Microbiota," *2nd Stockton International Geothermal Conference Proceedings*, Stockton, 1994, pp. 49-56. http://intraweb.stockton.edu/eyos/energy_studies/content/docs/proceedings?YORK

[58] V. M Fthenakis, H. C. Kim and E. Alsema, "Emissions from Photovoltaic Life Cycles," *Environmental Science & Technology*, Vol. 42, No. 6, 2008, pp. 2168-2174. doi:10.1021/es071763q

[59] H. Reiquam, "A Method for Optimizing Pollutant Emissions in an Airshed," *Atmospheric Environment*, Vol. 5, No. 1, 1971, pp. 57-64. doi:10.1016/0004-6981(71)90045-X

[60] J. H. Seinfeld, "Optimal Location of Pollutant Monitoring Situations in an Airshed," *Atmospheric Environment*, Vol. 6, 1972, pp. 847-858.

doi:10.1016/0004-6981(72)90056-X

[61] B. Snyder and M. J. Kaiser, "Ecological and Economic Cost-Benefit Analysis of Offshore Wind Energy," *Renewable Energy*, Vol. 34, No. 6, 2009, pp. 1567-1578. doi:10.1016/j.renene.2008.11.015

[62] M. R. Raupach, G. Marland, P. Ciais, C. Quere, J. G. Canadell, G. Klpper and C. B. Field, "Global and Regional Drivers of Accelerating CO_2 Emissions," *PNAS*, Vol. 104, 2007, pp. 10288-10293. doi:10.1073/pnas.0700609104

[63] EIA (US Energy Information Administration), "Energy-Related Carbon Dioxide Emissions," 2011. http://www.eia.gov/oiaf/ieo/emissions.html

[64] Bureau of Labor Statistics, "Labor Statistics," 2011. www.BLS.gov

[65] "NESCAUM. Northeast States Report," 2011. www.nescaum.org/documents/pr031104mercury.pdf.

[66] IOM (Institute of Medicine), "Seafood Safety," National Academy Press, Washington, 1991.

[67] IOM (Institute of Medicine), "Seafood Choices: Balancing Benefits and Risks," National Academy Press, Washington, 2006.

[68] R. L. Vengosh, G. S. Dwyer, H. Hsu-Kimand A. Deonaire, "Environmental Impacts of the Coal Ash Spill in Kingston, Tennessee: An 18 Months Survey," *Environmental Science & Technology*, Vol. 44, No. 24, 2010, pp. 9272-9278. doi:10.1021/es1026739

[69] I. Linkov, D. Loney, S. Cormier, F. K. Satterstrom and T. Bridges, "Weight-of-Evidence Evaluation in Environmental Assessment: Review of Qualitative and Quantitative Approaches," *Science of the Total Environment*, Vol. 407, No. 19, 2009, pp. 5199-5205. doi:10.1016/j.scitotenv.2009.05.004

10

Study on the Risk of Regional Energy Security Cooperation

Ying Shen, Xintong Yang, Xiaoli Guo[*]

Changchun University of Technology, Changchun, P.R.China

Email: shenying@mail.ccut.edu.cn

ABSTRACT

Energy security is an issue that many countries pay more attention to. Cooperation is a good way to solute it. According to the complex system theory, the regional energy cooperation means a state or process of balance system formed by the interaction and behavior coordination between agents. The features of the regional energy cooperation risk include: uncertainty, potential, fuzziness, diversity, relevance and particularity. And the regional energy cooperation risk can be divided into risk inside the system such as main body ability structure of the risk and main body decision-making risk, and the risk outside the system such as external environment factors and cooperation pattern factors.

Keywords: Energy Security; Regional Cooperation; Risk

1. Introduction

In the background of economic globalization and integration, with the growing trend of interconnect and interdependent between energy producers and consumers, as characters of International Energy Security, reciprocity, symbiosis and cooperative appear increasingly. It is a trend to carry out cooperation and participate in the international competition to develop the industry of each country. In international energy security cooperation, the economic environment, political environment, management terms are more complicated than in the domestic areas, while there are a great deal of advantages from participating in International Energy Cooperation. There are a lot of risks, which including not only general risk from opportunism, trust mechanisms, but also special risk because Energy as a commodity has the nature of political, economic and diplomatic. Therefore it is an important problem for cooperative behavior agents to strengthen risk prediction and prevention.

According to the complex system theory, the international energy cooperation risk can be divided into risk inside the system and the risk outside the system. The risk inside the system includes main body ability structure of the risk and main body decision-making risk. The risk outside the system includes external environment factors and cooperation pattern factors.

2. The Analysis of Regional Energy Cooperation Risk

2.1. Regional Energy Cooperation Risk Concepts and Features

"Energy cooperation" is still have no clear uniform definition, but the cognition about energy cooperation understanding of target is consistent: The aim of energy cooperation is to ensure energy security, cooperation is a kind of way to keep energy security, the essence of energy cooperation is take cooperation for safety. The concept of energy cooperation is that it is the unity of process and state, to prediction "static" state from the process of "move", and in turn reflects the process of "move" from the process of "static" state. The regional energy cooperation means a state or process of balance system formed by the interaction and behavior coordination between agents. Its connotation is: 1) behavior subject is governments; 2) cooperation method is dialogue and consultation; 3) the scope of cooperation involves the military, security, foreign, ecological and other areas of production consumption process; 4) goal is to achieve a win-win situation.

We can see that the regional energy cooperation is "set by set", it constitutes by the N (N$\geq$2 system elements, and each system elements is a small set. The unbalanced system is a chaotic system, and it is the original form of cooperation. Temporary balance system will become into the unbalanced system because of the changed conditions, and then become the next starting point of the conversion. The final balance system is an ideal state of cooperation, is also the highest state of cooperation, the performance of cooperation is the largest interests of the whole system.

---

[*]Corresponding author.

The regional energy cooperation contains both the balanced and disequilibrium system state and also contains the conversion process from disequilibrium state to balance state. The relationship between the body means the behavior interaction through the affection between each other in the outside world, the result of the interaction is contradictions and conflicts. Behavior coordination refers to the body through coordination behavior including means or measures to solve the contradictions and conflicts, such as pointing to a balanced "cooperation" and back to a balanced "betrayal", the purpose is to realize their own benefit maximization (cooperation benefits decrease the cost of the betrayal or cooperation income minus the betrayal costs). In particular space, when agents showed clear point to balance system behavior coordination, and began to have the organization function, so the cooperation began. The regional energy cooperation risk is due to the system internal and external potential factors, lead to regional energy cooperation deviation "equilibrium" the possibility and degree of state. This concept contains two meanings: one is the probability of imbalance happened; the other is the extent of the imbalance happened.

1) The uncertainty

The uncertainty means whether or not the risk is happening, whenever it occurs and the loss rate are not certain. This uncertainty is more evident in the regional energy cooperation. For example, some risk occur in the cooperation, some risk occur in the cooperation process produces, some occur in the end of cooperation. The state of risk and the influence degree of uncertainty change are random happen.

2) The potential

The potential is the existence of risk in basic form. In addition, the risk can exist in certain conditions as the prerequisite. With the change of external environment, the risk in regional energy cooperation can present different development trend.

3) The fuzziness

The internal and external environment is very complex in the regional energy cooperation. Some information is known, while some other information is unknown. The influence of risk factors is very difficult to quantify .In addition, risk factors can't accurately describe.

4) The diversity

The scope of the regional energy cooperation and energy of the economical, political and diplomatic triple attributes to the diversity of risk. The regional energy cooperation has much risk which includes economic, political and management.

5) The relevance

Risk factors in the regional energy cooperation can relate to each other .In addition, they can influence each other .In addition, a kind of risk factor may cause other

risk factors ensued, and even cause the disruption of cooperation .As a result, the influence of cooperation is integrity, such as political friction may bring economical contradiction and environmental protection contradiction, resulting in the risk of management techniques.

6) The particularity

Relative to the general cooperation, energy cooperation risk has its particularity. Energy is often as countries carry out the strategy of points, so the risk of energy cooperation has characteristics of political strategy.

2.2. The Risk Source of Regional Energy Cooperation

According to the definition, all the influence of factors which can destroy stable cooperation may result in the occurrence of cooperation risk.

To begin with, from benefits of the conflict, the regional energy cooperation is ultimately behavior between bodies of cooperation, the fundamental purpose of the behavior cooperation is benefits. If actions without expected benefits in the cooperation, it may choose to quit. Each behavior body can hope its own benefits are the largest. At the same time, they can hope to benefits from cooperation that shares the largest piece of. Once the behavior body do not get due reward, other agents from cooperation can get the far more profitable than him, the cooperative initiative will inevitably be hit, as soon as the other opportunities, very likely choose betrayal.

Secondly, the behavior of energy cooperation is spontaneously formed which has individual independence. In addition to contract or agreement, the behavior of energy cooperation is not coercion constraints or control. Agents to continue or not random cooperation, distrust and speculation will be moved cooperation. The characteristics of energy cooperation to cooperation risk is not completely avoided, so the only can do is is how to reduce the occurrence of possibility, cooperation is more likely.

Finally, the external environment of the regional energy cooperation is an important factor [1]. The regional energy cooperation not only considers the economical environment, but also considers the political environment. Cooperation of both sides is whether it helps to promote cooperation on international relations, domestic political environment is stable, policy and economic environment on energy cooperation is to encourage or restrictions.

3. The Risk of Subject Ability Structure

In recent years, some scholars study the influence of the ability structure to the regional cooperation. Cui Wei-guo (2004) [2] shows that when the ability is too big between regions, cooperation is almost have no chance. Yang Xian-ming (2005) [3] based on principal component analysis, choose the allocation of resources capacity, the transformation of the industrial structure ability, the technical development capability, trade development capacity and opening economy ability and these five indicators to evaluate the ability of structure level, and show the influence of the area ability structure to regional cooperation. Huang Ning (2008) [4] by using the coupling formula of district ability structure, explain two areas where the more similar the total capacity structure index and the bigger the ability structure of the coupling index, the greater likelihood the two areas to develop the economic cooperation. Proof that the two regions structure decides the ability to gain total benefits of cooperation size, the stronger the both part of cooperation ability structure, the greater total benefits the cooperation can achieve, and the part which have strong ability structure may obtain more benefit. These studies showed the main body ability structure decides the regional cooperation scope and cooperation efficiency.

This paper begin with the emphasize of the structure ability of main body, according to the characteristics of the international energy cooperation, Have the evaluation from the following four aspects: the allocation of resources capacity, technical development ability, the trade development ability and the ability to open economic, and get ability structure index through the calculation. This paper discusses the influence of the structure of the country's ability to the region's energy cooperation, "for short, body ability structure of the risk.

3.1. The Allocation of Resources Capacity

The index of resources capacity allocation mainly is constituted by the index of level of economic development, investment spending, and financial position of government, the degree of mercerization and allocation efficiency. It includes economy aggregate, population, social total retail sales of consumer goods, the total amount of investment and the level of industrial structure.

3.2. Trade Development Ability

Reflect economic extroverted degree, including total import and export, FDI absorbed dose and FDI absorbed dose per GDP.

3.3. The Economic Open Ability

Ensure that you return to the 'Els-body-text' style, the style that you will mainly be using for large blocks of text, when you have completed your bulleted list.

3.4. Technology Development Ability

Technology development capability index consists of several indicators including technology innovation, the technical transformation, technology absorb. It also In-

clude human input proportion, labor productivity and the percentage of R&D in GDP.

4. Subject Decision-making Risk

Behavior subject is the leader and participant of the energy cooperation, with the cooperation's generation and development, behavior between bodies from the start of the isolated affection to each other transformed into contact interaction. Behavior decision-making risk mainly shows the following respects in the game playing.

4.1. Trust Risk

Trust is a powerful guarantee to maintain the process of cooperation. Agents lack of mutual trust, which will increase wrong cognition and hostility between each other for cooperation intention, cooperation ability, increase conflict of interests in cooperation on both sides, influencing behavior body's decision-making. At present, psychology, sociology and economics, and other disciplines have relatively common understanding on trust production and the existence conditions, mainly refers to interdependence, communication and continuous contacts [5].

4.2. Opportunism Risk

Opportunistic is a form of behavior which pay attention to the interests of the short-term. The behavior body's subjective understanding and psychological anticipation can form different cooperation vision, and have different expectations about the future benefits. When a real income and ideal income have larger divide, the body behavior in the next cooperation may exit. Or in a game, earn temporarily interests, withdraw cooperation. If the actions in the cooperation opportunity in the body pursue socialist, seek temporary interest, and damage the interests of the partner, which will definitely affect the continuous and stable cooperation.

4.3. Information Asymmetry

The objective cause of the information asymmetry is that information is processed data, due to data collection staff, technical, the method, the purpose is different, after data processing the information upon which it is unified, and the more the processed process and levels and the errors to the more times amplified, the greater the information distortion. Subjective reason is that different economic individuals get information asymmetry of ability, and at the same time the provider are interested in news keep some important information or distribute some false information, and another on the market there are insider trading which will also lead to both trade information asymmetry. In the process of repeated game, information asymmetry often make the body to make the wrong decision behavior, the betrayal of a party in order to avoid suffered losses because of betrayal in the next cooperation.

5. The External Environment Factors

The international energy cooperation system's economic environment and the political environment is the main source of external cooperation risk [6]. The **Table 1** below shows the international energy cooperation risk environment factors.

6. Cooperation Pattern Factors

Cooperation pattern is to point to the energy consumer country realize the energy consumer energy security; take effective cooperation way to solve these problem [7]. According to the historical experience, the international energy cooperation pattern mainly has: the common goal cooperation, share information cooperation and common action cooperation and depend on each other cooperation [8]. The first three cooperation pattern is mainly used for the cooperation between consumer country and the last one is used for the cooperation between producers and consumers.

6.1. Common Goal Cooperation

This model only limited target, not set ways and the methods, the elastic is larger. According to their specific countries different measures have been taken, more

Table 1. The international energy cooperation risk environmental factors.

Environmental risk category	Risk environmental factors
Geopolitical environment	geopolitical
International relations	Political relations, economic relations
Economic environment	Economic level, inflation, and foreign exchange policy, economic risk
Cost environment	Exploration and development cost, profit standard, infrastructure, pipe facilities
Finance and tax environment	Currency stability, the terms of the contract, the tax system, the contractor earnings ratio, and cost recovery limit
Political environment	Political stability, political risk, political system
System environment	Availability of resources, system stability, policy available for foreigners, the mining right, the openness of resources cooperation, foreign cooperation policy

flexible, so, main body ability structure of the risk and the external environment factors have little influence. The target set by the all countries is consensus. Each country against the wishes of the cooperation is very small, speculative risk is not big. However, this pattern has not the effectiveness of enforcement, and set the common goal of general is long-term, profit is not sure in the long-term, which will affect cooperation faith and the enthusiasm in short term, trust risk and asymmetric information risk may produce during the cooperation. This model is used to solve "long-term supply" problems between consumer countries, set a unity of purpose. As for IEA is oil member in 1985 set 12 guiding principles about oil imports goals and energy policy.

6.2. Sharing Information Cooperation

Share information requirements agents to provide timely complete information to each other, can avoid the asymmetric information, to obtain more information from both parties, bring the long-term benefits, the risk of information asymmetry is not big. But, the parties will worry about the truth be understand by the external, thus creating the opportunity and conditions for the competitor or fear that the other party to provide false information and achieve unbalance benefit. So, it's easy to produce the trust risk and speculative risk. Second oil crisis, the IEA expand the member countries to share information cooperation, to the oil market, oil reserves, oil exploration information comprehensive monitoring, with important information released in time, avoid the market price, chaos and irrational decisions. Between member states set up promptly supply and demand in time, long-term supply and long-term needs of the comprehensive information sharing cooperation.

6.3. Common Action Cooperation

Common action cooperation pattern is a higher mandatory cooperation pattern, under this pattern the obligation of all parties burden is heavy, the cost of all countries and betrayal is higher, implement it step by step, depending on the behavior of the other countries to decide whether the next step cooperation is going on, this way effectively prevent betrayal behavior occurrence, beneficial to the long-term cooperation. But, this kind of cooperation pattern have the influence of international body ability structure is bigger, it is, the more urgent and mandatory the situation, the notable the ability structure risk. In addition, the cost and benefit from cooperation distribution difficulties, so trust risk and information asymmetry risk is great. IEA internal emergency oil sharing mechanism, IEA about national emergency oil sharing mechanism strategic petroleum reserve's number and the provisions of the common release, about the "minimum reserve price" regulation, in energy technology to research and belong to the category of common action contract mode.

6.4. Mutually Dependent Cooperation

Mutually dependent cooperation involves only two countries, by domestic situation great influence, if one party at a disadvantage, subject to the other party, cooperation uneven distribution of income, the main body of a large effect on the ability structure. Easy to cause the third party to misunderstanding or the third party conflicts of interest, the external environment is also great risk. So countries will not only take bilateral mutually dependent cooperation secure energy security, and will rely on each other and multilateral cooperation combined.

REFERENCES

[1] Energy Security, "Managing Risk in a Dynamic Legal and Regulatory Environment," www. World online. Oxford University Press, 2004.

[2] W. G. Cui and X. H. Liu, "Regional Economics," Bei Jing: Economic Science Press, 2004.

[3] X. M. Yang and Y. Li, "Study on the Pan Pearl River Region Cooperation Based on the Ability Structure," *Social Sciences in Guangdong,* Vol. 3, 2005, pp. 38-44.

[4] N. Huang, "Model of Relationship between Ability Structure and Economic Cooperation," *Contemporary Economics,* Vol. 10, 2008, pp. 108-110.

[5] P. Sztompak, "Trust-A Sociological Theory," Beijing: Zhonghua Book Company, 2005.

[6] H. Z. An, F. R. Chen and Y. L. Lei, "The Design of Energy Resource International Cooperation Environment Assessment Index System," *Reformation & Strategy,* Vol. 23, No. 11, 2007, pp. 45-49.

[7] O. Noreng, "Oil policy in 1980`s: Pattens of International Cooperation," New York: McGraw-Hill Book Company, 1978.

[8] Z. Z. Ye, "International Energy Cooperation Model and Strategic Choice of China," Doctoral Dissertation of China Foreign Affairs University, 2005.

11

Energy Simulation of PV Hybrid System for Remote Villages of Thailand

Jitiwat Yaungket[1], Tetsuo Tezuka[1], Boonyang Plangklang[2]
[1]Graduated School of Energy Science, Kyoto University, Kyoto, Japan
[2]Faculty of Engineering, Rajamangala University of Technology, Thanyaburi, Thailand
Email: jitiwaty@gmail.com

ABSTRACT

A surveys conducted in remote rural areas revealed that some people still do not have access to electricity. Subsequently, the Ministry of Interior was assigned to invite relevant agency to a meeting designed to assist those without access to electricity the opportunity to enjoy electricity comprehensively in all households. This paper presents the design simulation of PV Hybrid System for remote village. The design uses real data from field investigated for different regions in Thailand. The obtained data was compiled as load profile. The analysis results will describe for the design for given energy generation from PV hybrid system. The energy to be generated from the back up diesel generator when the power generated from renewable energy technologies fails to meet the energy demand. This information can use as guideline for future developing energy system planning in the area without the sources of electricity.

Keywords: Energy Simulation; Hybrid System; Remote Village

1. Introduction

Thai government was aware of the importance and necessity for promotion of popular participation, the government viewpoint was that Thai people in all areas should have the opportunity to be informed of the news [1]. However, surveys conducted in remote rural areas revealed that some people still do not have access to electricity. Subsequently, the Ministry of Interior was assigned to invite relevant agency to a meeting designed to assist those without access to electricity the opportunity to enjoy electricity comprehensively in all households, at the least the opportunity to view television for news information even though the investment is not economically worthwhile. The reason was that it was considered government duty to provide support for public equality. The goal set was for all households to have access to electric power within the years 2005-2006. The principle was that homes located in areas where power cable poles can be installed, the Provincial Electricity Authority (PEA) was to expedite operations. Meanwhile, homes located in remote areas, in wooded, mountainous, island areas, etc. where power cable poles cannot be installed, Solar Home System (SHS) was to be installed instead. The SHS standard kit, which is provided free of cost to the rural households is composed of a PV module of 120 Wp, 1 control unit circuit with inverter, 2 fluorescent lamps and a 12 V/125Ah battery, those locally produced. Private companies contracted by the Provincial Electricity Authority (PEA) are implementing the systems [2]. After systems installation has been carried out, maintenance work would be reassigned to local administration agencies at the sub-district (Tambon) level. In this way, we called Sub-district Administrative Organization (SAO). People will be given training and knowledge provided on systems usage and maintenance, etc. Accordingly, the SAO will be responsible for laying down administrative rules for further sustained operations in accordance with concerned projects. Target areas of the project are households still without access to electricity and located in remote areas where the PEA cannot provide service numbering around 290,716 households. The source of the operational budget is Local Administration Agency subsidy funds for Fiscal Years B.E. 2547 and 2548 (A.D. 2004-2005) totaling 7,631,295,000 baht[1] [3], which translates to an installation expense per household of 25,000 baht [1]. Power supplied by solar home system SHS is 120 Watts with Alternating Currency (AC) power used with two 10-Watt fluorescent lamps and one 14-inch television set. The project has

---

[1]30 Baht = 1 USD.

finalized operations at the end of 2006.

Energy community in remote villages of Thailand found themselve being equipped with solar PV systems provided by the government of Thailand as we mentioned above and diesel generators donated by the organizations in Thailand or agriculture diesel generator uses to generate electricity for support recovery efforts. However the electricity generating capacity the villagers could only benefit from an electricity supply for a few hours each day and besides this, important parts of the equipment failed. In addition, the conclusions of this study suggest that by using the existing workable components as a base, the creation of a hybrid energy system using diesel and renewable energy sources could provide a sustainable 24 hour-electricity supply for the villages. This is to be accomplished by showing data on real time system usage situations, simulation and guidelines on development of PV hybrid system to be the optimum electric power generation system. Supporting information comprise information directly from system survey questionnaires and interview with villagers was derived for four different regions in Thailand as follows are; Bann Bon Khao Kang Riang Village, Kanchanaburi Province (Western Thailand), Bann Koh Jik Island Chantaburi Province (Eastern Thailand), Bann Pa Ya Sai Village, Chiang Mai Province (Northern Thailand) and Bann Klong Rua Village, Chumporn Province(Southern Thailand) and secondary information from various agencies concerned [4]. The foremost hindrance to quality of life development on the study areas is the lack of national grid connected. It is unlikely that a grid extension could be possible in the near future as the current electricity demand on the study areas is low and not economically viable for Thailand's public utility, the Provincial Electricity Authority (PEA), to invest in the expensive underwater cable to connect the Island with the mainland's

grid system also in the mountainous areas. The planned hybrid system discussed in this study presents the possibility of expanding electricity supplies in ways that are controlled by the community itself.

2. Energy Demand Assessment

As mentioned, the principle design of this mobile PV hybrid system will be considered at the stability of power supply. The real time monitoring system is also included for data analysis. For this study, we selected an example of load profile as shown in **Figure 1**. The selected load profile is a typical load in remote areas which has the peak load in the evening. The load normally includes the daily life electrical load without air conditioning system.

After having the load profile, the capacity of the solar cell, P_{peak}, can be calculated by using a method [5] as followings.

$$Q = E_{el}/E_{th} \qquad (1)$$

$$E_{th} = n\text{Efficiency} \times I_{glob} \times APV_{array} \qquad (2)$$

$$P_{peak} = n\text{Efficiency} \times I_S \times APV_{array} \qquad (3)$$

$$E_{th} = P_{peak} \times \left(I_{glob}/I_S\right) \qquad (4)$$

$$Q = \left(E_{el}/\left(I_{glob}/P_{peak}\right)\right) \times I_S \qquad (5)$$

$$P_{peak} = \left(E_{el} \times I_S\right)/\left(I_{glob} \times Q\right) \qquad (6)$$

when: P_{peak} = peak power of the PV array under STC [kWp].

E_{el} = real electric output energy of the system [kWh/a].

I_S = incident solar radiation under S [1 kW/m²].

I_{glob} = annual global solar radiation [kWh/m²a].

Q = quality factor of the system.

E_{th} = theoretical output energy of the system [kWh/a].

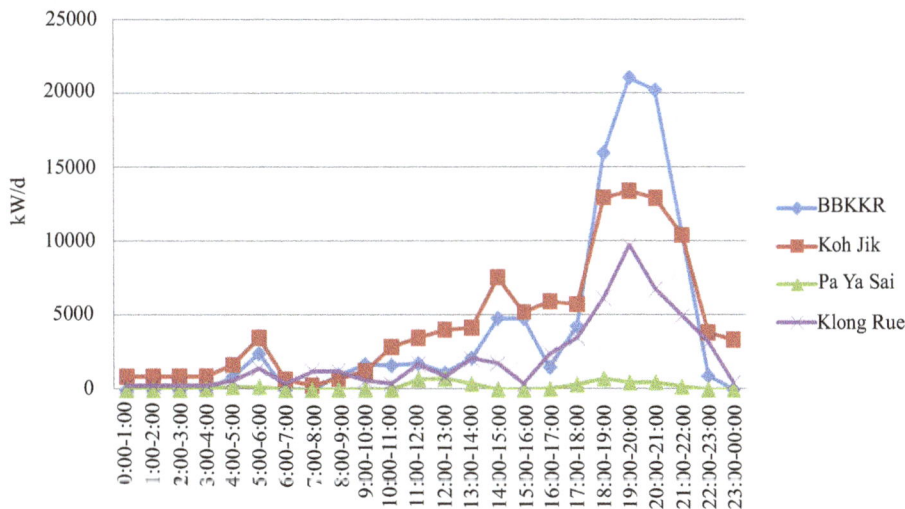

Figure 1. Daily load profile of areas study for system design.

n Efficiency = efficiency of the PV array [decimal].

APV_{array} = area of the PV array [m^2].

The battery capacity is calculated by (7).

$$CB = 10 \times P_{peak} \qquad (7)$$

when: P_{peak} = peak power of the PV array under STC [kWp].

CB = battery capacity [kWh].

Therefore the calculation can be done as following by having the E_{glob} (kWh/m^2) in Thailand and the load from the load profile (kWh/d) as shown the results in **Table 1**.

3. System Design and Selection of Operation System

The systems have 3 main operation modes. In the first mode where solar energy is sufficient for generate electricity and supply to the load directly, the excess energy will be charged into the battery, when the solar energy shortage battery will be discharge to supply the load. In the second mode where the solar and battery energy not sufficient the system will be switch to use the energy from the grid line until the energy level of solar and battery is possible to supply the load it will be switch back.

In the third mode where the energy from the system is into grid line by the grid connected inverter in case of battery full and no load demand.

The selection of operation system consists of system voltage and system operation.

- System voltage: for small hybrid system voltage should be 12 - 48 V, for medium size should be 120 - 240 V, and a big size is 480 - 600 V. Hence, the level of battery voltage can be selected depending on the level of load consumption [6]. For the hybrid system of Bann Pa Ya Sai is a very small size, there, the system voltage is 24 volts has selected. Therefore three villages larger than Bann Pa Ya Sai voltage level selected the voltage of battery 96 volts.
- The simulations for hybrid system of four villages are as shown in **Figures 2-5**.

Estimation of electricity generating system HOMER tool uses for estimate the electricity generating system. The capacity of the components of the hybrid system simulation in this paper is shown in **Table 2**.

Table 1. The calculation results of the capacity of PV and battery capacity.

Village	Peak power [kWp]	battery capacity [kWh]
BBKKR	30.3	303
Koh Jik	35	350
Pa Ya Sai	1.4	14
Klong Rue	16.3	163

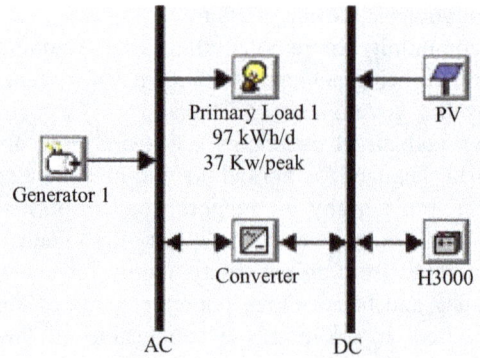

Figure 2. Simulation of BBKKR hybrid system.

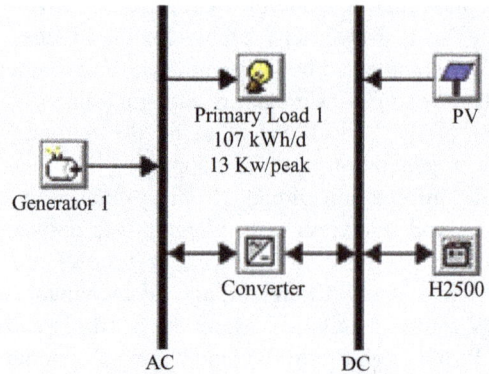

Figure 3. Simulation of Koh Jik hybrid system.

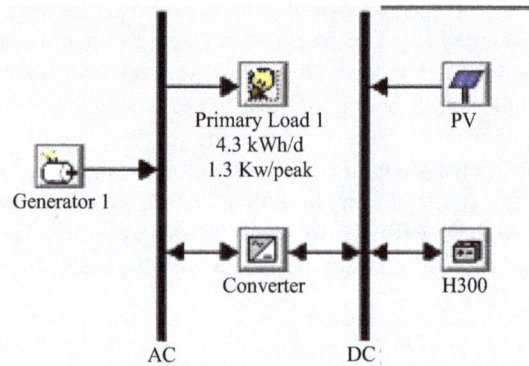

Figure 4. Simulation of Pa Ya Sai hybrid system.

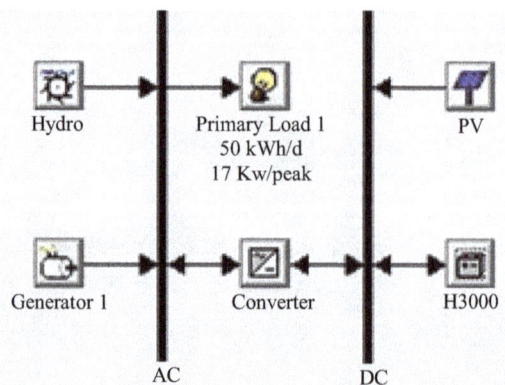

Figure 5. Simulation of Klong Rue hybrid system.

4. Results

The PV hybrid system simulated in the present investigation considered at the stability of power supply for different regions in Thailand. In order to show the ultimately PV hybrid power, the monthly average electric production for 4 villages is presented in **Figures 6-10**. The figures show distinctly of the different of energy sources to generate electricity. The Solar PV system is the single largest unit to supply the electricity in the north, west, east of Thailand whereas the south region of

Table 2. The capacity of the components.

Village	PV [kWp]	battery capacity [kWh]	Diesel Gen. [kW]	Hydro power [kW]
BBKKR	25	303	15	-
Koh Jik	20	350	15	-
Pa Ya Sai	1	15	2	-
Klong Rue_1	4	163	10	100
Klong Rue_2	7	163	20	w/o HP

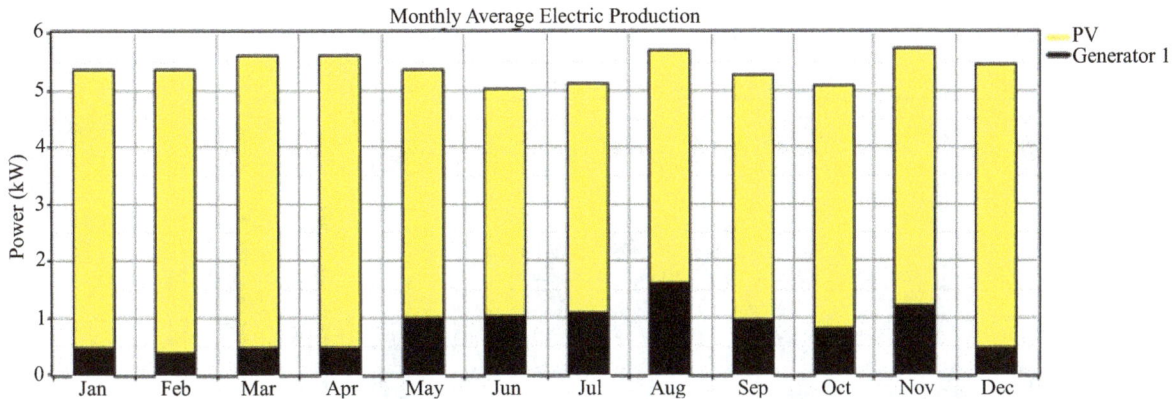

Figure 6. Monthly average electric production in BBKKR village.

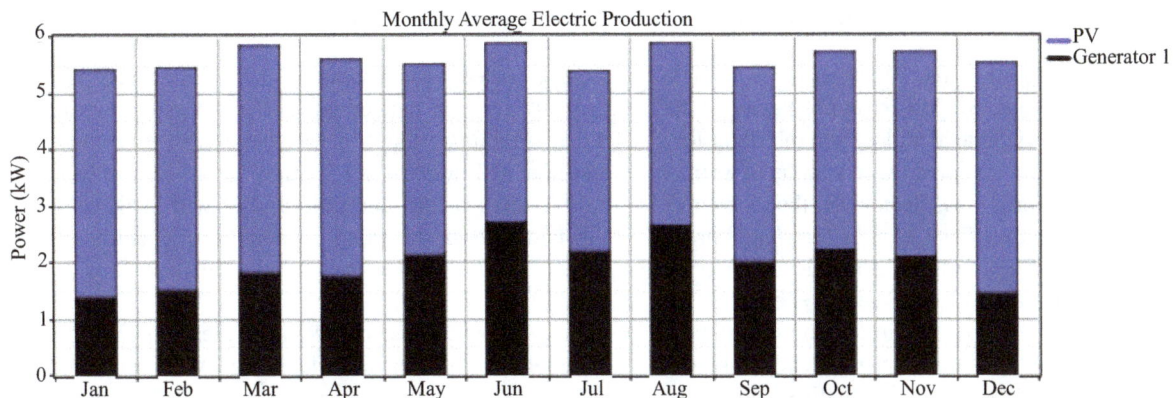

Figure 7. Monthly average electric production in Koh Jik village.

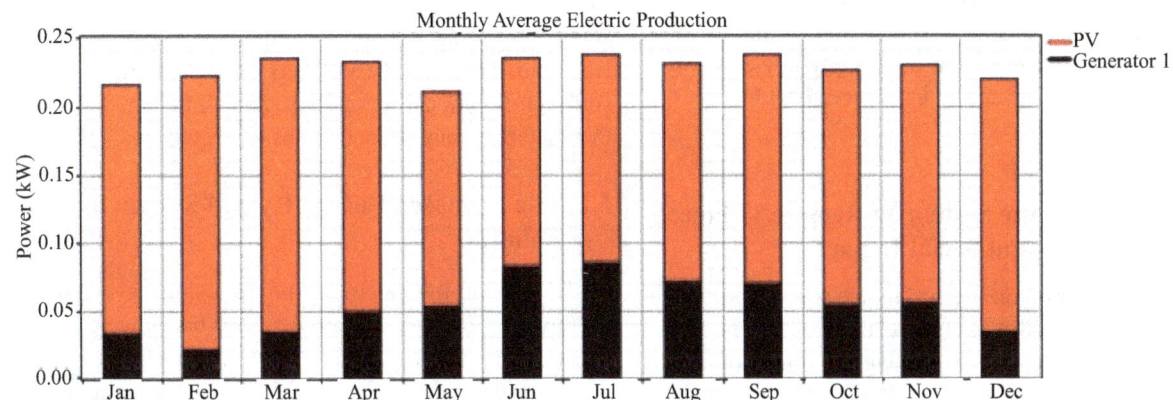

Figure 8. Monthly average electric production in Pa Ya Sai village.

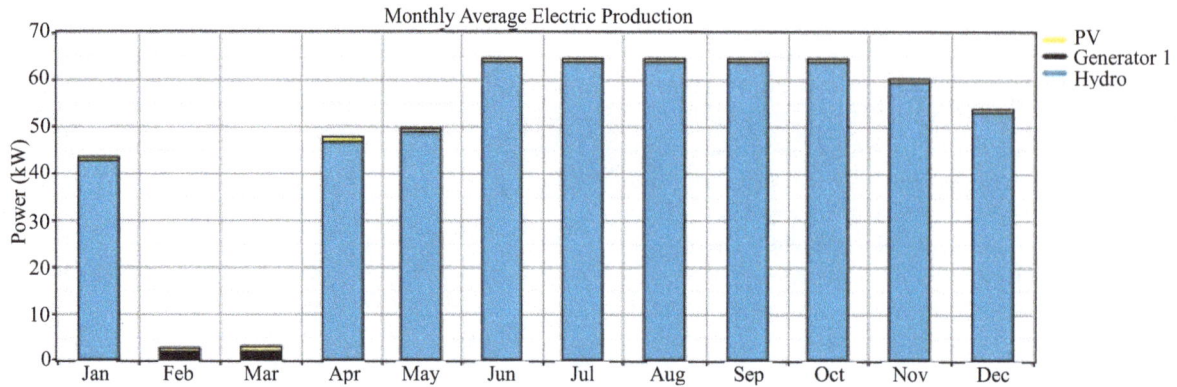

Figure 9. Monthly average electric production in Klong Rue village with hydro power.

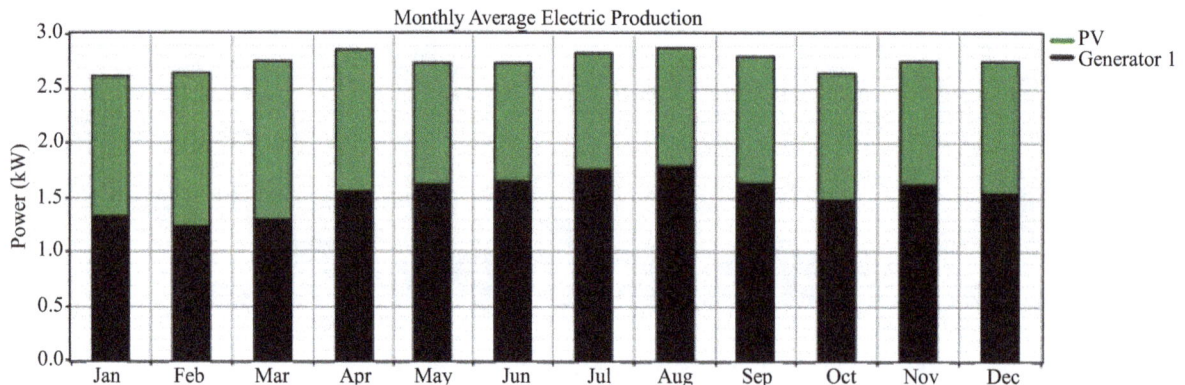

Figure 10. Monthly average electric production in Klong Rue village without hydro power.

Thailand, we found that the solar PV system was not able to fully satisfy their electric power supply needs. What this means is that the geographical location of Klong Rua village is situated in a regional with a rainy season 8 months per annum. Thus, it has been given the nickname: "*Eight Months Rainy and Four Months Sunshine*" province, where the solar PV system is unable to generate electricity while it is raining. Thus, diesel generator is the major share of the hybrid system in the south region of Thailand. It is due to these reasons that potency in southern areas regarding water is an alternative choice for electric power for this region with endless uses.

The figures below showed the results of the study areas from HOMER modeling. The results of simulation show that the designed system can supply power stably without energy shortage shown in **Figures 10-15**.

5. Discussions

5.1. Evaluation of the System on Remote Village: Sustainability and Social Impact

The most significant issues regarding the sustainability of the existing electricity arrangements are 1) ongoing technical sustainability of the system; 2) fuel expenses. From the **Figures 6-10** above we can say that so far the diesel generators have worked well, but there is limited local knowledge on operation and maintenance (O&M) and villagers are concerned that if the generators break down sometimes they do not know who to contact. In addition, PV systems are getting to fail in highly numbers of the systems, and failure rates will likely increase substantially in the rainy season (during summer and winter) because lower light levels will lead to a power deficit.

Environmental impacts from the diesel engine to generate electric power include noise. We mentioned as noise as an impact to concern in the Koh Jik village and the diesel engine appear closely from residential houses. Environmental impact from the PV system is a little small thing. The largest concern is the recycling of the lead-acid batteries. The villagers report that they already sold non-functional batteries several times in few years to a materials recycle that comes to the village.

5.2. The Proposed Future Hybrid System as Follows

- To provide the electricity power quality available to remote villagers, for instance; 24 hrs power, use of refrigerators, rice cooker, and so on electric appliances throughout the useful lifetime system operations.

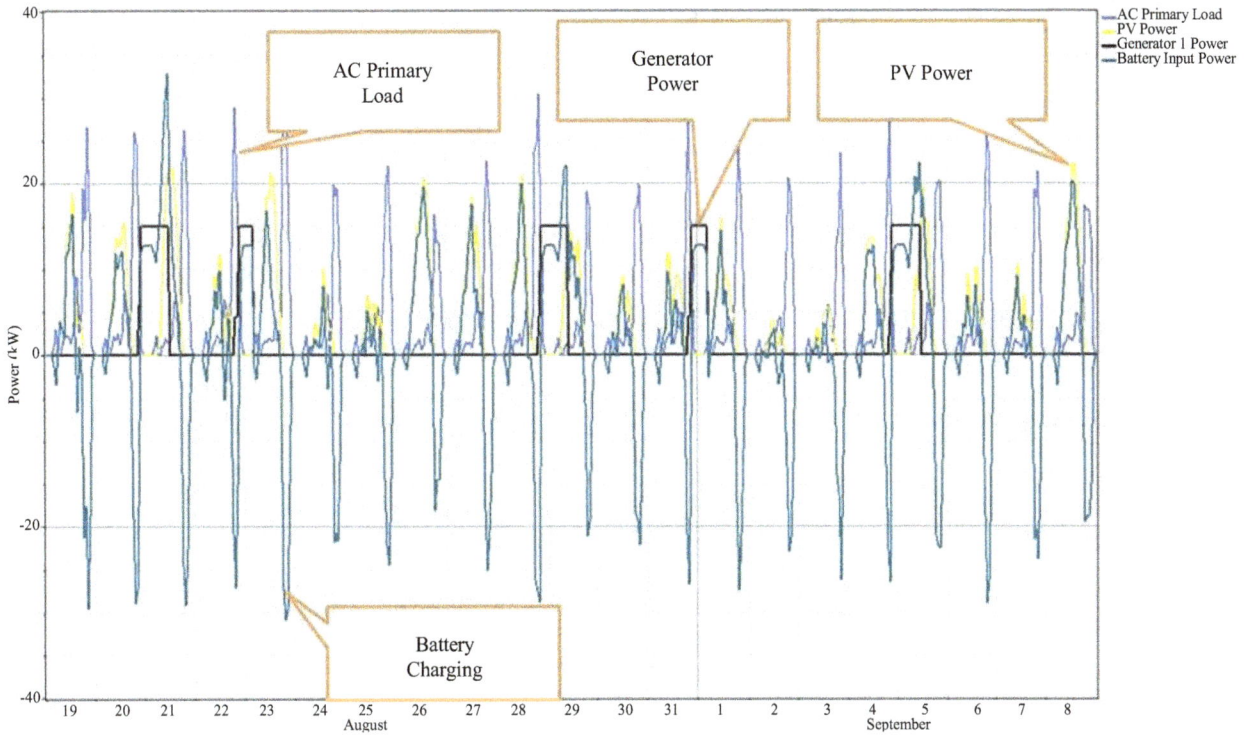

Figure 11. BBKKR simulation results.

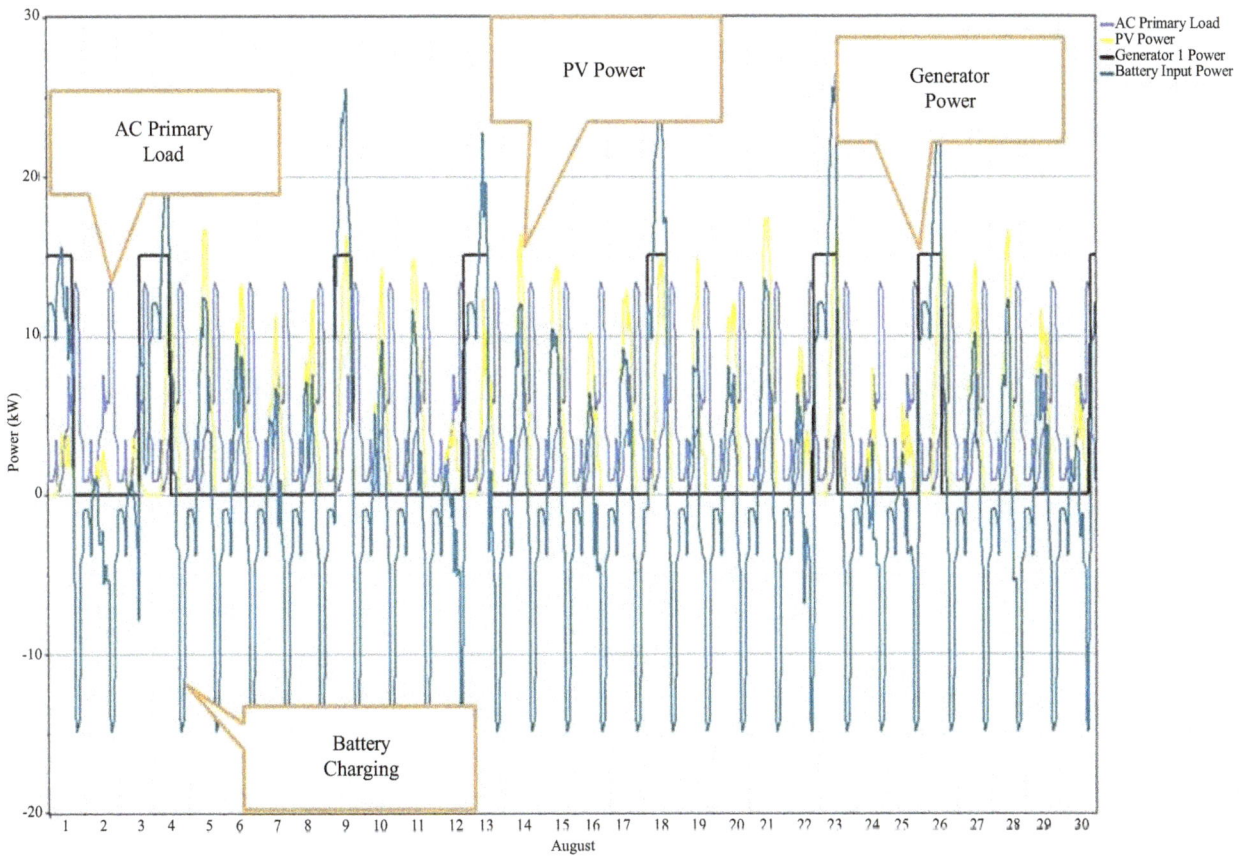

Figure 12. Koh Jik simulation results.

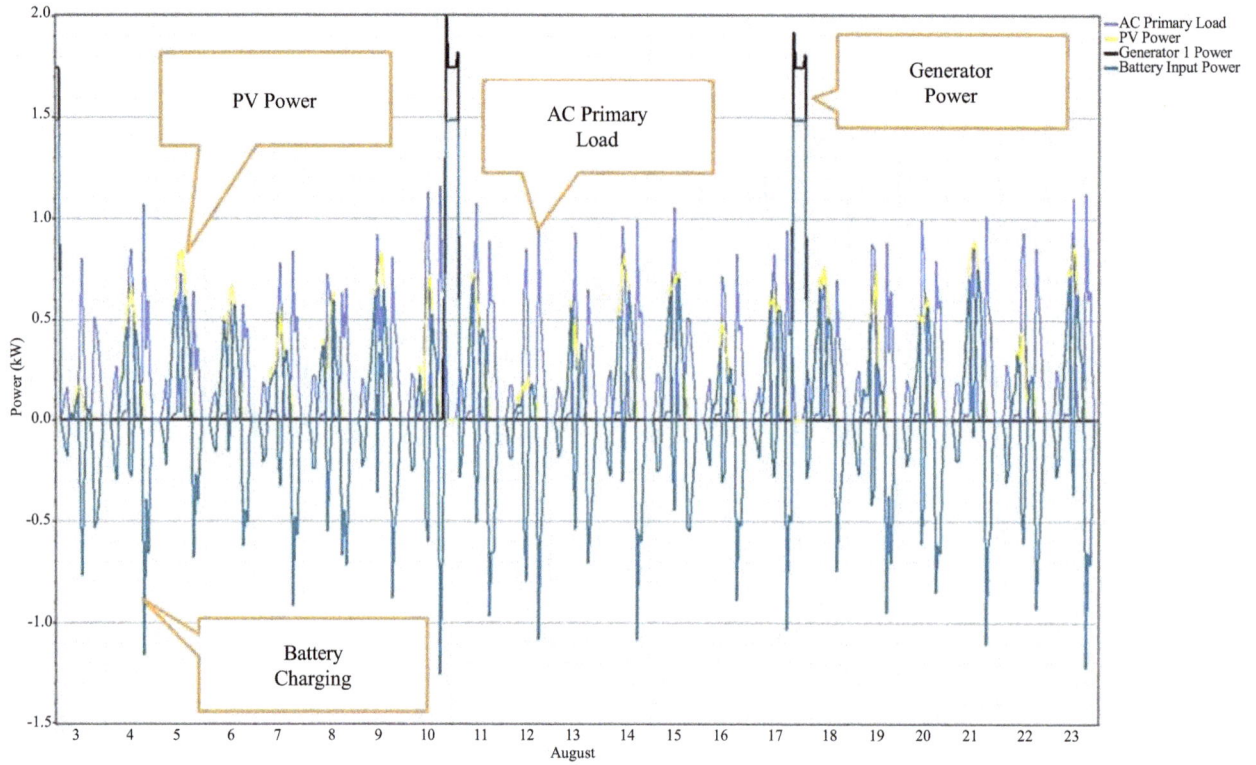

Figure 13. Bann Pa Ya Sai simulation result.

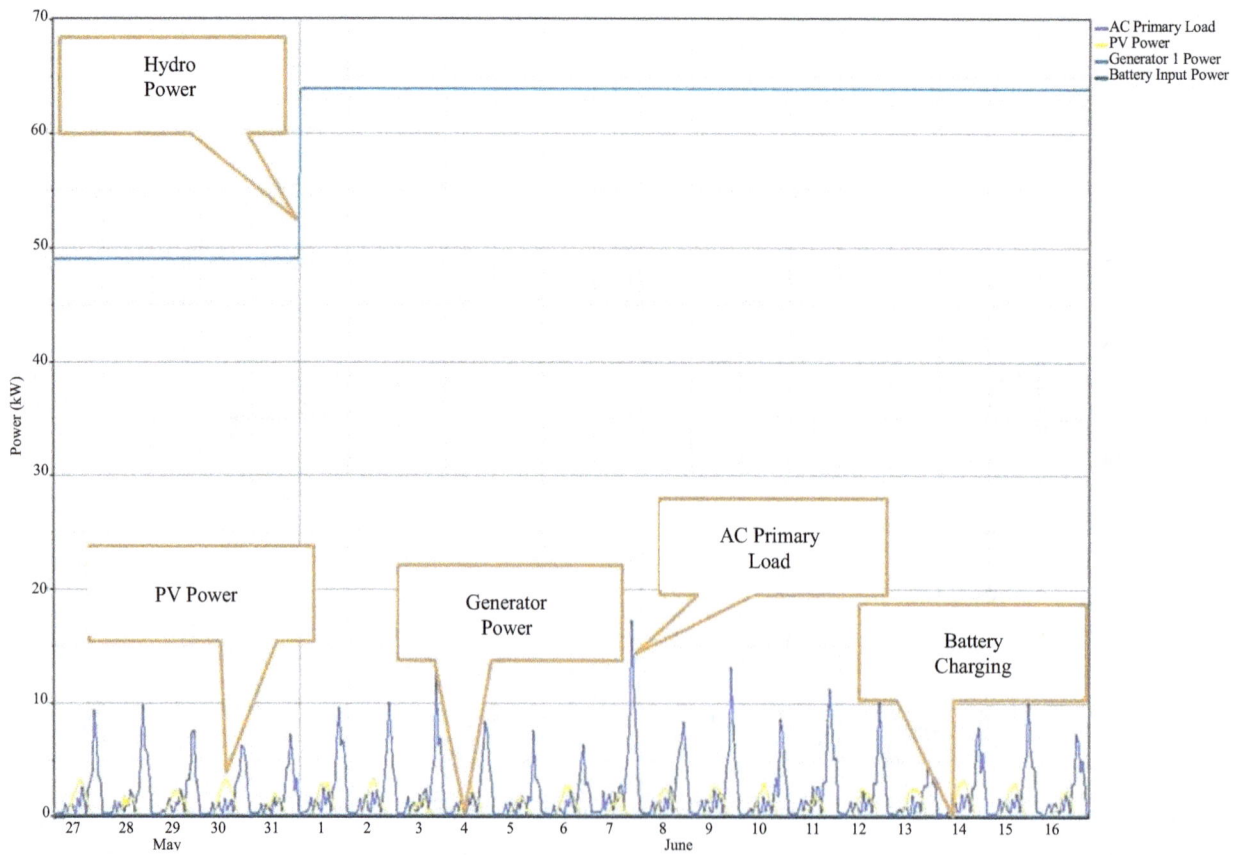

Figure 14. Bann Klong Rue simulation result with hydro power.

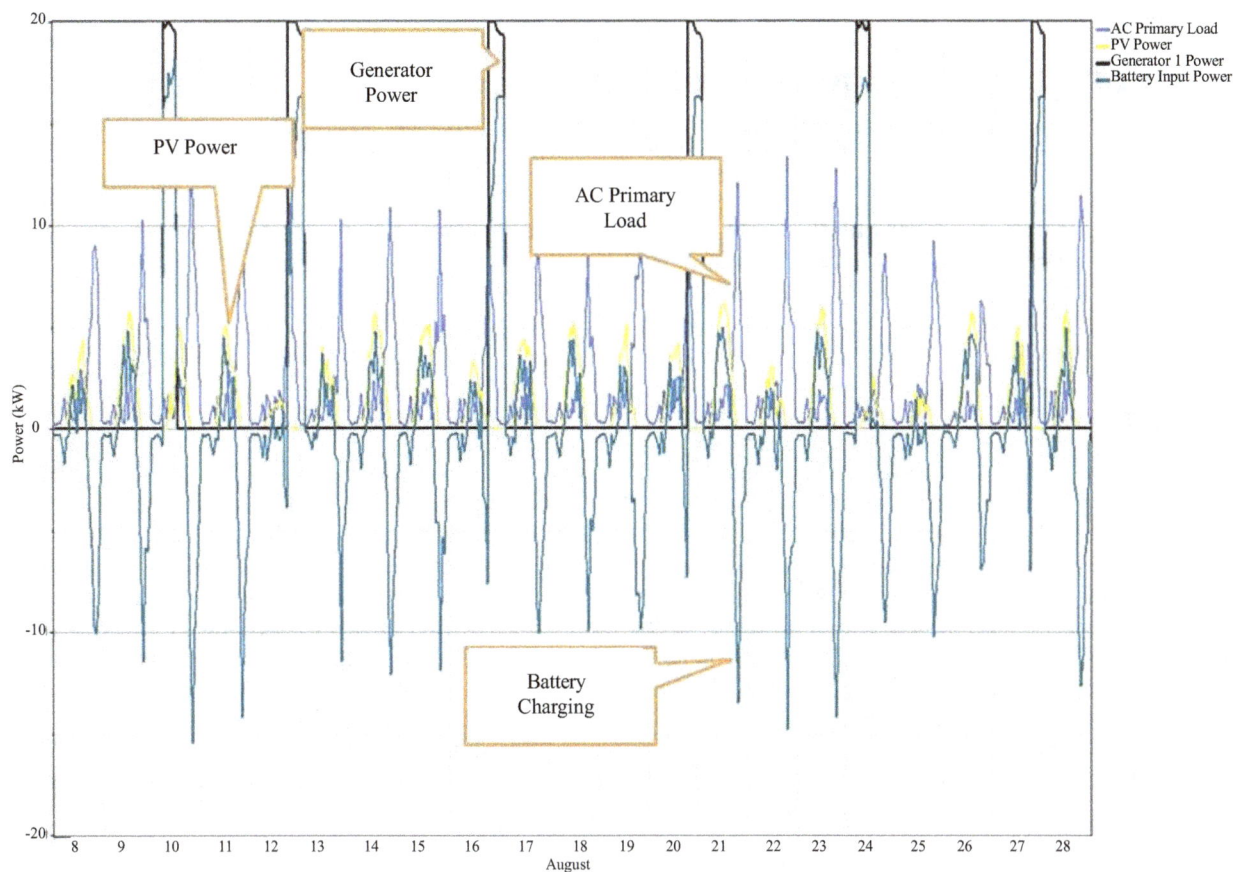

Figure 15. Bann Klong Rue simulation result without hydro power.

- To design in which renewable energy sources or combination of sources as hybrid renewable energy system for serving as a structured opportunity to build local technical.
- To gives the lowest costs to villagers through a more efficient use of the power generator.

6. Conclusion

The paper presents a method of design for PV hybrid system. The system design after the calculation is simulated by Homer program for system optimization. HOMER modeling results indicate that remote village electricity needs could be met at considerable available of resources with a hybrid (renewable energy/diesel) system compared with existing arrangements (separate diesel generator plus solar PV system).

REFERENCES

[1] Provincial Electricity Authority (PEA), "Solar Home System Project," 2005. http://www.pea.co.th/project/project_solar.htm

[2] R. Shrestha, S. Kumar, S. Martin and N. Limjeerajarus, "Role of Renewable Energy for Productive Uses in Rural Thailand," Global Network on Energy for Sustainable Development (GNESD), New York, 2006.

[3] P. Kruanpradit, S. Saengsrithorn and S. Sriphirom, "Electricity Supply by Solar Home System in Thailand," *Proceedings of* 14*th Technical Digest of the International PVSEC*, Bangkok, 2004, pp. 781-783.

[4] J. Yaungket and T. Tezuka, "A Survey of Remote Household Energy Use in Rural Thailand," Energy Procedia-Science Direct, 2013.

[5] B. Plangklang, "Photovoltaic Systems," 1st Edition, Triple Education Co., Ltd., Bangkok, 2011.

[6] B. Plangklang, R. Skunpong, K. Phumkittphich and S. Hiranvarodom, "Analysis of Energy Consumption and Behavior of Television in Resident Houses in Thailand," *The* 2*nd International Science, Social Science, Engineering and Energy Conferenc*e, 2010 *Procedia Engineering-Science Direct*, 2011, pp. 115-119.

12

Energy, Exergy and Economic Analyses of Energy Sourcing Pattern in a Nigerian Brewery

Wasiu Oyediran Adedeji, Ismaila Badmus[*]

Department of Mechanical Engineering, Yaba College of Technology, Lagos, Nigeria

Email: diranadedeji@yahoo.com, [*]ismail.badmus@gmail.com

ABSTRACT

Energy, exergy, and economic analyses of energy sourcing pattern in a Nigerian brewery have been carried out. The mean annual energy efficiencies have varied from 75.62% in 2004 to 81.71% in 2006, while the mean annual exergy efficiencies have varied from 42.66% in 2004 to 57.10% in 2005. Diesel fuel combustion, whether for local electricity generation via internal combustion engines or for process steam raising in boilers, has adversely affected the efficiencies of energy utilisation in the company. The negative effect of steam raising on efficient energy utilisation is more, although steam raising is unavoidable, due to the nature of the company under investigation. The annual mean energy unit costs have also varied from 27.86 USD per Giga-Joule in 2006 to 32.80 USD per Giga-Joule in 2004, confirming the inverse proportion of energy efficiency and costs. On the other hand, the annual mean exergy unit costs have varied from 40.19 USD per Giga-Joule in 2005 to 58.46 USD per Giga-Joule in 2004. The most efficient year has been 2006 energetically and 2005 exergetically. The difference in the two years lies in the proportions of generator diesel and boiler diesel utilised as the system exergy is most sensitive to boiler diesel use while the system energy is more sensitive to generator diesel utilisation due to their different device efficiencies.

Keywords: Exergy; Energy; Diesel; Electricity; Process Steam Boiler

1. Introduction

Historically, energy has been the pivot of economic development of most countries all over the world and this trend persists. It has brought great economic prosperity to nations and has been the centre for social and overall human development. Unfortunately, due to the way energy is sourced, produced and used historically, two major drawbacks have evolved. Firstly, the overall energy system has been very inefficient; and secondly, major local and global environmental, social and health problems have been associated with the energy system [1]. This throws up the twin challenge of energy conversion efficiency improvement and sustainable environmental management.

Nigeria is endowed with a vast amount of energy resources. According to the OPEC annual statistical bulletin [2], Nigeria proven crude oil reserves and natural gas are 37.2 billion barrels and 5292 trillion standard cubic metres, respectively.

Despite these huge resources which should have translated into cheap, affordable and reliably constant power supply, an estimated 60 million Nigerians now own power generating sets for their electricity, while the same number of people spend a staggering $10 billion to fuel them annually [3] quoted in ECN [4]. According to Oniwon [5], 15% of Nigeria's produced natural gas is still flared while only 12% is utilized locally between Industrial and power sectors.

Nigerian industrialists and other stakeholders have bitterly decried the situation of the Nigerian power sector. For instance, the manufacturers, who operate under different trade associations like the Manufacturers Association of Nigeria (MAN) and Nigeria Association of Small Scale Industries (NASSI), once said that the major problem facing the manufacturing sector was the lack of power, explaining that the volume of diesel consumed daily in Nigeria was currently put at between 12 million and 13 million litres [6] quoted in [4].

Activities in the company revolve round brewing of (non-alcoholic) malts and (alcoholic) lager beer. To carry out this production there are various processes involved and these include: decoration, bottle washing, filling, capping, pasteurization, cooling and so on. All these processes require steam, air, water, electricity, etc. Generators are used as alternative source of electricity when there is power outage from the national grid (the Power Holding Company of Nigeria, PHCN). But due to inefficient and unavailable electricity supply from the Power

---

[*]Corresponding author.

Holding Company of Nigeria, as testified to by many stakeholders like Iwayemi [7], most companies largely depend on generators. At this brewery, generators supply power for most of the running hours, while PHCN supply power for the rest. The boiler is used most of time because most of the processes require steam for washing, sterilizing, heating, pasteurizing, etc.

2. Theory

The principle of energy conversation is the first law of thermodynamics, which stipulates that under no circumstance can energy be destroyed. This law states that the amount of heat transferred into a system less the amount of work done by the system must be equal to the corresponding change in the system energy. In effect, heat and work are means by which systems exchange energy with one another. Mathematically,

$$_1Q_2 - {_1}W_2 = E_2 - E_1 \qquad (1)$$

Although electrical, mechanical and kinetic energies are all forms of energy which can be transformed into one other nearly completely, this is not so in the case of thermal energy as in case of internal combustion engine where the heat generation occurs in the cylinder but the useful component converted to mechanical use is less than 50%.

The first law of thermodynamics treats all energy forms in the same way. There are, however, certain types of energy that are more valuable than others.

Thus, we define the quality of the energy as the potential to produce useful work.

Exergy is the maximum work potential of a material or of a form of energy in relation to its environment. This work potential can be obtained by reversible processes. However, in reality there are only irreversible processes. Thus, we seek the work potential of a system, relative to the dead state, or reference environment.

The dead state of a system is the state in which it is in equilibrium with the environment. This means same temperature and pressure, no relative motion and same altitude.

For practical reasons a reference environment has been defined for the environment. The reference environment is considered to be so large, that its parameters are not affected by interaction with the system under consideration. In this work, the reference system as stated in Szargut et al. [8] and Kotas [9], both quoted in Cornelissen [10], has been used with a reference temperature (T_0) of 298.15 K and a reference pressure (P_0) of 1 atm. There are exergy transfers with work and heat transfers as well as material streams.

2.1. Exergy Transfer with Work Interaction

The exergy transfer with work interaction is associated with work transfer rate or shaft power. Because exergy is defined as the maximum work potential, work is equivalent to exergy in every respect.

2.2. Electrical Energy and Exergy

Electrical energy is not affected by ambient conditions and therefore is equivalent in work. In other words, electrical energy can be treated as totally convertible to work.

2.3. Exergy Transfer with Heat Interaction

The exergy transfer rate $(\dot{E})$ connected with the heat transfer rate $(\dot{Q})$ can be calculated using the following formula:

$$\dot{E} = \int_A \left(1 - \frac{T_0}{T}\right)\dot{Q}\,\mathrm{d}A \qquad (2)$$

where A is the heat transfer area, T_0 is the temperature of the environment, T is the temperature at which the heat transfer takes places. When there is a uniform temperature distribution, the expression becomes:

$$\dot{E} = \dot{Q}\left(1 - \frac{T_0}{T}\right) \qquad (3)$$

2.4. Exergy Transfer Associated with Material Streams

Chemical Exergy

One of the most common energy carriers is hydrocarbon/fossil/biomass fuels. The specific exergy of this class of thermodynamic systems is the chemical exergy. Chemical exergy is equal to the maximum amount of work obtainable when the substance under consideration is brought from the environmental state, defined by the parameters T_0 and P_0, to the reference state by processes involving heat transfer and exchange of substances only with the environment.

For many fuels the chemical structure is unknown. To overcome this problem the chemical exergy for these fuels can be estimated on the basis of the higher heating value (HHV). The relationship between the HHV and the chemical exergy is:

$$b_{ch} = \varphi \cdot HHV \qquad (4)$$

φ, the fuel chemical exergy factor can be calculated with formulae based on the atomic composition. For diesel [11], φ is 1.07 and for natural gas, it can be approximated as 0.94 [12] quoted in Hepbasli [13].

2.5. Energy and Exergy Efficiencies of the Processes

The expressions for energy efficiency (η) and exergy efficiency (ψ) for the main types of processes in this paper are as follows:

$$\eta = \frac{\text{Energy in products}}{\text{Total energy input}} \qquad (5)$$

$$\psi = \frac{\text{Exergy in products}}{\text{Total exergy input}} \qquad (6)$$

The particular efficiencies are as follows:
Boiler energy efficiency:

$$\eta_b = \frac{Q_{o,b}}{Q_{i,b}} \qquad (7)$$

Boiler exergy efficiency:

$$\psi_b = \frac{X_b}{\varphi Q_{i,b}} \qquad (8)$$

Generator energy efficiency:

$$\eta_g = \frac{W_{e,g}}{Q_g} \qquad (9)$$

Generator exergy efficiency:

$$\psi_g = \frac{X_{e,g}}{\varphi Q_g} \qquad (10)$$

Electrical energy efficiency:

$$\eta_e = \frac{W_{e,m} + \eta_g Q_g}{W_{e,m} + Q_g} \qquad (11)$$

Electrical exergy efficiency:

$$\psi_e = \frac{W_{e,m} + \psi_g \varphi Q_g}{W_{e,m} + \varphi Q_g} \qquad (12)$$

Total energy efficiency:

$$\eta_{\text{total}} = \frac{W_{e,m} + \eta_g Q_g + \eta_b Q_{i,b}}{W_{e,m} + Q_g + Q_{i,b}} \qquad (13)$$

Total exergy efficiency:

$$\psi_{\text{total}} = \frac{W_{e,m} + \psi_g \varphi Q_g + \psi_b \varphi Q_{i,b}}{W_{e,m} + \varphi Q_g + \varphi Q_{i,b}} \qquad (14)$$

In all cases,

$$Q = \text{fuel mass} \times \text{fuel heating value} \qquad (15)$$

2.6. Economic Analysis

2.6.1. Mains Electricity Tariff
Mains electricity tariff for the industrial sector from 1st Feb. 2002 to 30th June 2009, for power consumption above 20MVA is N8.50 per kWh [14]. This is equivalent to a unit cost, $C_{n,m}, (= C_{x,m} = C_m)$ of $15.74 per GJ of mains electricity, at N150 per US dollar.

2.6.2. Diesel Generator Output Electricity Unit Costs
Energy unit cost, $C_{n,g}$, is given by:

$$C_{n,g} = \frac{c_f}{\eta_g HV} \qquad (16)$$

Exergy unit cost, $C_{X,g}$, is given by:

$$C_{X,g} = \frac{c_f}{\varphi_g \psi_g HV} \qquad (17)$$

2.6.3. Boiler Steam Generation Unit Costs
Boiler energy unit cost, $C_{n,b}$, is given by:

$$C_{n,b} = \frac{c_f}{\eta_b HV} \qquad (18)$$

Boiler exergy unit cost, $C_{X,b}$, is given by:

$$C_{X,b} = \frac{c_f}{\varphi_b \psi_b HV} \qquad (19)$$

Since, in our case, both the boiler and the generator use the same fuel, $\varphi_b = \varphi_g = \varphi$

2.6.4. Mean Output Electricity Unit Costs
Mean electrical energy unit cost

$$\bar{C}_{n,e} = \frac{C_m \times W_{e,m} + C_{n,g} \times Q_g}{W_{e,m} + \eta_g Q_g} \qquad (20)$$

Mean electrical exergy unit cost

$$\bar{C}_{X,e} = \frac{C_m \times W_{e,m} + C_{X,g} \times \varphi Q_g}{W_{e,m} + \psi_g \varphi Q_g} \qquad (21)$$

2.6.5. Overall Mean Unit Costs
Overall mean energy unit cost

$$\bar{C}_n = \frac{C_m \times W_{e,m} + C_{n,g} \times Q_g + C_{n,b} \times Q_{i,b}}{W_{e,m} + \eta_g Q_g + \eta_b Q_{i,b}} \qquad (22)$$

Overall mean exergy unit cost

$$\bar{C}_X = \frac{C_m \times W_{e,m} + C_{X,g} \times \varphi Q_g + C_{X,b} \times \varphi Q_{i,b}}{W_{e,m} + \psi_g \varphi Q_g + \psi_b \varphi Q_{i,b}} \qquad (23)$$

3. Methodology

Due to difficulties in accessing the production process lines details; this work concentrates on the assessment of the company energy sourcing efficiencies rather than the end uses. Electricity has been sourced from both the national grid and diesel fuelled generators. Steam has been raised using diesel fuel alone to fire the boilers. Are there better options on ground for the company? This is the focus of this work.

To examine the energy utilization efficiency of the company, a five-year data (2004-2008) was collected from the company utilities section. The data collected covers the following areas:

i) Mains Electricity bill (PHCN)—kWh/$ values

ii) Energy value computed from volumes of fuel consumed for firing the boilers and running electrical generators on monthly basis (GJ).

Energy consumption in the factory affects the period costing and pricing directly. Considering energy value of diesel oil, one litre is equivalent to 39 MJ [15], using the higher heating value. The diesel engine power plant energy efficiency and exergy efficiency are taken to be 47% and 43.8% respectively [16]. Also, we are taking boiler energy efficiency to be 72.46% and its exergy efficiency to be 24.89% [17].

4. Results

The results in **Tables 1-5** were obtained for the years 2004-2008 respectively from available data.

5. Discussion of Results

Generally, the electrical energy and exergy efficiencies are very numerically close. This is because electrical energy and exergy values are thermodynamically equal. The small disparities that exist between the energy and exergy efficiency values in our case is due to the relatively low electrical energy generation efficiency of diesel powered internal combustion engines (47% for en-

ergy and 43.8% for exergy). Secondly, the total energy and exergy efficiencies are further brought down in value by the relatively low thermal efficiencies of the process steam boilers. The boiler energy efficiency (72.46%) and exergy efficiency (24.89%) have led to very wide gaps between the total energy efficiencies and total exergy efficiencies.

Considering the year 2004 (**Figure 1(a)**), all efficiencies except for total exergy one, record their lowest values in the month of January. This is due to the fact that the electricity supply is dominated by low efficiency diesel engine generated electric power for the month but the low percentage of boiler fuel in the overall fuel mix (**Figure 1(b)**) has led to the improvement of the overall exergy efficiency. February has high electrical energy and exergy efficiencies because the generator diesel portion of the total energy supply mix is only about 3.6%. In May, we have low overall energy efficiency since practically all the three energy supply types have equal quantities (**Figure 1(b)**). The month of June has the highest electrical energy and exergy efficiencies (97.06% and 96.68% respectively) for the year because its diesel generator supply (2.7%) is the least after the month of December (2.5%) but its boiler fuel supply (52%) is almost half of that of December (95.7%).

Table 1. Energy consumption pattern for the year 2004.

Month	Mains Electricity (GJ)	Generator Diesel Energy Value (GJ)	Generator Diesel Exergy Input (GJ)	Generator Elect. Energy Output (GJ)	Generator Elect. Exergy Output (GJ)	Total Elect. Energy Output (GJ)	Total Elect. Exergy Output (GJ)	Boiler Diesel Energy Value (GJ)	Boiler Diesel Exergy Input (GJ)	Boiler Energy Produced (GJ)	Boiler Exergy Produced (GJ)	Total Energy Produced (GJ)	Total Exergy Produced (GJ)
Jan.	255.4	682.4	730.168	320.728	319.8136	576.128	575.2136	546.9	585.183	396.2837	145.652	972.4117	720.8656
Feb.	447	39.2	41.944	18.424	18.37147	465.424	465.3715	614.7	657.729	445.4116	163.7087	910.8356	629.0802
Mar.	445.3	329	352.03	154.63	154.1891	599.93	599.4891	852.1	911.747	617.4317	226.9338	1217.362	826.423
April	748.9	214.8	229.836	100.956	100.6682	849.856	849.5682	771.9	825.933	559.3187	205.5747	1409.175	1055.143
May	1077.7	1008.7	1079.309	474.089	472.7373	1551.789	1550.437	1006.4	1076.848	729.2374	268.0275	2281.026	1818.465
June	897	52.7	56.389	24.769	24.69838	921.769	921.6984	1028.9	1100.923	745.5409	274.0197	1667.31	1195.718
July	900.4	195.8	209.506	92.026	91.76363	992.426	992.1636	1306.6	1398.062	946.7624	347.9776	1939.188	1340.141
Aug.	924.3	130.2	139.314	61.194	61.01953	985.494	985.3195	1612.6	1725.482	1168.49	429.4725	2153.984	1414.792
Sept.	1013.7	158.3	169.381	74.401	74.18888	1088.101	1087.889	1581.9	1692.633	1146.245	421.2964	2234.346	1509.185
Oct.	797.1	482.4	516.168	226.728	226.0816	1023.828	1023.182	2258.6	2416.702	1636.582	601.5171	2660.41	1624.699
Nov.	833.1	770	823.9	361.9	360.8682	1195	1193.968	1142.7	1222.689	828.0004	304.3273	2023	1498.295
Dec.	250.39	356	380.92	167.32	166.843	417.71	417.233	13.467	14409.69	9758.188	3586.572	10175.9	4003.805
Total	8.590	4.420	4728.865	2077.165	2071.243	10667.46	10661.53	26.190	28023.62	18977.49	6975.079	29644.95	17636.61
Mean	715.8575	368.2917	394.0721	173.0971	172.604	888.9546	888.4611	2182.525	2335.302	1581.458	581.257	2470.412	1469.718

Table 2. Energy consumption pattern for the year 2005.

Month	Mains Electricity (GJ)	Generator Diesel Energy Value (GJ)	Generator Diesel Exergy Input (GJ)	Generator Elect. Energy Output (GJ)	Generator Elect. Exergy Output (GJ)	Total Elect. Energy Output (GJ)	Total Elect. Exergy Output (GJ)	Boiler Diesel Energy Value (GJ)	Boiler Diesel Exergy Input (GJ)	Boiler Energy Produced (GJ)	Boiler Exergy Produced (GJ)	Total Energy Produced (GJ)	Total Exergy Produced (GJ)
Jan.	592.9	335.2	358.664	157.544	157.0948	750.444	749.9948	669.2	716.044	484.9023	178.2234	1235.346	928.2182
Feb.	575.2	225.4	241.178	105.938	105.636	681.138	680.836	928.1	993.067	672.5013	247.1744	1353.639	928.0103
Mar.	1068.1	208.9	223.523	98.183	97.90307	1166.283	1166.003	1054.9	1128.743	764.3805	280.9441	1930.664	1446.947
April	774.5	184.5	197.415	86.715	86.46777	861.215	860.9678	805.6	861.992	583.7378	214.5498	1444.953	1075.518
May	773.4	484.4	518.308	227.668	227.0189	1001.068	1000.419	1391.3	1488.691	1008.136	370.5352	2009.204	1370.954
June	688	117.8	126.046	55.366	55.20815	743.366	743.2081	92.3	98.761	66.88058	24.58161	810.2466	767.7898
July	773.4	117.8	126.046	55.366	55.20815	828.766	828.6081	923.3	987.931	669.0232	245.896	1497.789	1074.504
Aug.	643.18	87.6	93.732	41.172	41.05462	684.352	684.2346	1066.1	1140.727	772.4961	283.927	1456.848	968.1616
Sept.	914.8	169.3	181.151	79.571	79.34414	994.371	994.1441	39.6	42.372	28.69416	10.54639	1023.065	1004.691
Oct.	723.2	1274.8	1364.036	599.156	597.4478	1322.356	1320.648	1206.9	1291.383	874.5197	321.4252	2196.876	1642.073
Nov.	1032	1223.8	1309.466	575.186	573.5461	1607.186	1605.546	949.6	1016.072	688.0802	252.9003	2295.266	1858.446
Dec.	859.8	135	144.45	63.45	63.2691	923.25	923.0691	711.4	761.198	515.4804	189.4622	1438.73	1112.531
Total	9.418	4.565	4.884	2145.315	2139.199	11563.8	11557.68	9.838	10526.98	7128.832	2620.166	18692.63	14177.84
Mean	784.8733	380.375	407.001	178.7763	178.2664	963.6496	963.1397	819.8583	877.2484	594.0693	218.3471	1557.719	1181.487

Table 3. Energy consumption pattern for the year 2006.

Month	Mains Electricity (GJ)	Generator Diesel Energy Value (GJ)	Generator Diesel Exergy Input (GJ)	Generator Elect. Energy Output (GJ)	Generator Elect. Exergy Output (GJ)	Total Elect. Energy Output (GJ)	Total Elect. Exergy Output (GJ)	Boiler Diesel Energy Value (GJ)	Boiler Diesel Exergy Input (GJ)	Boiler Energy Produced (GJ)	Boiler Exergy Produced (GJ)	Total Energy Produced (GJ)	Total Exergy Produced (GJ)
Jan.	866.7	94.2	100.794	44.274	44.14777	910.974	910.8478	634.4	678.808	459.6862	168.9553	1370.66	1079.803
Feb.	611.8	237	253.59	111.39	111.0724	723.19	722.8724	1317.1	1409.297	954.3707	350.774	1677.561	1073.646
Mar.	783.2	348.3	372.681	163.701	163.2343	946.901	946.4343	1083.8	1159.666	785.3215	288.6409	1732.222	1235.075
April	782.5	276.2	295.534	129.814	129.4439	912.314	911.9439	1335.1	1428.557	967.4135	355.5678	1879.727	1267.512
May	973	176.1	188.427	82.767	82.53103	1055.767	1055.531	1370.1	1466.007	992.7745	364.8891	2048.541	1420.42
June	970.1	182.4	195.168	85.728	85.48358	1055.828	1055.584	927.1	991.997	671.7767	246.9081	1727.605	1302.492
July	862.4	174.3	186.501	81.921	81.68744	944.321	944.0874	1250.9	1338.463	906.4021	333.1434	1850.723	1277.231
Aug.	1159.4	13.3	14.231	6.251	6.233178	1165.651	1165.633	1496.1	1600.827	1084.074	398.4458	2249.725	1564.079
Sept.	1159.4	29.5	31.565	13.865	13.82547	1173.265	1173.225	1643	1758.01	1190.518	437.5687	2363.783	1610.794
Oct.	1169.1	13.1	14.017	6.157	6.139446	1175.257	1175.239	1122.3	1200.861	813.2186	298.8943	1988.476	1474.134
Nov.	929.5	249.5	266.965	117.265	116.9307	1046.765	1046.431	1905.3	2038.671	1380.58	507.4252	2427.345	1553.856
Dec.	1205.9	74.3	79.501	34.921	34.82144	1240.821	1240.721	1600.1	1712.107	1159.432	426.1434	2400.253	1666.865
Total	11473	1868.2	1998.974	878.054	875.5506	12351.054	12348.55	15685.3	16783.27	11365.57	4177.356	23716.62	16525.91
Mean	956.0833	155.6833	166.5812	73.17115	72.96257	1029.25445	1029.046	1307.108	1398.606	947.1307	348.113	1976.385	1377.159

Table 4. Energy consumption pattern for the year 2007.

Month	Mains Electricity (GJ)	Generator Diesel Energy Value (GJ)	Generator Diesel Exergy Input (GJ)	Generator Elect. Energy Output (GJ)	Generator Elect. Exergy Output (GJ)	Total Elect. Energy Output (GJ)	Total Elect. Exergy Output (GJ)	Boiler Diesel Energy Value (GJ)	Boiler Diesel Exergy Input (GJ)	Boiler Energy Produced (GJ)	Boiler Exergy Produced (GJ)	Total Energy Produced (GJ)	Total Exergy Produced (GJ)
Jan.	983.2	14	14.98	6.58	6.56124	989.78	989.7612	1608.6	1721.202	1165.592	428.4072	2155.372	1418.168
Feb.	1260.7	238.7	255.409	112.189	111.8691	1372.889	1372.569	1220.1	1305.507	884.0845	324.9407	2256.973	1697.51
Mar.	1034.5	153.9	164.673	72.333	72.12677	1106.833	1106.627	1808.2	1934.774	1310.222	481.5652	2417.055	1588.192
April	950	154	164.78	72.38	72.17364	1022.38	1022.174	1338.3	1431.981	969.7322	356.4201	1992.112	1378.594
May	1058.8	176.1	188.427	82.767	82.53103	1141.567	1141.331	1190	1273.3	862.274	316.9244	2003.841	1458.255
June	828.4	302.7	323.889	142.269	141.8634	970.669	970.2634	1493.3	1597.831	1082.045	397.7001	2052.714	1367.964
July	1070.8	186.2	199.234	87.514	87.26449	1158.314	1158.064	1824.7	1952.429	1322.178	485.9596	2480.492	1644.024
Aug.	1274.2	33.7	36.059	15.839	15.79384	1290.039	1289.994	1671.2	1788.184	1210.952	445.079	2500.991	1735.073
Sept.	1161.8	73.2	78.324	34.404	34.30591	1196.204	1196.106	12043.7	12886.76	8726.865	3207.514	9923.069	4403.62
Oct.	1248.2	61.6	65.912	28.952	28.86946	1277.152	1277.069	1957.3	2094.311	1418.26	521.274	2695.412	1798.343
Nov.	904.5	700.7	749.749	329.329	328.3901	1233.829	1232.89	1863.7	1994.159	1350.437	496.3462	2584.266	1729.236
Dec.	960.4	249.5	266.965	117.265	116.9307	1077.665	1077.331	1925.7	2060.499	1395.362	512.8582	2473.027	1590.189
Total	12.736	2.344	2508.401	1101.821	1098.68	13837.321	13834.18	29.945	32040.94	21698	7974.989	35535.32	21809.17
Mean	1061.292	195.3583	209.0334	91.8184	91.55662	1153.1101	1152.848	2495.4	2670.078	1808.167	664.5824	2961.277	1817.431

Table 5. Energy consumption pattern for the year 2008.

Month	Mains Electricity (GJ)	Generator Diesel Energy Value (GJ)	Generator Diesel Exergy Input (GJ)	Generator Elect. Energy Output (GJ)	Generator Elect. Exergy Output (GJ)	Total Elect. Energy Output (GJ)	Total Elect. Exergy Output (GJ)	Boiler Diesel Energy Value (GJ)	Boiler Diesel Exergy Input (GJ)	Boiler Energy Produced (GJ)	Boiler Exergy Produced (GJ)	Total Energy Produced (GJ)	Total Exergy Produced (GJ)
Jan.	321.7	509.4	545.058	239.418	238.7354	561.118	560.4354	1324.3	1417.001	959.58778	352.69155	1520.7058	913.12695
Feb.	92.3	29	31.03	13.63	13.59114	105.93	105.89114	870.5	931.435	630.7643	231.83417	736.6943	337.72531
Mar.	2	1.1	1.177	0.517	0.515526	2.517	2.515526	2055	2198.85	1489.053	547.29377	1491.57	549.80929
April	1234.4	81.6	87.312	38.352	38.242656	1272.752	1272.6427	2005.2	2145.564	1452.9679	534.03088	2725.7199	1806.6735
May	1122.6	174.1	186.287	81.827	81.593706	1204.427	1204.1937	1040	1112.8	753.584	276.97592	1958.011	1481.1696
June	635	1784.5	1909.415	838.715	836.32377	1473.715	1471.3238	1654.4	1770.208	1198.7782	440.60477	2672.4932	1911.9285
July	926.9	115.5	123.585	54.285	54.13023	981.185	981.03023	2403.9	2572.173	1741.8659	640.21386	2723.0509	1621.2441
Aug.	900.3	51	54.57	23.97	23.90166	924.27	924.20166	1973.1	2111.217	1429.7083	525.48191	2353.9783	1449.6836
Sept.	874.3	77	82.39	36.19	36.08682	910.49	910.38682	2124	2272.68	1539.0504	565.67005	2449.5404	1476.0569
Oct.	787	24.8	26.536	11.656	11.622768	798.656	798.62277	1692.3	1810.761	1226.2406	450.69841	2024.8966	1249.3212
Nov.	716.4	184.5	197.415	86.715	86.46777	803.115	802.86777	1983.8	2122.666	1437.4615	528.33157	2240.5765	1331.1993
Dec.	710.3	136.7	146.269	64.249	64.065822	774.549	774.36582	1789.3	1914.551	1296.5268	476.53174	2071.0758	1250.8976
Total	8323.2	3169.2	3391.044	1489.524	1485.2773	9812.724	9808.4773	20915.8	22379.906	15155.589	5570.3586	24968.313	15378.836
Mean	693.6	264.1	282.587	124.127	123.77311	817.727	817.37311	1742.983	1864.9921	1262.9657	464.19654	2080.6927	1281.5696

(a)

(b)

(c)

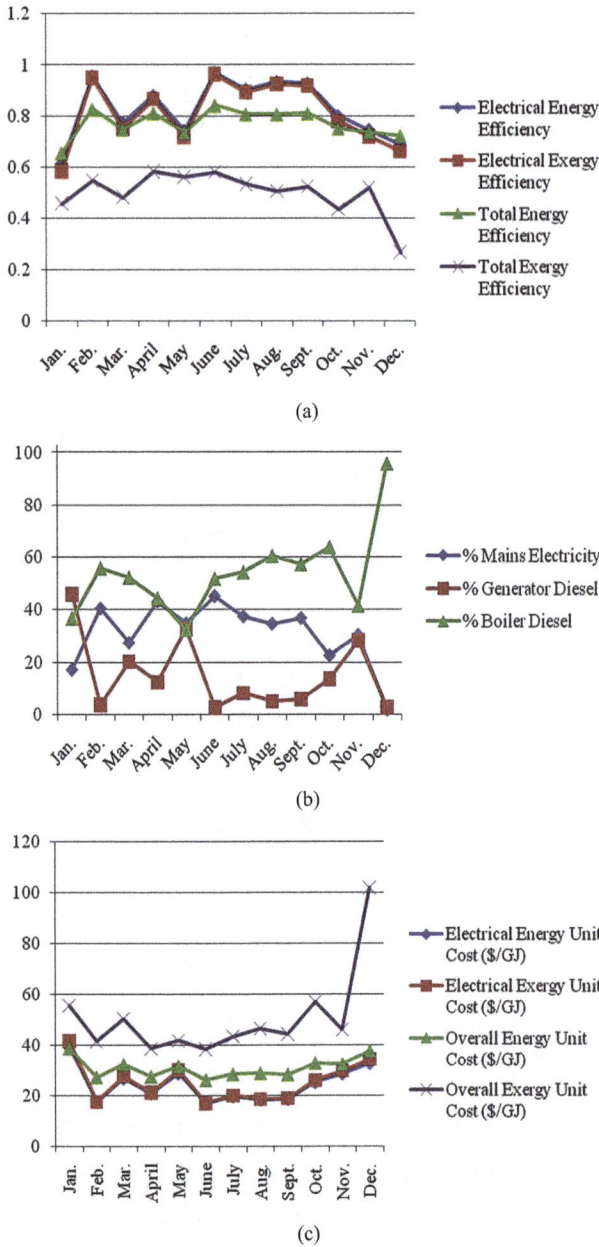

Figure 1. (a) Energy and exergy efficiencies for the year 2004; (b) Energy supply mix for the year 2004; (c) Energy and exergy unit costs for the year 2004.

The same line of argument applies to all the other years. For instance, for the year 2005, the months of June and September record high efficiencies (**Figure 2(a)**) and favourable energy mixes (**Figure 2(b)**) respectively. Similarly, the months of February, May, August and October have low efficiencies and unfavourable energy mixes respectively.

The year 2006 is with moderate efficiencies (**Figure 3(a)**) because although the boiler fuel consumption is generally high throughout the year, the electrical energy consumption from the mains is also generally high (**Fig-**

ure 3(b)). The year 2007 is similar to 2006 except for the month of September which has generally low electricity consumption (about 9.4% of total energy supply), with higher percentage share (8.75%) from the mains but very high boiler fuel consumption (90.7% of total energy supply), leading to the lowest total energy and exergy efficiencies for the year, despite high electrical energy and exergy efficiencies of 96.86% and 96.45%, respectively.

In the year 2008, the generator diesel consumption is generally low, except for the months of January (23.63%) and June (43.8%). However, the fact that June records

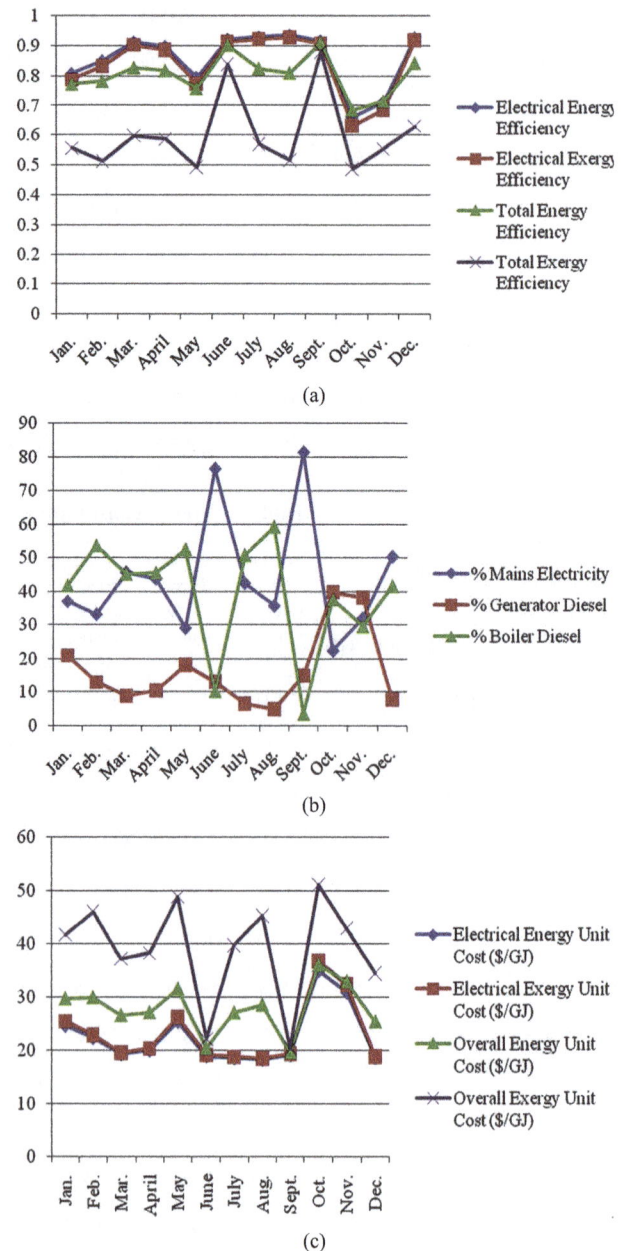

(a)

(b)

(c)

Figure 2. (a) Energy and exergy efficiencies for the year 2005; (b) Energy supply mix for the year 2005; (c) Energy and exergy unit costs for the year 2005.

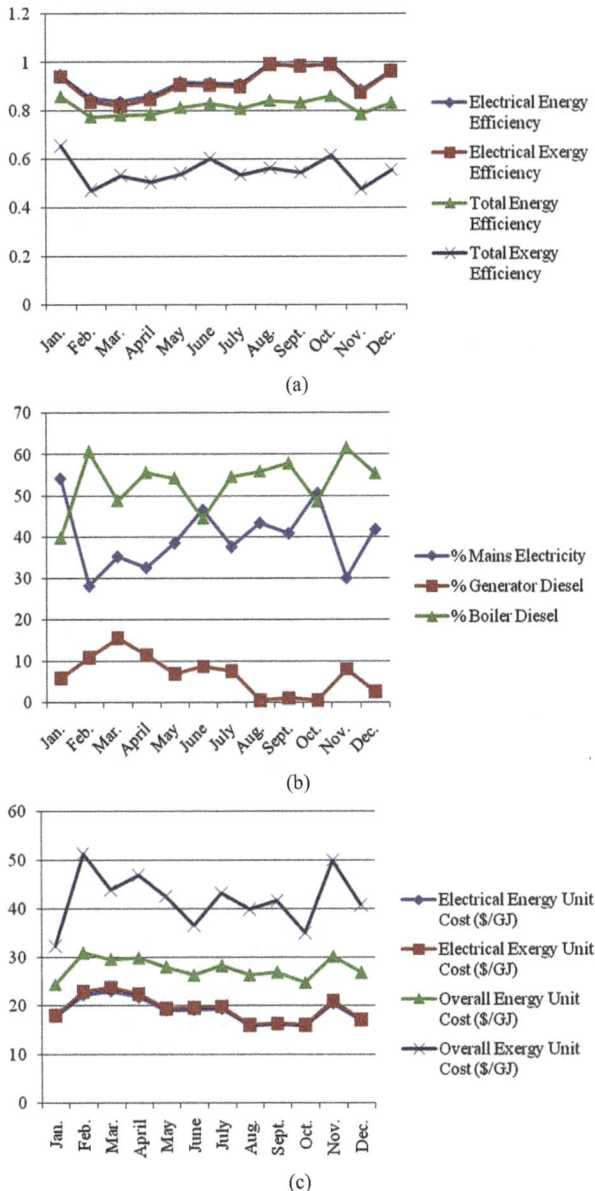

Figure 3. (a) Energy and exergy efficiencies for the year 2006; (b) Energy supply mix for the year 2006; (c) Energy and exergy unit costs for the year 2006.

the lowest boiler fuel consumption for the year has positively influenced the exergy efficiency for the month to make it fall within the average value. January boiler fuel consumption is very high (61.44%). This, coupled with low mains electricity, has resulted in low efficiencies. March has the lowest total exergy efficiency for the year due to the domination of its energy supply by boiler fuel (99.85%).

Finally, the years 2005 and 2006 are the most energy efficient years with exergy efficiencies of 57.1% and 54.6% respectively. The year 2006 has higher energy efficiency and lower exergy efficiency than the year 2005 despite practically equal percentages of mains elec-

tricity supply (39.54% for 2005 and 39.53% for 2006) due to the fact that 2006 has relatively low percentage of generator fuel consumption but high percentage of boiler fuel consumption while 2005 has relatively high percentage of generator fuel consumption and low percentage of boiler fuel consumption. This observation reinforces the fact that high generator fuel consumption largely affects the total energy efficiency adversely while high boiler fuel consumption negatively affects the total exergy efficiency.

For the economic analysis, comparisons of **Figures 1(a)** and **(c)**, **Figures 2(a)** and **(c)**, **Figures 3(a)** and **(c)**, **Figures 4(a)** and **(c)**, **Figures 5(a)** and **(c)** as well as **Figures 6(a)** and **(c)** show that graphical representations of energy and exergy efficiencies are practically mirror inverses of those of the energy unit costs. Months and years of minimum efficiencies correspond to months and years of maximum energy unit costs and vice versa. These comparisons inform us that where we have efficient energy utilisation the costs are reduced, resulting in corresponding economic gains and vice versa. Comparing **Figure 6(a)** with **Figure 7**, one discovers that with total switching to the mains supply for electricity sourcing, the mean annual electrical energy and exergy efficiencies both become 100%; the mean annual total energy efficiencies vary from 79.26% to 85.93% while the mean annual total exergy efficiencies now vary from 42.51% to 60.36%. This means that the optimised generator diesel/boiler diesel is zero (zero generator diesel) for electricity. Before the switching, the corresponding values were 82% - 92.58%; 80.05% - 91.66%; 75.62% - 81.71% and 42.66% - 57.1%, respectively. These imply at least 7% savings in electricity and at least about 3% savings in overall energy. Hence, as a first step, the company still needs to consider this power source switching. This suggests that the major energy challenge facing the brewery is in its boiler energy utilization.

6. Conclusions and Recommendations

6.1. Conclusions

Energy, exergy, and economic analyses of energy sourcing pattern in a Nigerian brewery have been carried out. The brewery has relied on electricity from both the national grid and diesel-powered electrical generators. It also utilises diesel fuel oil for process steam boiler firing. The mean annual energy efficiencies have varied from 75.62% in 2004 to 81.71% in 2006, while the mean annual exergy efficiencies have varied from 42.66% in 2004 to 57.10% in 2005. Diesel fuel combustion, whether for local electricity generation via internal combustion engines or for process steam raising in boilers, has adversely affected the efficiencies of energy utilization in the company. The negative effect of steam raising

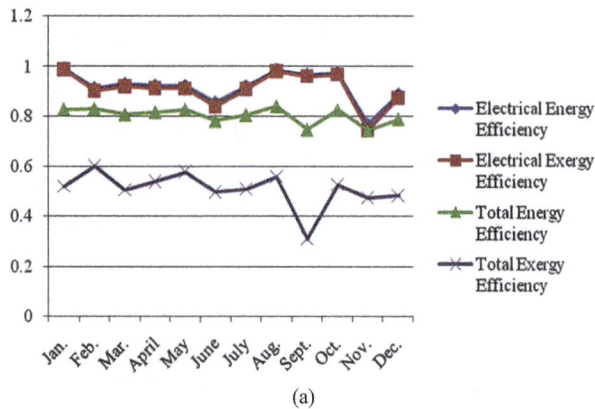

(a)

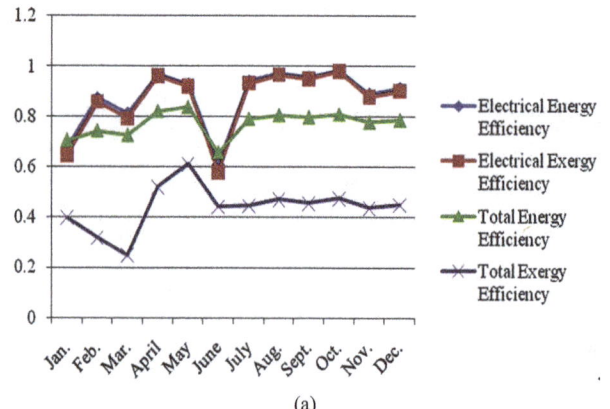

(b)

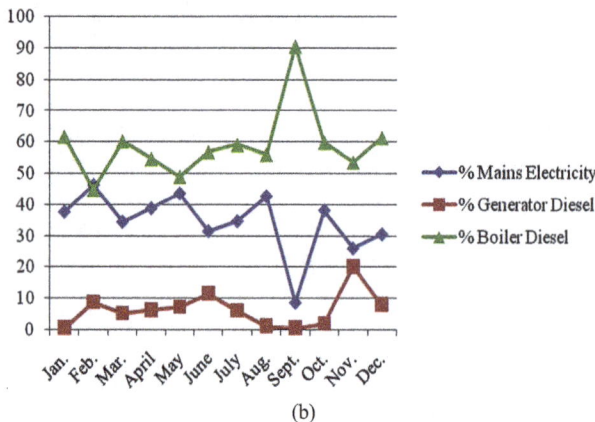

(c)

Figure 4. (a) Energy and exergy efficiencies for the year 2007; (b) Energy supply mix for the year 2007; (c) Energy and exergy unit costs for the year 2007.

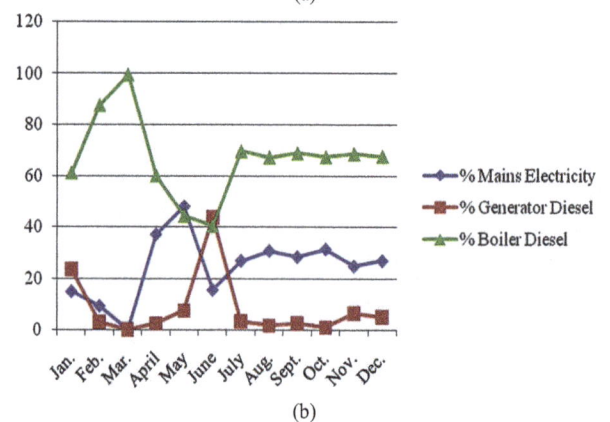

(a)

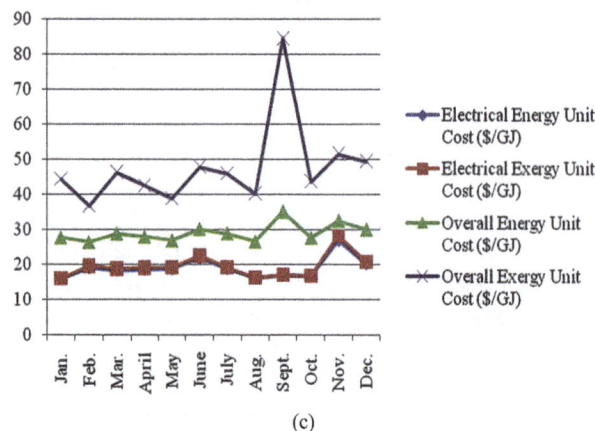

(b)

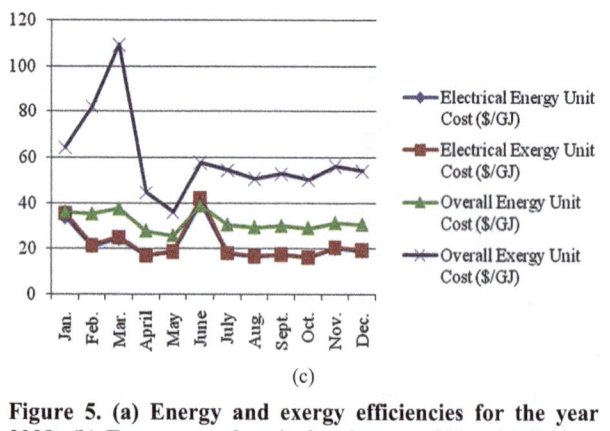

(c)

Figure 5. (a) Energy and exergy efficiencies for the year 2008; (b) Energy supply mix for the year 2008; (c) Energy and exergy unit costs for the year 2008.

on efficient energy utilisation is more, although steam raising is unavoidable, due to the nature of the company under investigation. The annual mean energy unit costs have also varied from 27.86 USD per Giga-Joule in 2006 to 32.80 USD per Giga-Joule in 2004, confirming the inverse proportion of energy efficiency and costs. On the other hand, the annual mean exergy unit costs have varied from 40.19 USD per Giga-Joule in 2005 to 58.46 USD per Giga-Joule in 2004. This also implies that year 2004 has been the worst year from all (energy, exergy and economic) points of view. The most efficient year

has been 2006 energetically and 2005 exergetically. The difference in the two years lies in the proportions of generator diesel and boiler diesel utilised as the system exergy is more sensitive to boiler diesel use while the system energy is more sensitive to generator diesel utilisation due to their different device efficiencies.

6.2. Recommendations

Based on the findings in this paper, it is necessary to avoid electricity generation from diesel powered generators as much as possible. Secondly, steam generation is an unavoidable but very expensive process. Hence, spe-

(a)

(b)

(c)

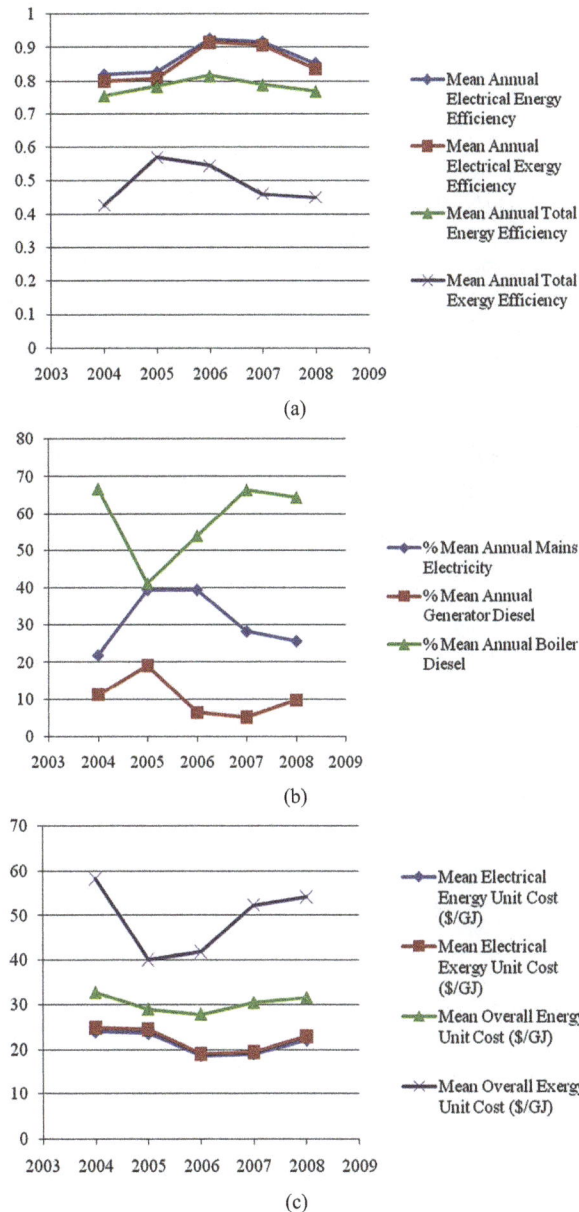

Figure 6. (a) Mean annual efficiencies; (b) Mean annual energy supply mixes for the years 2004-2008; (c) Annual mean energy unit costs.

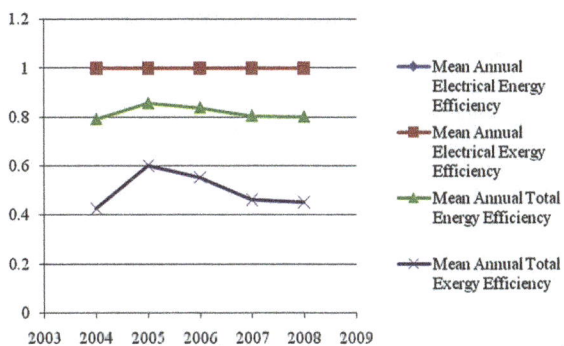

Figure 7. Mean annual efficiencies with zero generator diesel.

cial attention should be paid to process steam management to avoid unnecessary leakages and/or wastages. Lastly, in lieu of stable power supply from the national grid, a big company like this brewery should be able to consider a technically trusted energy conservation technique like cogeneration, since it needs both electrical power and process steam dearly.

For the boiler, the brewery may have to consider bigger size boilers. This is because authors like Pulkrabek [18] quoted in [16] have observed that general trend is that the greater is the plant size, the smaller is the specific fuel consumption. One reason for this is less heat loss due to the higher volume to surface area ratio of the combustion chamber

Secondly, a boiler fuel-switch from diesel to natural gas may be worthwhile, considering the fact that natural gas is a low-carbon fuel with a lower minimum allowable stack temperature than that of diesel oil [19] and Nigeria has the gas abundantly. A lower stack temperature would improve the boiler efficiency, reducing the company energy costs and bring down thermal pollution level, while the low carbon content would reduce CO_2, a green house gas, emission.

Finally, 2% - 8% boiler energy can be saved [17] by enhancing heat transfer rate of flue gases using nanofluids.

REFERENCES

[1]   O. Davidson, "Energy for Sustainable Development: South African Profile," Phase 1 Final Report, Energy Research Centre, University of Cape Town, South Africa, 2004.

[2]   Organisation of Petroleum Exporting Countries (OPEC), "Oil and Gas Data," *OPEC Annual Statistical Bulletin*, Vienna, 2009, pp. 22-23.

[3]   Manufacturers Association of Nigeria (MAN), "60 m Nigerians Now Own Power Generators," *Vanguard Newspaper*, 26 January 2009.

[4]   Energy Commission of Nigeria (ECN), 7 March 2012. http://www.energy.gov.ng/index.php?option=com_content&task=view&id=56&Itemid=58

[5]   A. Oniwon, "Oil and Gas in Nigeria's National Development: An Assessment," *A Presentation at the National Defence College*, Abuja, 9 November 2011.

[6]   Manufacturers Association of Nigeria (MAN), "Manufacturers Spend N1.8bn Weekly on Diesel," *Nigerian Tribune Newspaper*, 21 August 2009.

[7]   A. Iwayemi, "Nigeria's Dual Energy Problems: Policy Issues and Challenges," *Proceedings of the 31st International Association for Energy Economics International Conference*, Istanbul, 18-20 June 2008, pp. 17-21.

[8]   J. Szargut, D. R. Morris and F. R. Steward, "Exergy Analysis of Thermal, Chemical and Metallurgical Processes," Hemisphere Publishing Corporation, Washington DC, 1988.

[9]   T. J. Kotas, "The Exergy Method of Thermal Plant Ana-

lysis," 2nd Edition, Krieger Publishing Company, Melbourne, 1995.

[10] R. L. Cornelissen, "Thermodynamics and Sustainable Development," Febodruk, Enschede, 1997.

[11] J. O. Jaber, A. Al-Ghandoor and S. A. Sawalha, "Energy Analysis and Exergy Utilization in the Transportation Sector of Jordan," *Energy Policy*, Vol. 36, No. 8, 2008, pp. 2995-3000. doi:10.1016/j.enpol.2008.04.004

[12] G. Tsatsaronis and F. Cziesta, "Thermodynamics," Summer Course on Optimization of Energy Systems and Processes, Gliwice, 2003.

[13] A. Hepbasli, "A Study on Estimating the Energetic and Exergetic Prices of Various Residential Energy Sources," *Energy and Buildings*, Vol. 40, No. 3, 2008, pp. 308-315. doi:10.1016/j.enbuild.2007.01.023

[14] 10 March 2012. http://www.phcnonline.com/tariffNew.asp

[15] O. A. Fashade, "General Information for Engineering Students and Professionals," Fast Corp Publishers, Lagos, 1997.

[16] M. Kanoğlu, S. K. Işık and A. Abuşoğlu, "Performance Characteristics of a Diesel Engine Power Plant," *Energy Conversion and Management*, Vol. 46, No. 11-12, 2005, pp. 1692-1702. doi:10.1016/j.enconman.2004.10.005

[17] R. Saidur, J. U. Ahamed and H. H. Masjuki, "Energy, Exergy and Economic Analysis of Industrial Boilers," *Energy Policy*, Vol. 38, No. 5, 2010, pp. 2188-2197. doi:10.1016/j.enpol.2009.11.087

[18] W. W. Pulkrabek, "Engineering Fundamentals of Internal Combustion Engines," Prentice-Hall, Upper Saddle River, 1997.

[19] A. Bhatia, "Improving Energy Efficiency of Boiler Systems," Continuing Education and Development, Inc., Greyridge Farm Court Stony Point, New York, 2012. http://www.cedengineering.com

13

Solar Energy and Residential Building Integration Technology and Application

Ding Ma, Yi-bing Xue

Department of Architecture and Urban Planning.Shandong Jianzhu University, Jinan, China

Email: mading.1989@163.com

ABSTRACT

Building energy saving needs solar energy, but the promotion of solar energy has to be integrated with the constructions. Through analyzing the energy-saving significance of solar energy, and the status and features of it, this paper has discussed the solar energy and building integration technology and application in the residential building, and explored a new way and thinking for the close combination of the solar technology and residence.

Keywords: Solar Energy; Residential Building; Integration Technology and Application

1. Introduction

With the improvement of China's economic construction and people's living standards, the energy crisis and environmental degradation are also growing, we are facing the dual pressures of resources and environmental protection. Building energy consumption accounts for 25% to 40% of the total energy consumption, together with transportation and industry as the three major energy-consuming households, building energy efficiency plays an important role in the national energy conservation strategies. As the large construction number of residential buildings energy-saving or not, will no doubt have positive practical significance to realization of building energy conservation and environmental protection.

2. Status and Features of Solar Energy and Building Integration Technology

2.1. Status of Solar Energy and Building Integration Technology

As the building renewable energy, solar energy is clean, non-polluting and easy to get, more and more people of all ages. Solar energy resource in China is extremely rich, and the total annual solar radiation amount is more than 5.02 million KJ/m^2, annual sunshine hours over 2200h areas accounted for more than two-thirds of the land area, so there is great potential in the use of solar energy[1]. Currently, the use of solar energy in our country has made gratifying achievements, but the degree of realization of solar energy and building integration technologies is not high. Firstly, the development of solar thermal and photovoltaic is uneven, from the *Renewables* 2012 *global*

Status Report can see that China ranks the world's first in solar water heaters in 2011, while the development of solar photovoltaic is relatively slow[2]. Secondly, there exist many problems in the use of solar water heater, such as the water heaters rank highly on the roof, placed very messy, not only destroy the architectural aesthetics, but also affect the image of the city. The aim of implementing solar energy and building integrated technologies is to change the disjointed, fragmented status quo of each branch and link, incorporate the use of solar energy into the over all design of the environment, make architecture, technologies and aesthetics be in harmony as an organic whole, and make solar facilities to be part of the building.

2.2. Features of Solar Energy and Building Integration Technology

The main features of solar energy and building integration technology are:

- Involving a wide range

To achieve the application of solar energy and building integration technology need the coordination and joint effort of several departments, such as national policies and regulations department, component construction department, solar manufacturers, property developers, designing institution and construction enterprises.

- High technical content

Solar energy and building integration technology is a comprehensive technical which combines multiple disciplines such as optical, thermal, electronic, fluid mechanics and architecture, etc.

- Complicated construction

Solar energy and building integration technology also requires the integration of construction process and technology, in addition to the conventional construction, there also need to conduct the complex construction of waterway and circuit and to complete the installation and debugging tasks of solar equipments .

● High initial investment

Due to the increased solar equipments, pipeline and appliances, as well as the corresponding structural and construction detail handing, the initial investment of solar energy and building integration technology is high, and the construction cost is also high[3]. But with the use of solar energy, the economic benefits that generate from its energy conservation, environmental protection, safe and efficient will become increasingly apparent.

3. Application of Solar Energy and Building Integration Technology in Residential Building

The application of solar energy and building integration technology in residential buildings, mainly has three aspect: solar thermal technology, solar photovoltaic technology, solar optical technology, and mainly set on roofs, balconies, exterior walls and somewhere with ample sunshine.

3.1. Solar Thermal Technology

Solar thermal is mainly used to supply domestic hot water, heating and refrigeration. In designing the integration of solar hot water system and residential buildings, not only need to consider the layout of solar hot water system, but need to further improve the form of the system itself. Traditional solar hot water system with vacuum tube can not meet the needs of the ever-changing layout and style of the residential buildings, beyond that, it have other deficiencies, such as the installation is very difficult, easy to destroy the waterproof layer of the roof, have security risks if the lightning protection and drought exclusion device not in place, vacuum tube belongs to quick-wear part and the maintenance ratio is high, water pipes are exposed to the outdoor cause large heat loss, etc. In short, the traditional solar hot water system with vacuum tube can not meet the need of integration of solar energy and building either in quality or in performance. Now, the flat plate solar collector system is gradually replacing the solar hot water system with vacuum tube, for it has higher adaptability, and the installation of it can better achieve the perfect combination with the construction[4]. Solar collector system mainly operate on the split double-cycle under pressure, the hot water tank can located in the basement, attic, staircase, balcony and other hidden parts, and not occupy the indoor space,

avoiding the load-bearing of roofs, balconies and exterior walls; water tank can use single tank, double tank and even a multi-tank, so as to achieve a larger holding tank capacity, when the tank capacity is increased, the installation area is correspondingly increased to meet the hot water needs; hot water is not just use for bath, but also used for heating and supplying domestic water, the water quality should keep clean to meet the drinking water standards.

Integrating solar collector with the roofs, balcony rails of the south façade, bay windows and walls, can make the appearance of residential buildings be overall unified, and have rich hierarchies[5] (**Figure 1**). When installed on

(a) combined with the roof

(b) combined with the balcony

(c) combined with the exterior wall

Figure 1. Ways of the solar collector combined with the residential building.

the sloping roof, the solar collector can be embedded in the roof like a sunroof or flat out on the roof, integrating with the construction to increase the building beauty. When installed on the flat roof, the flat-plate solar collector can act as roof covering or insulation layer, not only conforms to the residential modeling requirements, but also avoids the repeated investment and reduce the cost. In addition, the flat-plate solar collector can be combined with balconies, bay windows, outside walls of residential buildings, to maximize the use of solar energy and provide new ways and means to the residential façade design, and achieve the aim of multi-purpose as well (**Figure 2**).

3.2. Solar Photovoltaic Technology

Solar photovoltaic technology applying in residential buildings is mainly used for photovoltaic conversion and lighting. BIPV (Building Integrated Photovoltaic) is a

new concept for the application of solar power, in short, installing the solar photovoltaic phalanx on the surface of the maintenance structure of the building to provide electricity[6]. Photovoltaic arrays do not take up additional floor space when integrate with the construction, and is the best installation way of photovoltaic generation system, thus attracting much attention.

BIPV can be divided into two categories according to the forms that photovoltaic array integrated with the buildings[7]. One is the combination of photovoltaic array with building, installing the PV array on the building, and the building play a supporting role as a photovoltaic carrier (**Figure 3**). The other is the integration of photovoltaic array with building, PV modules appear as the building material, and the photovoltaic array become the integral part of the construction, such as photoelectric tile roof, photoelectric curtain wall and photoelectric lighting roof, etc (**Figure 4**).

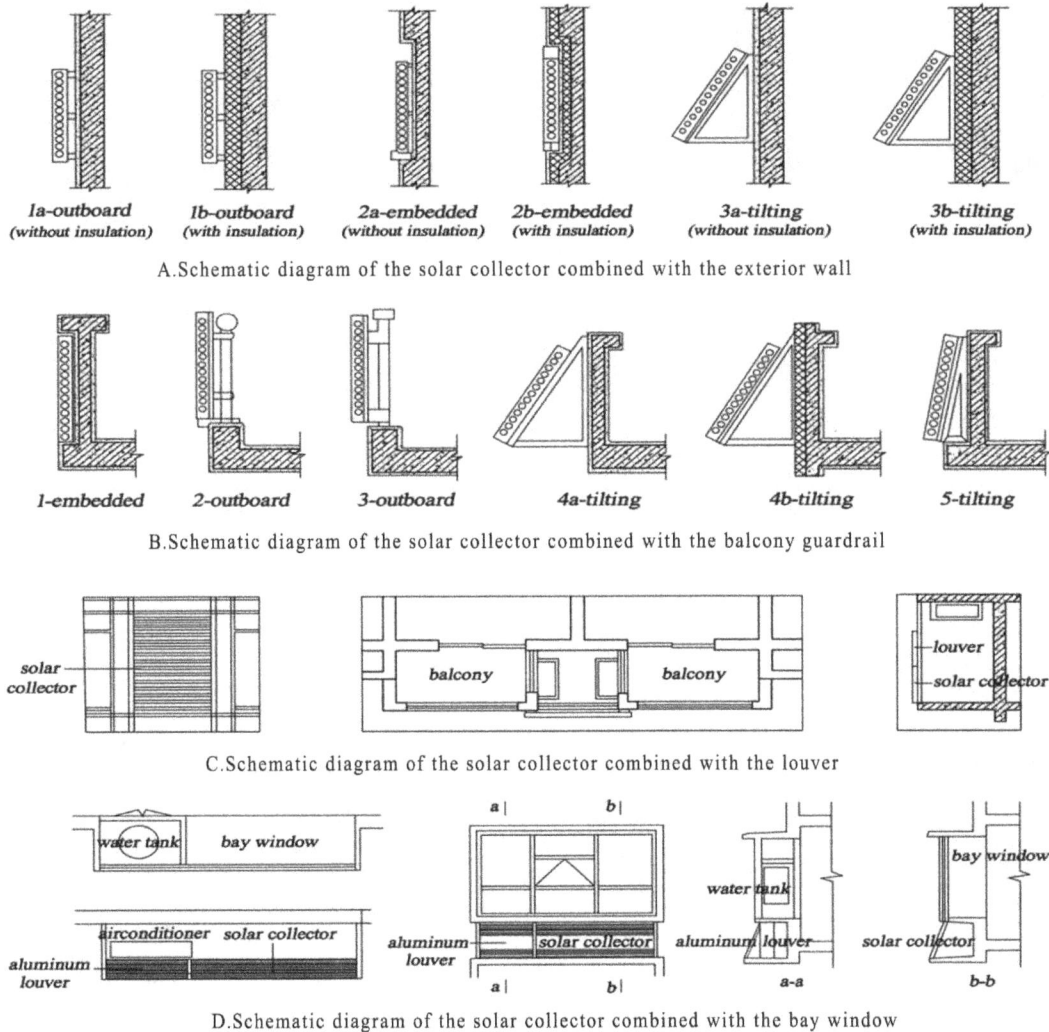

1a-outboard (without insulation)   1b-outboard (with insulation)   2a-embedded (without insulation)   2b-embedded (with insulation)   3a-tilting (without insulation)   3b-tilting (with insulation)

A.Schematic diagram of the solar collector combined with the exterior wall

1-embedded   2-outboard   3-outboard   4a-tilting   4b-tilting   5-tilting

B.Schematic diagram of the solar collector combined with the balcony guardrail

C.Schematic diagram of the solar collector combined with the louver

D.Schematic diagram of the solar collector combined with the bay window

Figure 2. Construction node design of the solar collector combined with the residential building.

(a) combined with the roof   (b) combined with the façade

Figure 3. Ways of the solar PV arrays combined with the residential building.

(a) photoelectric tile roof   (b) photoelectric lighting roof

Figure 4. Ways of the solar PV arrays integrated with the residential building.

3.3. Solar Optical Technology

The main use of the solar optical technology in residential buildings is for lighting, natural light can enter into the function rooms through the light guide tube, thus improve the indoor daylighting situation, such as underground garage, equipment room and storage room. Due to the utilization of the solar optical system is also subjected to the impact of the climate, it is suitable for the regions that have abundant natural light and less cloudy sky[8]. The light guide tube is mainly composed of three parts: a light collector for collecting the daylight; tubing portion for transmitting light; the light exit portion for controlling the distribution of the light in the room. Using the light guide tube on the roof must ensure that there have no obstructions, and well water treatment to avoid leaking during the installation (**Figure 5**). Moreover, as the instability of the natural light, the light guild tubes must in combination with the adjustable artificial light, so as to be an effective supplement when the daylight is insufficient.

4. Conclusions

Solar energy and residential building integration technology has broad application prospects, despite the many

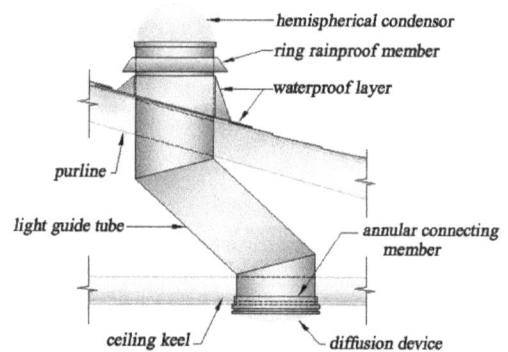

(a) cross-section drawing

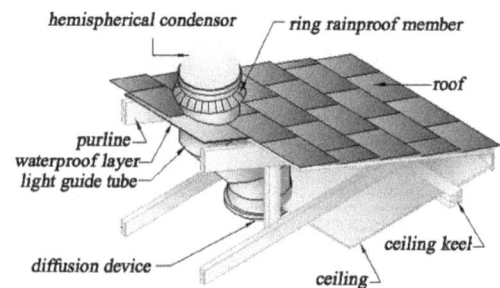

(b) axonometric drawing

Figure 5. Solar optical system on the roof.

problems, such as the integration degree of solar water heating system and building is not high, solar photovoltaic industry is lack of technological breakthroughs, and the production cost is high, etc. But with the introduction of the "Renewable Energy Law" and the detrusion of a series of policies and measures that encouraging the use of renewable energy, is bound to promote people's enthusiasm toward the use of renewable energy, increase the technology innovation and development efforts, make the building integrated solar technologies become more mature, and make the solar energy and building integration technology more closely combined with residential and broader development prospects.

5. Acknowledgements

This research was partially co-funded by the National Natural Science Foundation (51078223), We would like to thank the Key Laboratory of Renewable Energy Utilization Technologies in Buildings of the National Education Ministry and the Key Laboratory of energy-saving technology of Shandong Province for providing us with historical data of solar energy, integration technology and application etc. We also thank Professor Wang Chong-jie, a specialist on the solar energy and building integration technology, for his valuable advice.

REFERENCES

[1] C. J. Wang and Y. B. Xue, "Solar Building Design," Beijing: China Architecture & Building Press, 1st Edition, 2007, pp. 87-90.

[2] S. N. Wang, "The Development of Solar Building Technology in Domestic and Foreign," *Journal of New Building Materials*, No. 10, 2008, pp. 44-46.

[3] T. S. Xin and C. M. Yang, "Application of Solar Energy Technology to Congregated House," *Journal of Architecture*, No. 8, 2006, pp. 22-25.

[4] Y. H. Wang, "Design of solar Residential Building Integration," *Journal of Building Energy Efficiency*, Vol. 38, No. 1, 2010, pp.53-55.

[5] C. H. Xu and M. L. Qin, "The Application of Solar Energy and Building Integrated Multi-technology," *Journal of Construction Science and Technology*, No. 5, 2012, pp.69-71.

[6] L. R. Zhang, "The Application of Motor Control Unit in the Cement," *Journal of Equipment Manufacturing Technology*, No. 8, 2011, pp. 208-209.

[7] W. Jin, "Application of Building Integrated Photovoltaic (BIPV) in Green Buildings," *Journal of Architecture Technology*, Vol. 42, No. 10, 2011, pp. 907-908.

[8] A. Y. Wang and G. Shi, "New Process of Daylighting Technology," *Journal of Architecture*, No. 3, 2003, pp. 64-66.

14

Power Analysis for Piezoelectric Energy Harvester

Wahied G. Ali, Sutrisno W. Ibrahim

Electrical Engineering Department, King Saud University, Riyadh, KSA

Email: wahied@ksu.edu.sa, suibrahim@ksu.edu.sa

ABSTRACT

Piezoelectric energy harvesting technology is used to design battery less microelectronic devices such as wireless sensor nodes. This paper investigates the necessary conditions to enhance the extracted AC electrical power from exciting vibrations energy using piezoelectric materials. The effect of tip masses and their mounting positions are investigated to enhance the system performance. The optimal resistive load is estimated to maximize the power output. Different capacitive loads are tested to store the output energy. The experimental results validated the theoretical analysis and highlighted remarks in the paper.

Keywords: Vibration; Energy Harvesting; Piezoelectric Materials; Power Analysis; Resonant Frequency

1. Introduction

Energy harvesting technology is used to generate electrical power from natural (green) energy sources. The concept of energy harvesting generally relates to the process of using ambient energy, which is converted into electrical energy. Research on energy harvesting technology became progressively larger over the last decade to design self-powered microelectronic devices. With the advances being made in wireless technology and low power microelectronics, wireless sensors are being developed and can be placed almost anywhere. Wireless sensor networks are progressively used in many applications such as: structure health monitoring, automation, robotics swarm, and military applications. However, these wireless sensors require their own power supply which in most cases is the conventional electrochemical battery. Once these finite power supplies are discharged, the sensor battery has to be replaced. The task of replacing the battery is tedious and can be very expensive when the sensor is placed in a remote location. These issues can be potentially alleviated through the use of power harvesting devices.

One of typical wasted energy is an ambient vibration that presents around most of machines and biological systems. This source of energy is ideal for the use of piezoelectric materials, which have the ability to convert mechanical strain energy into electrical energy and vice versa [1]. In general, there are three techniques to harvest the energy from the vibration: electrostatic, electromagnetic, and piezoelectric. Piezoelectric materials have a superior performance to be used for energy harvesting from ambient vibrations, because they can efficiently convert mechanical strain to an electric charge without any additional power and have a simple structure for real time applications [2,3].

In general, a piezoelectric energy harvesting can be represented as shown in **Figure 1**. The mechanical energy (e.g., applied external force or acceleration) is converted into mechanical energy in the host structure. Then, this energy is converted into electrical energy by the use of piezoelectric material, and is finally transferred into electrical form to a load and/or a storage stage [4]. Therefore, three basic processes are performed: conversion of the input energy (vibration) into mechanical energy (strain) using a cantilever structure, electromechanical conversion using piezoelectric material, and electrical

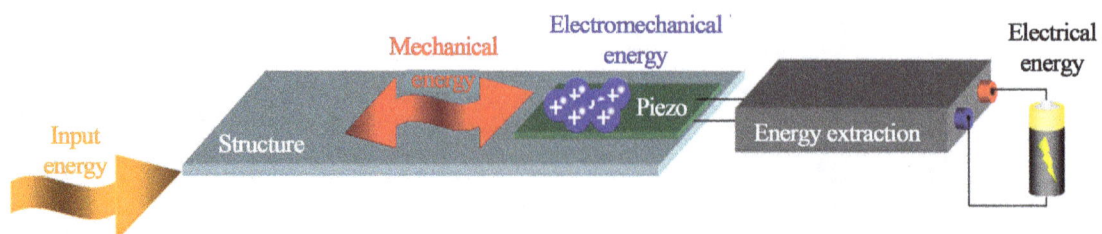

Figure 1. Schematic diagram of piezoelectric energy harvester [4].

energy transfer. In the literature, several books are recently published in this research domain [5-8]. Several review papers are also published in all different aspects concerning energy harvesting technologies [9-15].

This paper investigates the necessary conditions to enhance the extracted AC electrical power from the exciting vibration energy using piezoelectric material. This paper is organized as follows: the next section presents theoretical background to maximize the generated electrical power. Section 3 describes the experimental setup to carry out the real time measurements. Section 4 discusses the experimental results; while the last section concludes the paper and highlights the future research directions in this field.

2. Theoretical Background

The piezoelectric effect is a direct transformation of mechanical energy into electrical energy. Piezoelectricity was discovered by Jacques Curie and Pierre Curie in 1880 [16-18]. They observed that certain crystals respond to pressure by separating electrical charges on opposing faces and named the phenomenon as piezoelectricity. In the literature, several design parameters have been investigated to maximize the generated power from mechanical vibrations to electrical output using piezoelectric material. These parameters are summarized as follows [1,3]:

Material type as PZT, PVFD, Quick-Pack, and PVFD. Material with high quality factor (Q-factor) produces more energy and recently piezoelectric micro fiber composite (MFC) has more efficiency to generate electrical power up to 65% of the input mechanical energy.

Geometry, tapered form produces more energy while the strip form is commercially available.

Thickness, thin layers produce more energy.

Structure, bimorph structure doubles the output than unimorph structure.

Loading mode, d_{31} produces large strain and more energy for small applied forces.

Resonant frequency has to be matched with the fundamental vibration frequency.

Electrical connection, parallel (to increase the output current) and series (to increase the output voltage source).

Fixation, cantilever produces more strain than simple beam.

Load impedance, it has to be matched with the piezoelectric impedance at the operating frequency.

In this paper, the development is focused on the necessary conditions to maximize the AC power output from the piezoelectric harvester using MFC material with fixed dimensions. The piezoelectric harvester is a cantilever with an effective length (I_b) extends from the clamped end to the free end with tip mass (m) as shown in **Figure 2**. Electrodes must be plated onto the piezo-

electric material to collect the generated charges as the beam flexes.

Williams and Yates developed a generic model based on inertial kinetic energy [19]. The model is a lumped-parameter second order dynamic system which relates the input vibration $y(t)$ to the output relative displacement $z(t)$ (see **Figure 3**). By applying D'Almbert's law, the dynamic equation for this system is given as:

$$m\ddot{z} + d\dot{z} + kz = -m\ddot{y} \qquad (1)$$

where $y(t)$ is the input vibration, $\ddot{y}(t)$ is the input acceleration, $z(t)$ is the relative displacement of the mass with respect to the vibrating cantilever, k is the spring constant (device stiffness), d is the total damping (parasitic damping and electrical damping), and m is the effective mass of cantilever. The transfer function $G(s)$ for this system is given as:

$$G(s) = \frac{-m}{ms^2 + ds + k} \qquad (2)$$

Thus natural frequency ω_n is given by

$$\omega_n = \sqrt{\frac{k}{m}} = \sqrt{\frac{Ywh^3}{4l^3\left(m_t + 0.24m_c\right)}} \qquad (3)$$

where Y is the young's modulus, w, h, and l are the width,

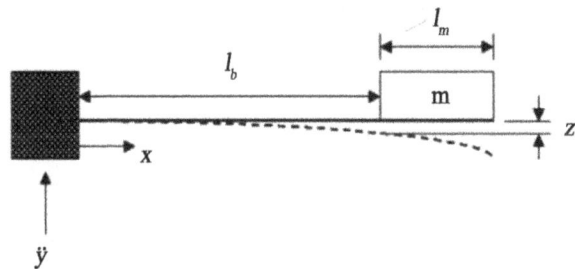

Figure 2. Piezoelectric cantilsever [3].

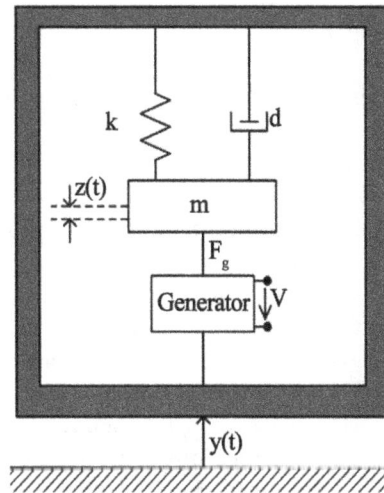

Figure 3. Dynamic model of vibration harvester [19].

the thickness, and the length of the cantilever respectively, m_t is the tip mass, and m_c is the cantilever's mass. The total damping factor η (mechanical and electrical) is given by

$$\eta = \frac{d}{2m\omega_n} \qquad (4)$$

The total quality factor is given by [3,6]

$$Q = \frac{1}{2\eta} = \frac{f_n}{f_{bw}} \qquad (5)$$

The quality factor is inversely proportional to the damping factor and can be computed as the ratio between the resonant frequency (f_n) and the frequency bandwidth (f_{bw}). For a sinusoidal vibration signal ($y(t) = A\sin(\omega t)$) the instantaneous dissipated power (P) within the damper equals the product of the velocity and the damping force. This power is computed using the following formula [6]:

$$P = \frac{m\eta A^2 \left(\dfrac{\omega}{\omega_n}\right)^3 \omega^3}{\left[1 - \left(\dfrac{\omega}{\omega_n}\right)^2\right]^2 + \left[2\eta\dfrac{\omega}{\omega_n}\right]^2} \qquad (6)$$

where; A is the amplitude of vibration. The maximum power dissipated in the damper occurs at the natural frequency and can be calculated by the following formula:

$$P = \frac{mA^2\omega_n^3}{4\eta} \qquad (7)$$

where; the peak of input acceleration (a) is given by $A\omega^2$. Maximum power conversion to electrical domain occurs when the mechanical damping equals the electrical damping. Therefore, the maximum electrical output power is equal to half the value in the Equation (7).

$$P_e(\max) = \frac{mA^2\omega_n^3}{8\eta} = \frac{ma^2}{8\eta\omega_n} = \frac{ma^2 Q}{4\omega_n} \qquad (8)$$

The above equation indicates that the maximum power is directly proportional to the effective mass, the input acceleration and the quality factor. Whatever, it is inversely proportional with the natural frequency and the total damping. The piezoelectric cantilever itself can generate output voltage due to the (mechanical strain) relative displacement $z(t)$. The strain effect utilizes the deformation of the piezoelectric material to generate positive and negative charges on both sides. The equivalent electrical circuit can be considered as shown in **Figure 4**. The piezoelectric harvester has high resistive impedance in mega Ohm and capacitive value in nano Farad. At resonance, the current source is I_{piezo} equal to $mA\omega_n^2$.

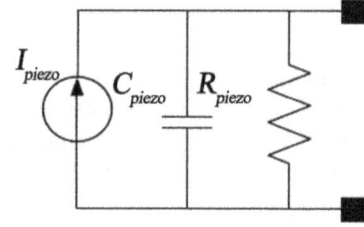

Figure 4. Equivalent circuit at resonance [6].

The harvester's resistive value could be ignored due to its too high value in mega Ohm, therefore its effective impedance is a capacitive type. This impedance Z_i can be computed using the following equation

$$Z_i = \frac{1}{\omega_n C_{piezo}} = \frac{1}{2\pi f_n C_{piezo}} \qquad (9)$$

The maximum power transfer to the external load occurs when the value of this load is close to the harvester's internal impedance. To store the energy in a capacitive load, the extracted AC power has to be rectified using a 4 bridge rectifier. The stored energy in the capacitor is given as:

$$W(t) = \frac{1}{2}C\big[V(t)\big]^2 \qquad (10)$$

where $W(t)$ is the stored energy in Joule at instant time t, C is the capacitive value in Farad, and $V(t)$ is the measured voltage across the capacitor at instant time t. The accumulated energy has to be enough to run the load for a pre-defined time. The application run time depends on the consumed current, the output voltage to the load, and the stored energy in the capacitor. The run time of a certain application can be computed by

$$T = \frac{W}{VI_{app}} \qquad (11)$$

where W is the stored energy in Joule, V the voltage across the capacitor to the load, and I_{app} is the consumed current by application.

3. Experimental Setup

The experimental set-up consists of (see **Figure 5**):
- Amplifier module to generate the vibration signal at different levels and different frequencies ranged from 5 Hz to 60 Hz with incremental step of 1 Hz;
- Desktop shaker to generate mechanical vibrations;
- Harvester module using a Micro Fiber Composite (MFC) material (M-8528-P2, *Smart Material*) with dimensions of 8.5 cm × 2.8 cm;
- Variable impedance module (VIM, *Smart Material*);
- Different capacitors, 22 µF, 470 µF, and 0.1 F;
- NI-PXIe system with DAQ (Data Acquisition) card as a hardware platform;

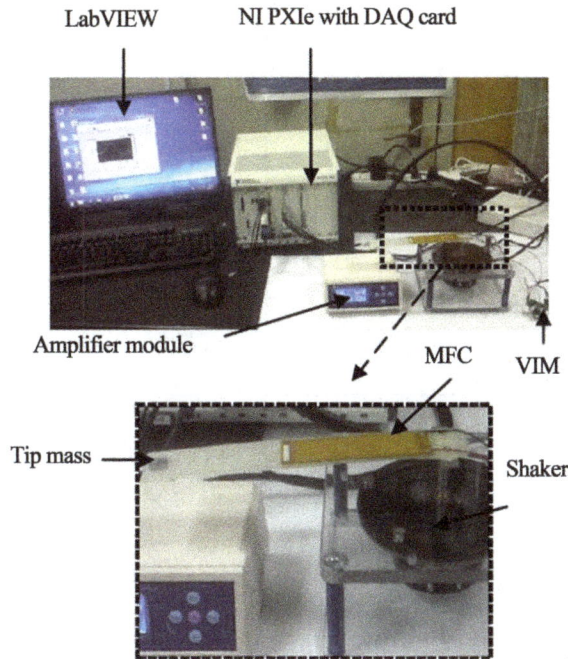

Figure 5. Experimental set-up.

- NI-LabVIEW software for monitoring and analyzing the acquired signals.

The MFC is glued on a flex cantilever substrate that is mounted on a shaker. The dimension of the cantilever is 22 cm × 3.5 cm × 0.1 cm. The shaker is excited by an amplifier module to generate vibrations. Under exciting vibration, the piezoelectric harvester produces AC electrical output. Then, the output signal from the harvester is connected to a variable resistive load or a bridge rectifier with a capacitive load. NI-PXIe with DAQ card is used to perform real time measurements.

The vibration frequency and its excitation level (amplitude) are varied to test the performance of the harvester. The peak-to-peak voltage or maximum power is used to evaluate the system performance. The real time monitoring and analysis for the harvested electrical signal is performed using LabVIEW software. The harvesting cantilever has to be selected with high Q-factor to increase the power output. Higher Q-factor indicates a lower rate of energy loss relative to the stored energy. However, high Q-factor also means narrow operational frequency bandwidth. The harvester produces significant power when it works under excitation frequency that closing to its resonant frequency. The resonant frequency of the harvesting cantilever under a given set conditions is identified experimentally by monitoring the peak of power output. The resonant frequency is changed by adding or removing tip masses. Different tip masses are used to investigate their effect to the resonant frequency. The effect of different mounting positions for this tip mass is also investigated.

The maximum power transfer happens when the load impedance is closely matched with the harvester's internal impedance. The harvester's internal impedance is a capacitor type and consequently not only depends on MFC characteristics but also the excitation frequency. The harvester resonant frequency will be investigated firstly to achieve optimal frequency for excitation. Such experiment using maximum power point tracking (MPPT) method will be performed to determine maximum power transfer and its corresponding optimal load. A variable impedance module (VIM) is used as variable resistive load (potentiometer) to change the electrical load easily.

4. Results and Discussion

4.1. Resonant Frequency without Tip Mass

The generated AC electrical power from the piezoelectric harvester has been monitored in real time using Lab-VIEW software as shown in **Figure 6**.

The resonant frequency under fixed excitation level (level 2) is identified by monitoring the maximum output power as a function of excitation frequency. The excitation frequency from the amplifier module is varied from 11 to 30 Hz. In this experiment, no tip mass is mounted on the cantilever and the load is fixed to be 50 kΩ. **Figure 7** shows the generated power from the harvesting system without tip mass.

It is observed that the resonant frequency of the harvester is 18 Hz (maximum peak of the extracted power). Under this excitation frequency; the power output reached to its maximum value equal to 250 µW. Another local peak is also observed at 23 Hz with lower power output. This phenomenon is commonly known for physical systems, in which such systems have multiple resonant frequency modes due to its flexible structure.

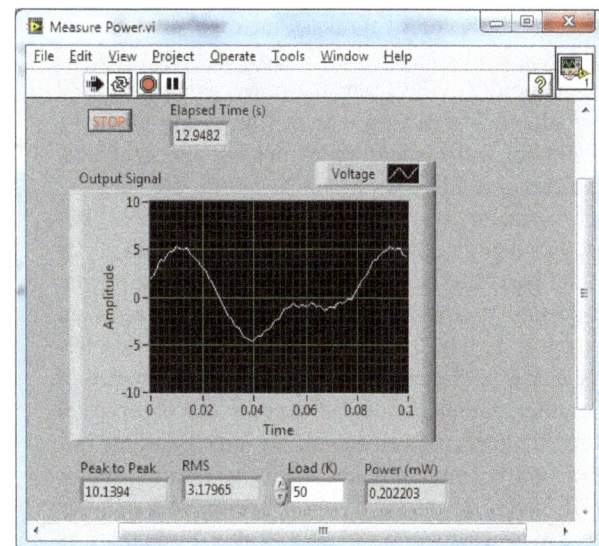

Figure 6. LabVIEW front panel.

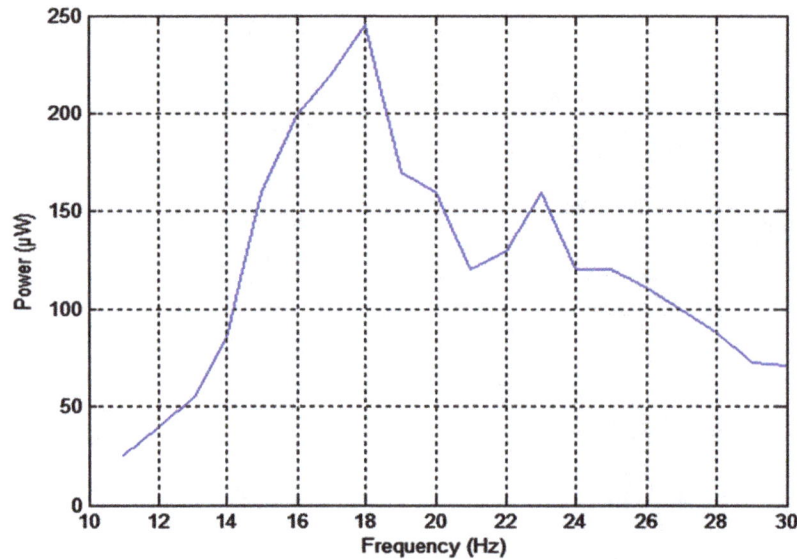

Figure 7. Generated power without tip mass.

4.2. Effect of Tip Masses

From Equation (3), the resonant frequency is inversely proportional to square root of effective mass. Theoretically, increasing cantilever mass will decrease the resonant frequency (nonlinear characteristics). This theoretical result has been validated by experimental work as shown in **Figure 8**.

The additional masses are varied from 1 to 5 grams were mounted at fixed position 1 cm from the free end of the cantilever with vibration excitation level 2. The resonant frequency is shifted to 17 Hz, 15 Hz, 14 Hz, 13 Hz, and 13 Hz respectively for 1, 2, 3, 4 and 5 gram additional tip mass. In real time applications, this method could be used effectively to tune the resonant frequency if the fundamental frequency of vibration source is pre-determined. It is observed also that the power output increases as the value of tip mass increases that means; the Q-factor is also increased. This result was expected from Equation (8). The generated power is equal to 380 μW by adding 5 grams tip masses (increasing factor by 60%). While, without adding tip masses; the generated power is equal to 245 μW.

High Q-factor also means narrow operational frequency bandwidth. Although, with 5 gram additional tip mass; the harvester is able to produce maximum power, its frequency bandwidth is too narrow. With resonant frequency at 13 Hz, a slightly deviation from this frequency causes significant degradation in performance of the harvester. Another secondary peak is also observed at 27 Hz with lower power output. As noted earlier, this phenomenon is commonly known for physical systems, in which such systems have multiple resonant frequency modes due to its flexible structure. **Table 1** summarizes the complete results for these experiments.

4.3. Effect of Mounting Positions for Tip Mass

Resonant frequency of a cantilever depends on its effective mass rather than its total mass. Two cantilevers with similar material and mass but with different shape or mass density distribution will have distinct resonant frequency. The closer the center of gravity to the free end of the cantilever, the greater the effective mass is achieved, and hence the lower resonant frequency. **Figure 9** shows the experiment results using 3 grams tip masses at different mounting positions on the cantilever with vibration excitation level 2.

The mounting position of tip mass is varied from 1 to 7 cm from the free end of the cantilever. The resonant frequencies are 14, 15, 15, and 16.5 Hz respectively for 1, 3, 5, and 7 cm mounting positions. **Table 2** summarizes the complete results for these experiments.

4.4. Optimal Resistive Load

As noted earlier, the maximum power transfer can be extracted when the load impedance is closed to the harvester's internal impedance. Due to high resistive value for piezoelectric material, the internal impedance could be simplified as capacitive type. From the datasheet, the internal capacitance for the MFC is equal to 172 nF; this value has been also checked using a high sensitivity capacitance meter. To investigate optimal load for the harvester, no mass is mounted on the cantilever and the harvester is excited at its resonant frequency (18 Hz). By using Equation (9), the internal impedance (Z_i) is computed as follows:

$$Z_i = \frac{1}{(2\pi)\, x\, (18)\, x\, (170 \times 10^{-9})} = 52.47 \text{ k}\Omega$$

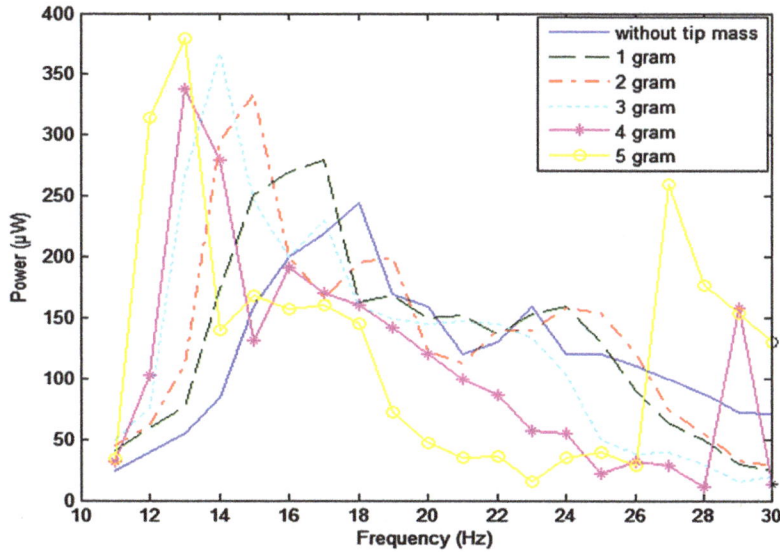

Figure 8. Generated power with different tip masses.

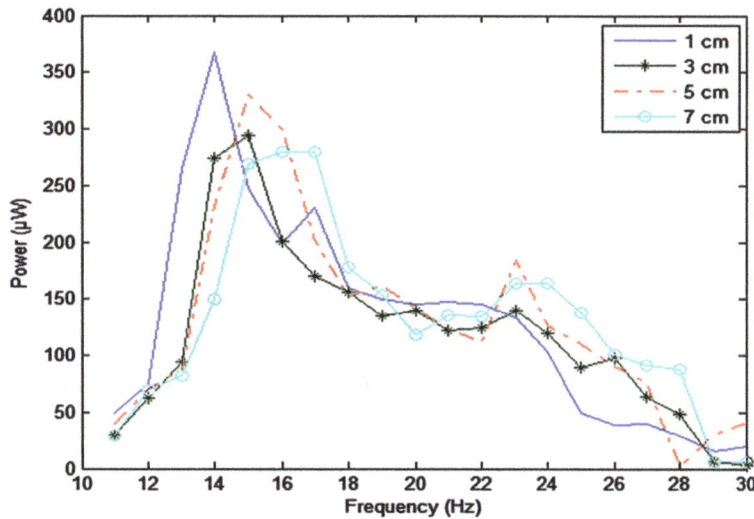

Figure 9. Generated power with different mounting positions.

Table 1. Experiment for different tip masses.

EXP.	P_{max} (μW)	f (Hz)	f_{BW} (Hz)	Q-factor
Without tip mass	245	18	11	1.64
+ 1 gram	280	17	11	1.55
+ 2 gram	334	15	6	2.5
+ 3 gram	368	14	5	2.8
+ 4 gram	338	13	2	6.5
+ 5 gram	380	13	2	6.5

Table 2. Experiment results for mounting positions.

EXP.	P_{max} (μW)	f_n (Hz)	f_{BW} (Hz)	Q-factor
1 cm	368	14	5	2.8
3 cm	294	15	5	3
5 cm	330	15	4	3.75
7 cm	280	16.5	6	2.75

in **Figure 10**. The maximum power is obtained at 50 kΩ, which is close to the harvester's internal impedance.

The upper curve in **Figure 10** represents the AC power from the piezoelectric material; while the lower curve represents the rectified power using diode-bridge. The difference between the two curves is due to the power loss in the rectifier circuit. The maximum power for the

The harvester is predicted to generate maximum output power when the (external) load is closed to the above value. The variable impedance module (VIM) is varied from 0 Ω to 500 kΩ, the extracted power output is shown both curves is obtained at the same value of the resistive load.

4.5. Capacitive Load to Store the Energy

After rectifying the AC voltage, the extracted power is stored in a capacitor. Three different capacitors are tested for charging characteristics: 22 µF (10 V maximum voltage), 470 µF (25 V maximum voltage), and 0.1F (5.5 V maximum voltage). **Figures 11-16** show the measured output voltage and the corresponding stored energy for each capacitor.

The charging process for small capacitors (22 µF and 470 µF) was very fast. The output voltage was increased to more than 3.3 V which is a practical value for low power microcontroller applications. The third capacitor is too large (0.1 Farad), it takes long time for charging process but it can store more energy which is more practical to run the real time application. The chattering effect in **Figures 15** and **16** is due to the large time con-

stant with respect to the periodic variations in the bridge circuit. It can be easily eliminated by stabilizing the output voltage from the bridge circuit.

5. Conclusions and Future Directions

Piezoelectric energy harvesting is a promising avenue of research to develop self powered microelectronic devices. Wireless remote monitoring of mechanical structures, low power wireless sensors, and biomedical sensors are strongly candidates for piezoelectric energy harvesting applications. The piezoelectric energy harvester has a limited power and the optimization to extract maximum power in the whole stages is needed to enhance the device performance. The maximum (mechanical/electrical) power transfer depends on piezoelectric material properties and other matching operating conditions. In this paper, the experimental results validated the theoretical analysis to enhance the system. The experimental results highlighted the following points:

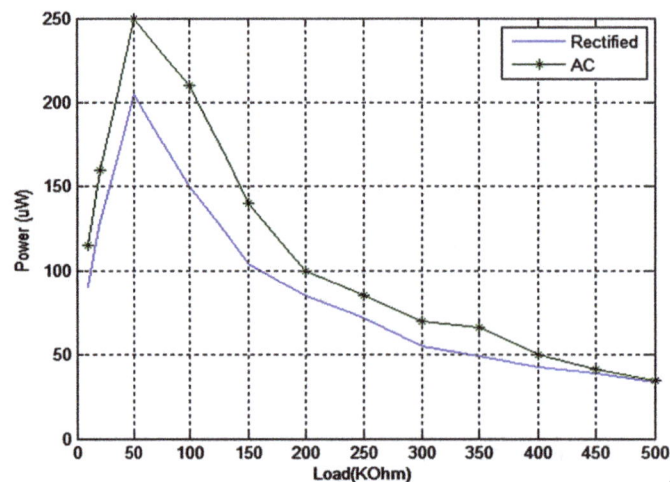

Figure 10. Output power with different external load.

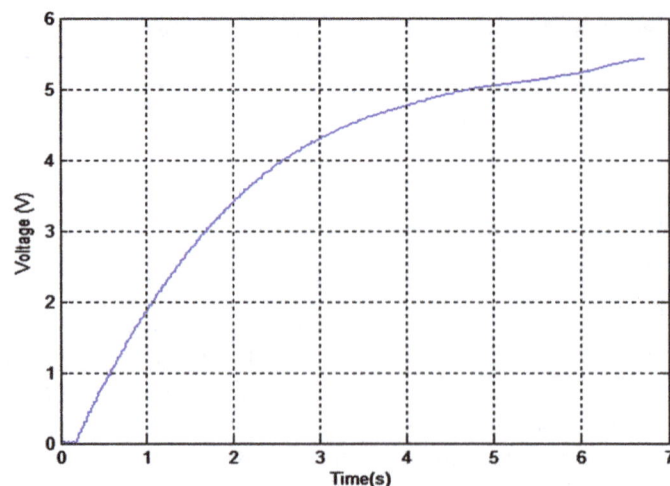

Figure 11. Output voltage using 22 µF capacitor.

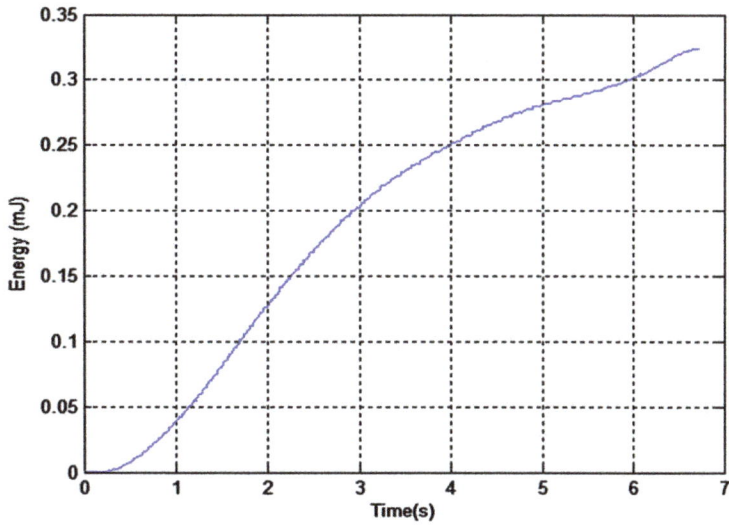

Figure 12. Stored energy using 22 μF capacitor.

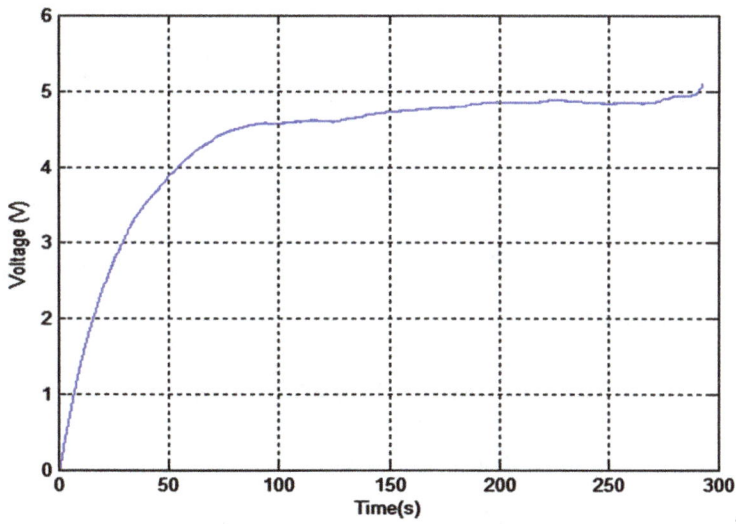

Figure 13. Output voltage using 470 μF capacitor.

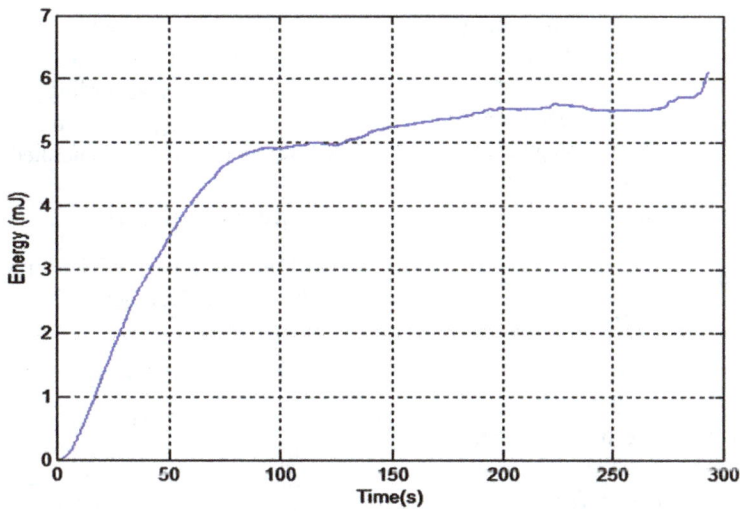

Figure 14. Stored energy using 470 μF capacitor.

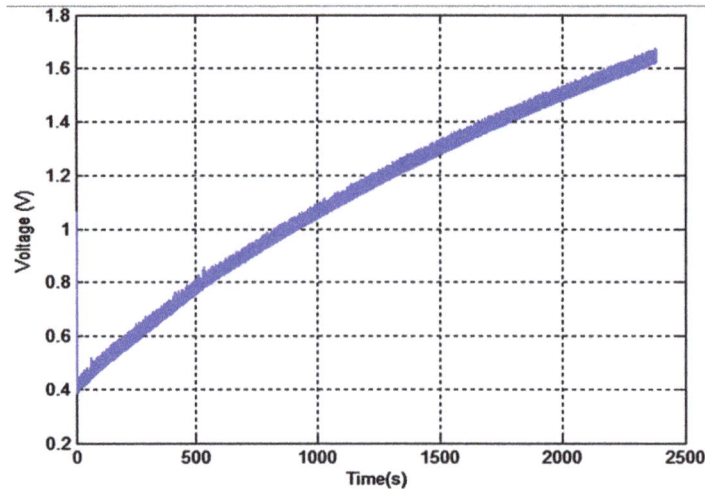

Figure 15. Output voltage using 0.1 F capacitor.

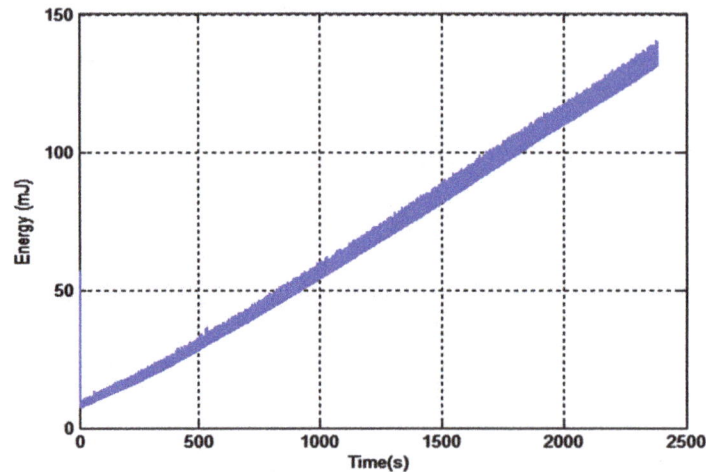

Figure 16. Stored energy using 0.1 F capacitor.

- Resonant frequency of the harvester can be identified experimentally by tracking the maximum extracted electrical power.
- Increasing tip mass decreases the resonant frequency.
- Output power increases as the value of tip mass increases that means; the Q-factor is also increased.
- After certain limit; increasing tip mass decreases the Q-factor due to the increasing of the damping effect.
- The position of tip mass has a great effect on the effective mass of the harvesting cantilever and also its resonant frequency.
- Piezoelectric harvester has an effective internal impedance as capacitive type which depends on the operating frequency.
- Maximum power transfer occurs when harvester's internal impedance matches the resistive load.
- For a fixed level of excitation, the output power is inversely proportional to the natural frequency of a harvesting structure and hence it is preferable to operate at the first harmonic or the fundamental fre-

quency of the vibrating structure.

In the future, the design of autonomous frequency tuning will be investigated using these parameters to enhance the system performance. In this case, the piezoelectric harvester will be smarter to autonomously adapt its resonant frequency according to the fundamental frequency in the exciting vibration signal. Super-capacitors will be studied to store the generated energy.

6. Acknowledgements

This work is supported by NPST program by King Saud University, Project Number 10-NAN1036-02.

REFERENCES

[1] H. A. Sodano and D. J. Inman, "Comparison of Piezoelectric Energy Harvesting Devices for Recharging Batteries," *Journal of Intelligent Material Systems and Structures*, Vol. 16, No. 10, 2005, pp. 799-807. doi:10.1177/1045389X05056681

[2] C. Lee, "Hybrid Energy Harvesters Could Power Hand-held Electronics," SPIE Digital Library, Newsroom 18, October 2010.
http://spie.org/x42029.xml?ArticleID=x42029

[3] K. A. Cook-Chennault, N. Thambi and A. M. Sastry, "Powering MEMS Portable Devices—A Review of Non-Regenerative and Regenerative Power Supply Systems with Special Emphasis on Piezoelectric Energy Harvesting Systems," *Smart Materials and Structures*, Vol. 17, No. 4, 2008. doi:10.1088/0964-1726/17/4/043001

[4] D. Guyomar and M. Lallart, "Recent Progress in Piezoelectric Conversion and Energy Harvesting Using Non-Linear Electronic Interfaces and Issues in Small Scale Implementation," *Micromachines*, Vol. 2, No. 2, 2011, pp. 274-294. doi:10.3390/mi2020274

[5] A. Erturk and D. J. Inman, "Piezoelectric Energy Harvesting," John Wiley & Sons, Hoboken, 2011. doi:10.1002/9781119991151

[6] T. J. Kazmierski and S. Beeby, "Energy Harvesting Systems: Principles, Modeling and Applications," Springer Science, Berlin, Heidelberg, 2011.

[7] S. Priya and D. J. Inman, "Energy Harvesting Technologies," Springer Science, Berlin, Heidelberg, 2010.

[8] S. Beeby and N. White, "Energy Harvesting for Autonomous Systems," Artech House, London, 2010.

[9] A. Harb, "Energy Harvesting: State-of-the-Art," *Renewable Energy* 36 (*Elsevier*), Vol. 36, No. 10, 2011, pp. 2641-2654.

[10] S. Saadon and O. Sidek, "A Review of Vibration-Based MEMS Piezoelectric Energy Harvesters," *Energy Conversion and Management*, Vol. 52, No. 1, 2011, pp. 500-504. doi:10.1016/j.enconman.2010.07.024

[11] J. Paulo and P. D. Gaspar, "Review and Future Trend of Energy Harvesting Methods for Portable Medical Devices," *Proceedings of the World Congress on Engineering WCE* 2010, London, 30 June-2 July 2010, pp. 909-914.

[12] A. Khaligh, P. Zeng and C. Zheng, "Kinetic Energy Harvesting Using Piezoelectric and Electromagnetic Technologies—State of the Art," *IEEE Transactions on Industrial Electronics*, Vol. 57, No. 3, 2010, pp. 850-860. doi:10.1109/TIE.2009.2024652

[13] D. Jia and J. Liu, "Human Power-Based Energy Harvesting Strategies for Mobile Electronic Devices," *Energy Power Engineering in China*, Vol. 3, No. 1, 2009, pp. 27-46. doi:10.1007/s11708-009-0002-4

[14] R. Bogue, "Energy Harvesting and Wireless Sensors: A Review of Recent Developments," *Sensor Review*, Vol. 29, No. 3, 2009, pp. 194-199. doi:10.1108/02602280910967594

[15] S. Chalasani and J. M. Conrad, "A Survey of Energy Harvesting Sources for Embedded Systems," 2008 *Southeast Conference*, Huntsville, 3-6 April 2008, pp. 442-447.

[16] A. Safari and A. E. Korary, "Piezoelectric and Acoustic Materials for Transducer Applications," Springer, Berlin, Heidelberg, 2010.

[17] W. Heywang, K. Lubitz and W. Wersing, "Piezoelectricity Evolution and Future of a Technology," Springer, Berlin, Heidelberg, 2008.

[18] A. Arnau, "Piezoelectric Transducers and Applications," Springer, Berlin, Heidelberg, 2008.

[19] C. B. Williams and R. B. Yates, "Analysis of Micro-Electric Generator for Microsystems," *Sensors and Actuators A: Physical*, Vol. 52, No. 1-3, 1996, pp. 8-11. doi:10.1016/0924-4247(96)80118-X

15

Contribution of the Energy Sector towards Global Warming in Malawi

Gregory E. T. Gamula[1], Liu Hui[1], Wuyuan Peng[2*]
[1]School of Environmental Studies, China University of Geosciences, Wuhan, China
[2]School of Economics and Management, China University of Geosciences, Wuhan, China
Email: *pengwy@cug.edu.cn

ABSTRACT

This paper presents the energy demand projection for Malawi considering implementation of two energy development strategies. The strategies are Malawi Biomass Energy Strategy (BEST) and Malawi electricity investment plan. Long-range Energy Alternatives Planning System (LEAP) software was used as the simulation tool. Environmental effects of the energy sector towards global warming by the energy sector as a result of implementing the strategies have been investigated. Three scenarios were developed, the first one to reflect on business as usual, the second one depicting implementation of Malawi BEST and the third one depicting implementation of the Malawi Electricity Investment Plan. A fourth scenario was developed to depict implementation of both strategies. 2008 was used as the base year with energy mix of consisting of biomass consumption decrease in all the scenarios due to better efficiency in the utilization of biomass and change in life style by people as a direct response to available energy alternatives. Implementing both the Malawi BEST and Malawi Electricity Investment Plan for energy sector development would be better in terms of both energy supply and global warming effects.

Keywords: Energy Demand; Energy Development Strategies; Scenarios; Emissions

1. Introduction

Energy demand projection is an important tool in having a foresight of the direction that the energy sector is headed depending on the prevailing circumstances. Development of any country depends upon a stable, sufficient and reliable energy supply to drive the different sectors that contribute towards economic development. Malawi has been facing energy supply challenges for some time and this has been one of major setbacks in attaining certain economic goals as set out in government's agenda. Department of Energy in the Ministry of Natural Resources, Energy and Environment is responsible for energy data collection. Malawi's energy supply is dominated by biomass, constituting 88.5% of energy supply for Malawi in 2008. Petroleum products contributed 6.4%, electricity contributed 2.8%, coal contributed 2.4% and renewable energy sources were just starting to appear in the energy mix, although as a percentage of the total energy it was still negligible. This paper looks at simulation results depicting implementation of two energy development strategies separately and also com-

bined. Scenarios according to the strategies were developed and comparison of projected energy demand and resulting global warming effects has been made.

Energy development has been one of the government's main priority areas in efforts to move towards better economic development. A number of government's policies such as the Malawi Poverty Reduction Strategy (MPRS), the Malawi Growth and Development Strategy (MGDS) and Vision 2020 have energy generation and supply as a key focus area. Another area is that of environmental protection which is affected by operations in energy generation and supply. This study has been carried out to project energy demand, energy mix and resulting environmental effects as a result of implementing Malawi Biomass Energy Strategy (BEST) and Malawi Electricity Investment Plan. 2008 was chosen as the base year for simulations because data was available on energy demand by the four sub-sectors of household, industry, service and transport and also the energy mix by the various energy supply sources.

Energy transformation and energy utilization result in emission of different types of pollutants and therefore brings about the need to examine effects that energy has

---

*Corresponding author.

on the environment. Energy related activities contribute both directly and indirectly to the emission of carbon dioxide and other greenhouse gases such as methane and nitrous oxide. In 2005, the energy sector contributed about 68% of global greenhouse gases emissions [1]. Other pollutants from energy utilization and transformation are carbon monoxide, nitrogen dioxide, sulphur dioxide and particulate matter.

Malawi Energy Situation

2008 statistics indicated that about 90% of Malawi's population used wood for firewood and charcoal production, accounting for about 88.5% of the country's energy requirements. The balance of energy was supplied by petroleum products, electricity, and coal in the proportions of 6.4%, 2.8% and 2.4 respectively. Only a negligible 0.06% was supplied from renewable energy sources. Distribution of energy demand by sub-sectors showed that households accounted for 83% of all energy consumption, with industry accounting for 12%, transport 4% and the service sector accounting for only 1%. A majority of the population (85.7%) use paraffin in hurricane and pressure lamps for lighting, 7.2% use electricity, 2.2 use candles, 2.6% use firewood and 1.4% use other means of alternatives for lighting [2]. For cooking, 88% of the population use firewood, 8% use charcoal, 2% use electricity, 1% use paraffin and 1% use other means such as crop residues, animal dung and other sources not mentioned here.

Malawi has got a small electricity supply system of 302 MW when compared to her neighbors: Mozambique, Tanzania and Zambia whose electrical power capacities are 2483 MW, 1186 MW and 1737 MW respectively (installed capacities as of 2008) [3]. In addition to local resources, there are plans to connect Malawi to the Southern African Power Pool (SAPP) in the next few years. Electricity from photovoltaic modules is still insignificant when looking at the overall picture, but it is increasing in utilization, finding applications in households, telecommunications, lighting, refrigeration and water pumping.

Malawi has no refineries for petroleum products and therefore imports about 97% of its refined oil products. The balance is contributed by locally-produced ethanol which is sold directly to the oil companies for blending with gasoline on a maximum 20:80 ratio of ethanol:gasoline. The mixing ratio is in practice usually 12:88 because ethanol production is inadequate. The blending of ethanol and gasoline started in 1982 but has not been expanding much such that to date only a small proportion of gasoline is blended with ethanol. The internal storage capacity of oil for the country to avert supply disruptions by natural or man made emergencies is supposed to be 30 days but this is not the case on ground due to a number of economical and logistical challenges. The transport sector relies heavily on the oil imported supply which means that its pricing is heavily influenced by trends on the international oil market. Government of Malawi is undertaking oil explorations in Lake Malawi. Extensive Strategic Environmental Assessment will be carried out before any type of work can be done. The assessment will be presented to Ministry of Natural Resources, Energy and Environment for approval before exploration data acquisition can be started.

Biomass energy in Malawi is currently not sustainable which has resulted in wood resource base diminishing mainly because woodlands and trees in agricultural areas are being cleared up to start new farming land. Statistics show that between 1991 and 2008 about 669,000 hectares of woodlands were converted to farmland [4]. Diminishing standing stock is leading into gradual reduction of biomass that can be harvested. Although Malawi is heavily dependent on biomass fuels, the national energy policy has little information on biomass energy supply. There was therefore a need to look into the biomass side of energy supply which resulted into a strategy being formulated for this supply source in 2009.

Estimated coal reserves in Malawi are between 80 million tons and 1 billion tons. Quality of the coal varies with energy values ranging from 17 to 29 MJ/kg. There are four coal fields in Malawi, three in the northern part of the country and one in the southern part of the country. Currently only two fields are being mined and the coal produced from these sites is not enough to suffice the country's industrial needs consequently the balance is imported from Mozambique. Coal is used in various industries for heating but is currently not generally used as a domestic energy supply which means that its use in households is negligible.

Malawi is endowed with a number of renewable energy resources yet utilization of these resources is still in infancy stages. There are quite high levels of solar energy in the range of 1200 W/m^2 in the warm months and 900 W/m^2 in the cool months which means that both photovoltaic systems and solar thermal systems perform well. There are quite a good number of areas in the country with mean wind speeds above 5 meters per second almost throughout the year which would enable the use of wind energy systems. Malawi lies along the Great Rift Valley and therefore traces of geo-thermal reservoirs have been said to exist. There have been a number of initiatives to enhance utilization of renewable energy sources in various sub-sectors.

Malawi has about 63,000 tons of proven reserves of uranium at two sites. Mining of uranium started in 2008 but currently all the uranium that is mined is exported. Energy that is currently being used in mining of uranium is supplied from generators operating on diesel.

2. Investigating Energy Development Strategies

The energy supply sector in Malawi is categorized into five key components and these are biomass supply, coal supply, electricity supply, liquid fuels and gas supply and other renewable energy sources [5]. A sixth possible component is uranium for nuclear energy supply which at the moment is being mined for export only. Only two supply sectors have well defined development strategies and these are the biomass sector and the electricity sector.

2.1. Malawi Biomass Energy Strategy (BEST), 2009

The main objective of the Malawi BEST is to ensure sustainable of woodfuel through increased supply, improved efficiency and creation of institutional capacity to manage the sector. Sustainability of supply of woodfuels would be achieved through:

- Establishing a master plan for woodfuels supply;
- Designing and implementing woodfuel management plans at district level;
- Improving charcoal flow monitoring and control;
- Promoting production of alternative fuels which are also affordable [4].

The strategy has been able to clearly define steps to be taken to improve the supply of firewood and charcoal but is vague on the other fuels in the biomass category. The strategy has mostly been dedicated to traditional utilization of biomass leaving out modern biomass applications which would have a greater impact in economic development for the country through provision of better energy forms such as electricity.

Malawi has 4 urban areas namely Blantyre and Zomba in the southern region, Lilongwe in the central region and Mzuzu in the northern region. **Figure 1** shows the 4 urban centers of Malawi and the surrounding areas that form part of the catchment areas. Firewood is mainly used for cooking (about 76%), while for water heating accounts for about 22% and only about 2% is used for space heating. A high proportion (about 90%) of firewood users cook using traditional three-stone stoves with efficiencies ranging from 10 % to 14%. Most of charcoal users (over 80%) use ceramic charcoal stoves with and efficiency of about 30% while the rest use metal stoves with efficiency of about 20%.

2.2. Malawi Electricity Investment Plan, 2010

Electrical Energy Supply is provided by a government owned company called Electricity Supply Corporation of

Figure 1. A map showing the three catchment areas, covering the four main urban centers (*source*: *Malawi BEST*, 2009).

Malawi (ESCOM) Limited which was originally established by an Act of Parliament in 1957 and was revised in 1963 and then 1998. Total installed capacity of electrical power by ESCOM is about 302 MW, of which 94% is generated by hydropower and the remaining 6% is thermal. All ESCOM's hydro generation stations except one are located in the Southern region of Malawi along Shire river and the exception is on Wovwe river in the northern region with a capacity of 4.5 MW [6]. The map of **Figure 2** shows the location of all the hydro power stations in Malawi.

The current demand for electrical power is more than what is available by the installed plants. Some interventions have been planned to address the situation and these are in three categories, namely short term, medium term and long term interventions. Short term interventions have the potential of increasing the generation capacity by 234 MW while medium term interventions have a potential of adding a generation capacity of 1240 MW and long term interventions have the potential of adding a generation capacity of 770 MW.

Short term plans are those that are to be implemented within 5 years and they involve developing 3 new hydro power plants, 1 on Ruo river, 1 on Shire river and 1 on Lunyina river; also implementation of demand side Management techniques, upgrading of Nkula A power station from 24 MW to 50 MW; development of hydro matrix power plants and installation of diesel powered stand-by generators. Medium term plans are to be implemented between 5 to 10 years and involve construction of a coal fired power plant, 8 additional hydro power plants, 3 biomass fired power plants, and wind generating systems. The coal generating plant will have a capacity of

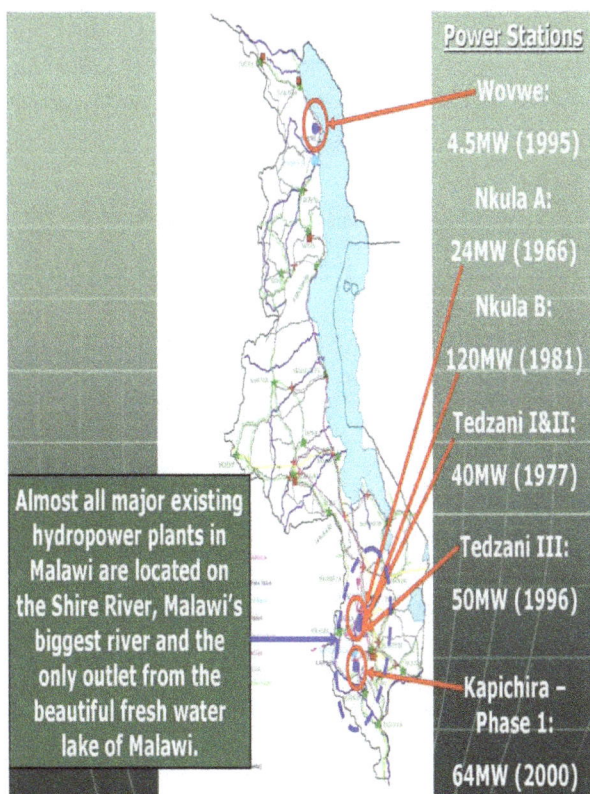

Figure 2. Location of hydro power plants in Malawi, capacity and commissioning year (*source*: *Malawi Electricity Investment Plan*, 2010).

300 MW which will require an annual coal supply of about 1 million metric tonnes. The sites for hydro power plants are: 1 site on Songwe river, 4 sites on Bua river, 1 site on Shire river at Mpatamanga, 1 site on North Rukuru river and 1 site on South Rukuru river. Long term plans are those to be implemented after 10 years and include 3 more hydro plants and a 150 MW modular nuclear power plant. Sites for the hydro power stations are on Dwambazi river, Luweya river in Nkhatabay and Kholombidzo on Shire river [7].

3. Developing Scenarios

Four scenarios were developed, energy demand projections were evaluated and the negative environmental effects were simulated. These scenarios are as follows:

- Business as usual scenario which will be used as the reference in which it will be assumed that the current trends are allowed to continue without any mitigating interventions in the energy sector and associated environment sector except those already in place.
- Biomass energy scenario in which new technologies are enhanced in the biomass energy supply sector through use of better efficiency stoves for both wood and charcoal and also use of modern biomass technologies. Developing the biomass sector in terms of better efficiency for the current wood and charcoal stoves; better forest management programs; increase the production of ethanol and other energy products from biomass like the growing of bio-energy plants; use of biogas for cooking.
- Electricity scenario which will examine a number of options which are:
  - ♦ Developing additional hydro electrical generating power stations along rivers where previous studies have already shown that such schemes are viable;
  - ♦ Developing thermal electrical generating plants operating on coal;
  - ♦ Developing renewable energy technologies to generate electricity.
- Combined scenario in which both the biomass scenario and the electricity scenario are implemented.

3.1. Scenario Assumptions

Malawi had a population of 13,077,160 in 2008 and the average population growth in the previous decade was 2.8%. About 84.5% of the population in this year lived in the rural areas. The population growth rate projections are slightly lower [8] which led to the use of an average annual growth rate of 2.6% for the simulations. From 1988 to 2008, the average annual growth rate for GDP was 3.7 but looking at the later years the GDP growth rate was as shown in **Figure 3** [9].

3.2. Key Assumptions in LEAP

- Population for the country is 13.1 million.
- Population growth rate is 2.5%.
- Average household size is 4.57.
- Income per capita is USD 800.
- Average annual GDP growth rate is 3.5%.
- Current urbanization is 15.5% and end year urbanization is 40%.
- End year of simulation is 2050.

These are common for all the scenarios under consideration. Energy data collection is done by the Department of Energy in the Ministry of Natural Resources, Energy and Environment [5]. Department of Energy coordinates with Malawi Energy Regulatory Authority (MERA), ESCOM and Petroleum Importers Limited (PIL) for demand details in the respective supply industries. Energy demand in various sub-sectors for energy utilization in 2008 is given in **Table 1** in which energy demand is grouped according to fuel type. This is the information that was used for the base year energy demand when doing simulation.

4. Demand Projections Results

The results of energy demand projections after simulations for the four different scenarios are shown in **Figure 4**, while the actual values of energy demand are as given in **Table 2**.

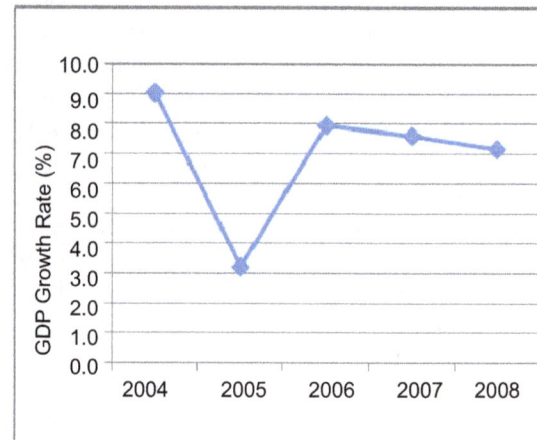

Figure 3. Annual GDP growth rate from 2004 to 2008.

Table 1. Total energy demand in sub-sectors by fuel type in Malawi in 2008 (Adapted from Malawi Biomass Energy Strategy, 2009).

Sector	Energy demand by fuel type (TJ)				
	Biomass	Coal	Electricity	Petroleum	Renewables
Household	127,574	5	1798	672	
Industry	10,004	3481	2010	3130	
Transport	270	15	35	5640	
Service	452	174	477	558	
Sub-Total (% of total)	**138,300 (88.48%)**	**3,675 (2.37%)**	**4,320 (2.8%)**	**10,000 (6.4%)**	**0.09 (0%)**
Total			**156,295**		

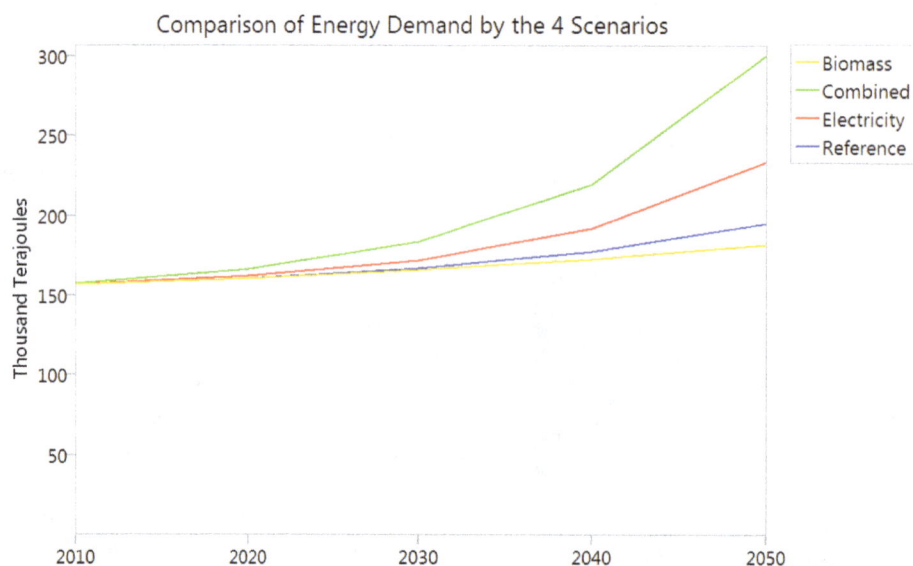

Figure 4. A graph showing energy demand projections comparing the four scenarios.

Table 2. Comparison of demand projections for the four scenarios.

	2008	2010	2020	2030	2040	2050
Reference (Thousand TJ)	156.3	156.9	160.8	167.2	177.7	195.4
Biomass (Thousand TJ)	156.3	156.9	160.8	166	172.9	182
Electricity (Thousand TJ)	156.3	157	162	172	192.2	234.2
Combined (Thousand TJ)	156.3	157.5	166.4	183.8	220	300.8

It can be observed that Biomass scenario has the lowest energy demand projections followed by the reference scenario then Electricity scenario while the combined scenario has the highest energy demand projections.

5. Greenhouse Gases (GHGs) Emissions

The GHGs that are considered in this study are carbon dioxide (CO_2), methane (CH_4) and nitrous oxide (N_2O). LEAP software divides GHGs emissions into two: emissions due to energy utilization and emissions due to transformation of energy from one form into another. Combination of the two components gives us the overall emissions. **Figures 5-8** show us the emissions trend for each one of the GHGs considering each scenario in thousand metric tonnes CO_2 equivalent. These figures were done using Microsoft Office Excel from the results that were imported from LEAP software. Total GHGs emission for Malawi in 2008 was 6,900 thousand metric tonnes CO_2 equivalent, [10] of this 2,025,500 tonnes was from the energy sector.

All the figures have a common trend in which there is a rapid increase of carbon dioxide emissions, a decrease in methane emissions and a slight decrease of nitrous oxide emissions. In reference scenario, carbon dioxide increased from 769.3 in 2008 to 5860.6 thousand metric tonnes in 2050; methane decreased from 1060 in 2008 to 892.7 thousand metric tonnes in 2050 and nitrous oxide remained almost constant, 196.1 in 2008 and 204.5 thousand metric tonnes in 2050. In biomass scenario, carbon dioxide increased from 769.3 in 2008 to 5,932.3 thousand metric tonnes in 2050; methane decreased from 1060 in 2008 to 831 thousand metric tonnes in 2050 and nitrous oxide remained almost constant, 196.1 in 2008 and 187.9 thousand metric tonnes in 2050. In electricity scenario, carbon dioxide increased from 769.3 in 2008 to 10,319.9 thousand metric tonnes in 2050; methane decreased from 1060 in 2008 to 720.9 thousand metric tonnes in 2050 and nitrous oxide remained almost constant, 196.1 in 2008 and 185.4 thousand metric tonnes in 2050. In combined scenario, carbon dioxide increased from 769.3 in 2008 to 10533.9 thousand metric tonnes in 2050; methane decreased from 1060 in 2008 to 577.6 thousand metric tonnes in 2050 and nitrous oxide remained almost constant, decreasing slightly from 196.1

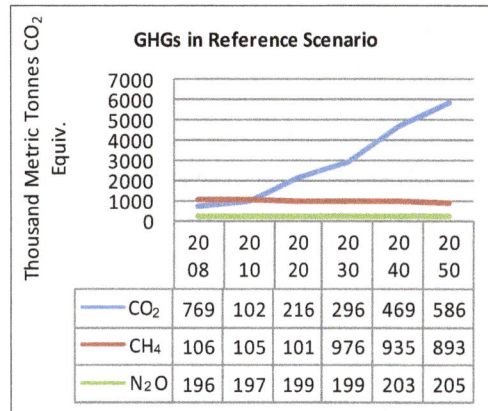

Figure 5. GHGs emissions in reference scenario.

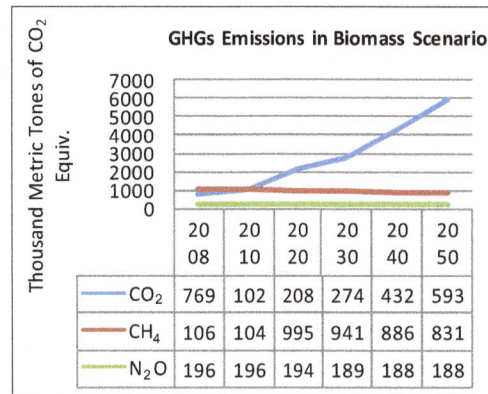

Figure 6. GHGs emissions in biomass scenario.

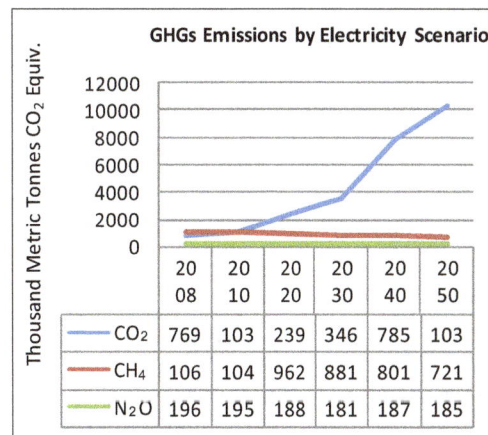

Figure 7. GHGs emissions in electricity scenario.

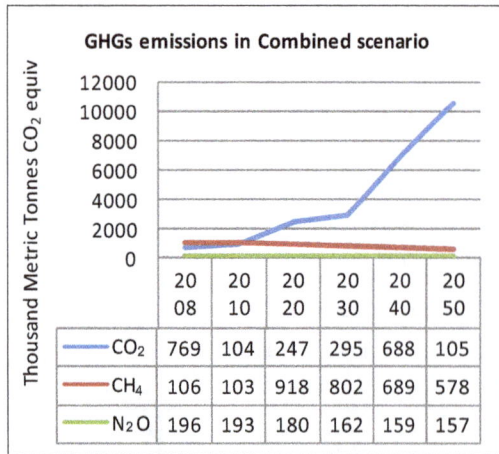

Figure 8. GHGs emissions in combined scenario.

in 2008 to 156.7 thousand metric tonnes in 2050.

The overall GHGs emissions have been expressed in CO_2 equivalent and the results for all the scenarios are as shown in **Figure 9** as global warming potential by each one of the scenarios.

It can be observed when looking at **Table 2** that the scenarios reference, biomass, electricity and combined registered increases of 25%, 16.4%, 49.8% and 92.5% in energy demand respectively. In the final year of simulation, energy demand and energy mix are as shown in **Tables 3-6** for all scenarios.

6. Conclusion

Implementing both the Malawi BEST and Malawi Electricity Investment Plan would be the best course of action

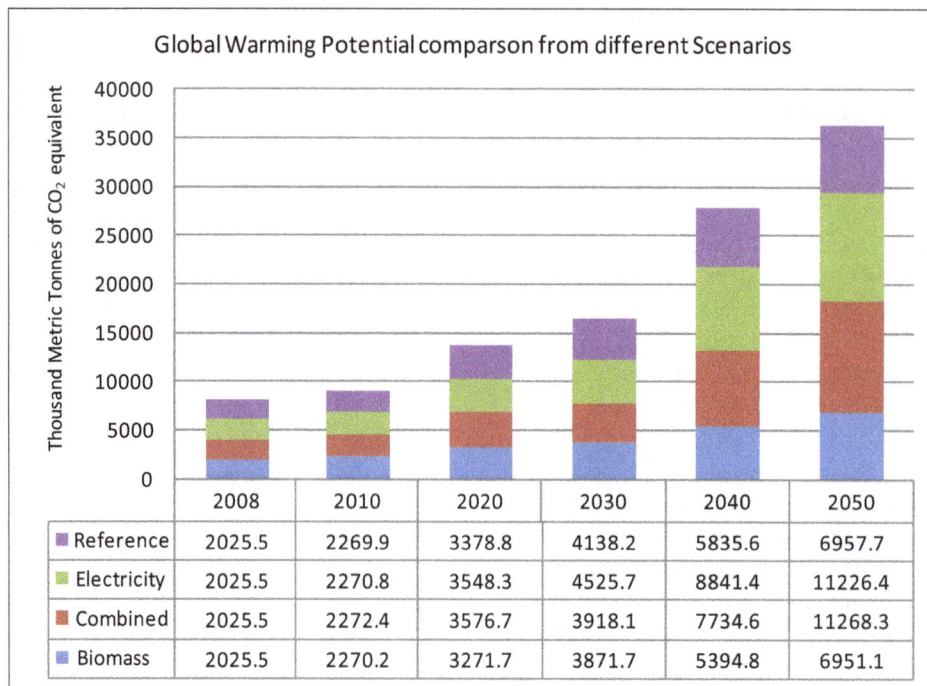

Figure 9. Comparison of GHGs emissions considered in the study in CO_2 equivalent.

Table 3. Summary of energy demand by sub-sector and energy mix for the reference scenario in 2050.

Sector	Energy demand by fuel type (TJ)				
	Biomass	Coal	Electricity	Petroleum	Renewables
Household	59,714	0	35,007	9915	19,305
Industry	15,194	12,047	23,230	3,150	31
Transport	0	0	724	8328	0
Service	845	555	6717	570	38
Sub-Total (% of total)	75,753 (38.8%)	12,602 (6.5%)	65,678 (33.6%)	21,691 (11.1%)	19,374 (9.9%)
Total			195,369		

Table 4. Summary of energy demand by sub-sector and energy mix for the biomass scenario in 2050.

Sector	Energy demand by fuel type (TJ)				
	Biomass	Coal	Electricity	Petroleum	Renewables
Household	60,053	0	26,869	13,458	23,561
Industry	24,952	6506	8525	3150	30
Transport	1	0	891	9913	0
Service	1038	491	1651	570	39
Sub-Total (% of total)	86,044 (47.3%)	6,996 (3.8%)	37,936 (20.8%)	27,091 (14.9%)	23,630 (13%)
Total			182,031		

Table 5. Summary of energy demand by sub-sector and energy mix for the electricity scenario in 2050.

Sector	Energy demand by fuel type (TJ)				
	Biomass	Coal	Electricity	Petroleum	Renewables
Household	50,688	0	41,192	11,517	20,542
Industry	12,335	12,047	50,932	3150	42
Transport	50	10	968	10,768	0
Service	1127	904	38,202	570	45
Sub-Total (% of total)	63,709 (27.2%)	12,649 (5.4%)	110,893 (47.3%)	26,004 (11.1%)	20,629 (8.8%)
Total			234,247		

Table 6. Summary of energy demand by sub-sector and energy mix for the combined scenario in 2050.

Sector	Energy demand by fuel type (TJ)				
	Biomass	Coal	Electricity	Petroleum	Renewables
Household	38,983	30	42,794	11,693	30,471
Industry	27,088	18,076	75,008	3161	65
Transport	75	0	1009	11,221	0
Service	1052	174	477	558	90
Sub-Total (% of total)	67,198 (22.3%)	18,980 (6.3%)	157,013 (52.2%)	26,636 (8.9%)	30,626 (10.2%)
Total			300,831		

for the country because it offers the highest supply of energy to meet the demand. It has the advantage of increasing electrical energy supply as well as making biomass energy sustainable. Implementing only the Malawi BEST on its own would be the worst even when comparing to business as usual scenario. This is so because the strategy concentrates on traditional use of biomass hence electrical power supply ends up being minimal. The way to make this strategy better is to include modern use of biomass including bio-energy. Combined scenario has the highest global warming potential but it is only a little higher than the electricity scenario. Looking at the increase in projected demand, it can be seen that scenarios reference, biomass, electricity and combined registered increases of 25%, 16.4%, 49.8% and 92.5% in energy demand respectively. Implementing both the Malawi BEST and Malawi Electricity Investment Plan would be the best since this combination almost doubles the energy demand, yet the global warming effect is not much different from that of implementing only the Malawi Electricity Investment Plan. Nuclear power supply has not been included in the simulations because uranium reserves in the country have not been fully ascertained and there is a need to verify that there would still be enough reserves to support a nuclear power plant towards the end of the simulation period.

REFERENCES

[1] T. Nakata, D. Silva and M. Rodionov, "Application of Energy System Models for Designing a Low-Carbon Society," *Progress in Energy and Combustion Science*, Vol. 8, No. 2, 2010, pp. 1-41.

[2] Malawi Government, Census Main Report, National Statistical Office, Zomba, 2009.

[3] http://www.sapp.co.zw/docs/

[4] Malawi Government, "Malawi Biomass Energy Strategy," Ministry of Natural Resources, Energy and Environment, Lilongwe, 2009.

[5] Malawi Government, "National Energy Policy," Department of Energy Affairs, Lilongwe, 2003.

[6] Escom & Electricity, Power Generation. http://www.escom.mw

[7] Malawi Government, "Malawi Electricity Investment Plan," Ministry of Natural Resources, Energy and Environment, Lilongwe, 2010.

[8] Malawi Government, "Population Projections of Malawi," National Statistical Office, Zomba, 2009.

[9] Malawi Government, "Annual Economic Report," Ministry of Economic Planning and Development, Lilongwe, 2009.

[10] Millenium Development Goals Indicators. http://mdgs.un.org/unsd/mdg/SeriesDetail.aspx?srid=749

16

An Overview of the Energy Sector in Malawi

Gregory E. T. Gamula[1], Liu Hui[1], Wuyuan Peng[2*]
[1]School of Environmental Studies, China University of Geosciences, Wuhan, China
[2]School of Economics and Management, China University of Geosciences, Wuhan, China
Email: *pengwy@yahoo.cn

ABSTRACT

This paper presents the status of the energy sector in Malawi which is not effectively contributing to the national economic development because it is unreliable and insufficient hence not able to meet the energy demand. About 83% of Malawi's population live in rural areas and rely on fuel wood for energy supply. High reliance on biomass has had negative environmental impacts through indoor pollution, deforestation, soil erosion leading to high soil sediment loads to water bodies resulting in poor water quality in rivers and lakes. Electricity supply is much less than demand resulting in deficient and unreliable supply. Malawi is endowed with a number of renewable energy resources yet utilization of these resources is still a major challenge. Little progress has been made in improving the energy supply situation due to a number of reasons, main ones being poverty, lack of political will and wrong approaches in addressing the energy problem. Most current approaches aim at improving the current situation but our opinion would be to develop new strategies to shift the energy situation from its present condition to the desired status. The paper starts with an introduction, then status of the energy sector, then goes on to discuss energy development efforts according to sectors and finally gives a conclusion.

Keywords: Energy Supply; Energy Development; Energy Resources; Electrical Energy

1. Introduction

Malawi is a land locked country located in Southern Africa, between latitudes 9°22'S and 17°3'S and longitudes 33°40'E and 35°55'E. The country is about 900 km long and 80 - 161 km wide, with a total area of 118,484 km^2 (11.8 million ha), of which 80% is land. The remaining 20% is covered by water, mainly comprising Lake Malawi, which is about 586 km long and 16 - 80 km wide. The rest of the water area is made up of the following lakes: Lake Chilwa, Lake Malombe and Lake Chiuta and there are also rivers a majority of them flowing into Lake Malawi. The country is divided into three administrative regions namely Northern Region, Central Region and Southern Region with population distribution, population density and energy resources as shown in **Table 1**.

Malawi is one of the poorest countries in the world (in the bottom 10%) with gross domestic product based on purchasing-power-parity (PPP) per capita GDP of about USD 900 in 2010 [1]. GDP composition by sector is 35.5% agriculture, 19.9% industry and 44.6% services. The population of Malawi is currently about 15 million and recently, it has been increasing at a rate of about 2.8% per annum. About 83% of the population lives in rural areas and about 75% of the population carries out

farming as smallholders on fragmented customary land [2]. Crops that contribute significantly to the economy of the country are tobacco, tea, sugarcane and cotton with these crops accounting for about 75% of total exports for the country and tobacco alone contributing about 52%. Urban growth is increasing at a rate of about 6.7%, of this 60% - 70% live in traditional housing areas and unplanned settlement areas. Poor planning has resulted in extreme urban squalor and deprivation, poor sanitation, and the rapid spread of communicable waterborne diseases. Improper disposal of wastes, agro-chemicals (fertilizers and pesticides) and effluent from industries, hospitals and other institutions are major urban problems with only 77% of proper disposed waste and most of the waste water enters the river systems that provide drinking water for downstream communities as raw sewerage. There is little environmental impact from mining and industries due to the minor economic contribution from these sectors. There is an increase in extraction of construction materials like sand and clay for bricks (with a high demand for fuel wood for curing of the bricks), lime for cement and quarry stones for concrete. Sand and gravel extraction leave large holes, which provide breeding environments for disease vectors and waterborne pathogens. Another important mining industry activity is making of cement, which is the second greatest contributor

---

*Corresponding author.

Table 1. Malawi population, land area and energy resources by region in 2008 (adapted from Malawi biomass strategy, 2009).

	Northern Region	Central Region	Southern Region	National
Rural population ('000)	1343	4814	5147	11,304
Urban population ('000)	251	959	1116	2326
Total population ('000)	1594	5773	6263	13,630
Land area (sq. km)	27,200	35,600	31,200	94,000
Population density (people/ha)	0.59	1.62	2.01	1.45
Available energy resources	Biomass, coal, hydro, uranium	Biomass, hydro	Biomass, coal, hydro	Biomass, coal, hydro, uranium
Current electric generating capacity (MW) and source by ESCOM	2.15-Diesel 4.5-Hydro	-	15-Gas turbine 279.3-Hydro	2.15-Diesel 15-Gas turbine 283.8-Hydro

to greenhouse gases in Malawi, after agricultural-related processes. The industry sector is a great contributor to creation of noise, dust, air pollution from furnaces and effluent by-products, however, these impacts are currently quite low.

Economic Development

Alleviation of poverty in Malawi has been a point of focus for government policies for over a decade. This endeavor has been spearheaded by two documents, namely the Malawi Poverty Reduction Strategy (MPRS, 2002) and the Malawi Growth and Development Strategy (MGDS, 2005). The MGDS has been built upon the foundations of MPRS and incorporated lessons that were learnt in the implementation of MPSR. MGDS recognizes that strong and sustainable economic growth is a key to poverty reduction and it focuses on:

- Agriculture and food security;
- Irrigation and water development;
- Transport infrastructure development;
- Energy generation and supply;
- Integrated rural development;
- Prevention and management of nutrition disorders and HIV/AIDS.

The first phase of MGDS was implemented between 2006 and 2011 and now it is being extended for another five years. In the MDGS under the energy generation and supply theme, Malawi is aiming to increase access to reliable electricity and reduce reliance on biomass fuels. Improving energy supply is clearly a vital input to the development of productive sectors of any country's development. There have been quite some challenges in focusing on the priority areas that have been set out in the MGDS in that it seems there is no order in which the priority areas are being addressed. As we are approaching the first phase of MGDS, food security has been attained but what remains is the sustainability aspect. The energy sector has been receiving some attention in doing

feasibility studies for alternative energy supply sources but no course of action has been undertaken to implement long term plans for the sector. National budget allocation to the energy and mining sector has remained well below 1% of the total national budget for the past 15 years.

2. Energy Situation

One of the challenges the country faces is being able to meet the energy needs of the various sectors in the country. Energy supply deficiencies are common which result in interruptions to processes that require energy as an input. A prominent example is the national electrical energy system which is accessible to less than 1% of the rural population and is unreliable. From 2008 statistics, about 90% of Malawi's population use wood for fuel and charcoal production, accounting for about 88.5% of the country's energy requirements, 6.4% comes from petroleum, 2.8% from electricity and 2.4% from coal. Households account for 83% of all energy consumption, with industry taking 12%, transport taking 4% and the service sector taking 1%. Statistics show that 85.7% of the population use paraffin in hurricane and pressure lamps for lighting, 7.2% use electricity, 2.2% use candles, 2.6% use firewood and 1.4% use other means of alternatives for lighting. For cooking, 88% of the population use firewood, 8% use charcoal, 2% use electricity, 1% use paraffin and 1% use other means such as crop residues, animal dung and those not mentioned above [3]. Looking more closely at the various energy sources for lighting and cooking for both rural and urban areas gives us results as summarized in **Table 2**.

The Government of Malawi has embarked on quite a number of programs and projects to improve the standard of living for the rural masses which should be able to eventually result in energy utilization switch. Even with such programs being carried out, less than 1% of the rural population has access to electricity.

In 2004, four energy laws were created to help operations in the energy sector in Malawi. These are four Energy Acts aimed at addressing various aspects of the energy sector which are: Act 20, the Energy Regulation Act which established Malawi Energy.

Regulatory Authority (MERA); Act 21, the Rural Electrification Act which laid the foundation for the formation of Rural Electrification Management Committee and Rural Electrification Fund; Act 22, the Electricity Act which deals with electricity issues in terms of licensing, tariffs, generation, transmission, distribution, sales contracts and related issues; Act 23, the Liquid Fuels and Gas (Production and Supply) Act which handles issues related to liquid fuels and gas production in terms of licensing, safety, pricing, taxation, strategic reserves and any other related issues. Following some surveys that were carried out, energy consumption by sector and by fuel type in 2008 was as summarized in **Table 3**.

Total energy produced in Malawi is less than energy that is consumed hence a need to import some energy products in form of oil products and coal. **Figure 1** shows the trend of energy produced compared to energy consumed for Malawi over the past recent years.

It can be observed that the gap between energy produced and that consumed is increasing which translates into increasing energy imports.

It is common knowledge that as the environment de-

grades, there are consequences that are observed in energy systems, whether it be the supply, transportation or utilization aspects. A simple example of rural communities depending on biomass as their main source of energy will face energy challenges because of deforestation. For Malawi which relies mainly on hydro power generation, there has been a growing impact of power generation from environmental degradation. Population growth and the pressures associated with it have resulted in most of the hills being laid bare in most of the catchment areas of the rivers. People have been opening up gardens in areas previously protected and use a lot of fertilizers to compensate for low yields and in addition the use of firewood and charcoal both as a household energy source and for business has depleted the forests. This has resulted in more soils being prone to erosion. When raining, a lot of soils are eroded into the river tributaries and later to the Shire River, which is the biggest river in Malawi and where electrical generation is mainly being done. The soils being eroded are full of nutrients from the use of artificial fertilizers and so when these soils and their nutrients are deposited into the river they provide necessary nutrients to the aquatic plants and they then grow and multiply [4]. Also the soil being eroded into the river cause problems at the intakes of the water which goes into the power stations.

Table 2. Population distribution by source of energy for lighting and cooking (source: Malawi census main report, 2009).

Source of energy	Cooking (% of population)			Lighting (% of population)		
	National	Rural	Urban	National	Rural	Urban
Charcoal	8	1.7	43.4	0	0	0
Electricity	2	0.4	13.6	7.2	1.9	37.5
Firewood	88	95.7	41.8	2.6	2.9	0.4
Gas	0	0	0.1	0	0	0
Paraffin	1	1.2	0.7	85.7	92.5	46.5
Others	1	1	0.5	1.3	1.6	0.3
Candles	0	0	0	3.2	1.1	15.3

Table 3. Total energy demand by sector by fuel in Malawi in 2008 (source: national biomass energy strategy document, 2009).

Sector	Energy demand by fuel type (TJ)				
	Biomass	Coal	Electricity	Petroleum	Total (% of total)
Household	127,574	5	1798	672	**130,049 (83.2%)**
Industry	10,004	3481	2010	3130	**18,625 (11.9%)**
Transport	270	15	35	5640	**5960 (3.8%)**
Service	452	174	477	558	**1661 (1.1%)**
Total (% of total)	**138,300 (88.5%)**	**3675 (2.4%)**	**4320 (2.8%)**	**10,000 (6.4%)**	**156,295 (100%)**

Energy production and consumptionn

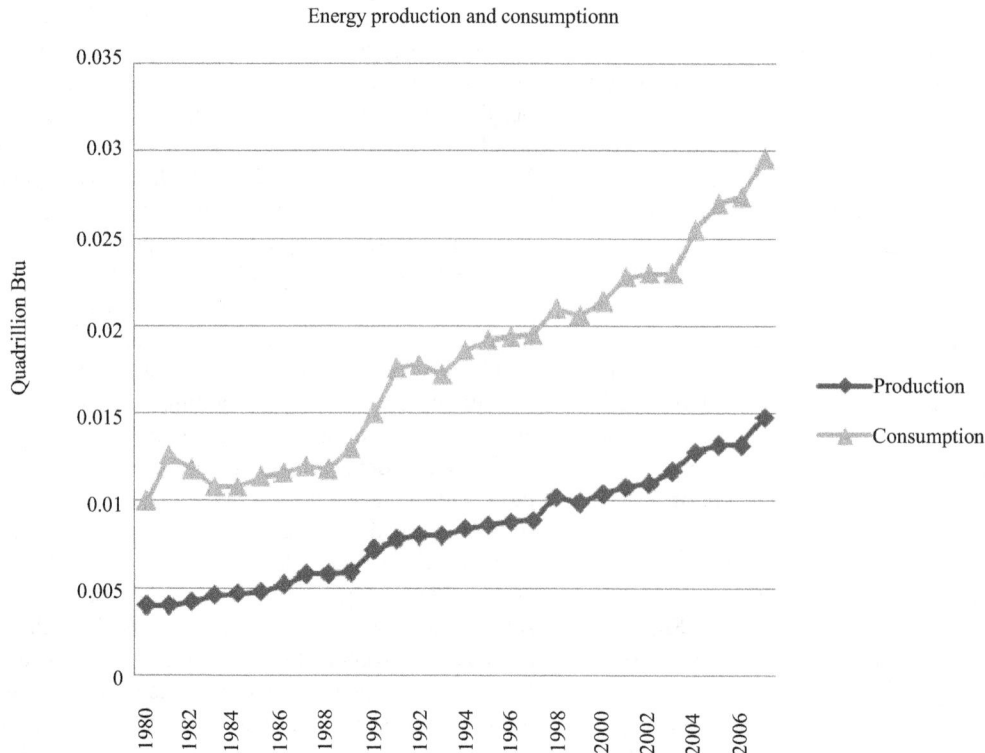

Figure 1. A graph showing energy produced and energy consumed (adapted from IAEA international energy statistics).

3. Energy Development

The energy sector in Malawi has not been able to meet all the energy requirements across the other sectors and this has been a major challenge for some time and has considerably contributed to the slow development of the various sectors in the country. Many studies that have been done have already shown that a correlation exists between energy and economic development, with various aspects of developmental trends being exposed depending on the area of study. At household level, it has been observed that total energy use increases with income. Commercial energy intensity follows a different path in that at low levels of economic development, commercial energy use is negligible which means that commercial energy intensity approaches zero. When there is economic development, industry booms and commercial energy intensities increase as well [5]. Similar patterns are evident in other economic activities linked to natural resources utilization such as local air pollution. It has also been established from studies in a number of countries that economic prosperity leads to more energy consumption [6,7]. Also income growth is associated with an increasing trend to use modern energy services. Since it has been observed that energy consumption increases with economic prosperity, it means that rising GDP is generally accompanied by increased energy consumption per capita [8]. The increasing trend in energy consump-

tion with increasing economy is decreasing in the recent times because of improved efficiencies and utilization of modern energy technologies. Generally total primary energy use rises with income but the two do not change in lockstep. At early stages of economic development, the dominance of inefficient fuels and technologies results in large inputs of energy but as economic development proceeds and more modern, efficient fuels and technologies are adopted, energy intensity input per unit of economic output begins to decline. The general downward trend in energy intensity may be interrupted as a country enters the first stages of industrialization and material-intensive manufacturing becomes the dominant economic driver [9].

Elements of achieving national energy security are included in other government's key policy and strategy documents such as the Vision 2020, the Millennium Development Goals and the National Energy Policy. There is a need for periodic policy and strategic reviews to ensure realization of energy sufficiency for the country. One of the major challenges facing the energy sector is the financial allocation (a negligible fraction of the national budget) that is not in line with the global energy trends and this has to change if any meaningful progress is to be made in achieving the government's agenda in improving energy supply. The financial spending figures for Malawi between 1996 and 2010 show that Energy and Mining Services sector had the highest allocation of the total national budget in 2010 which was 0.53% [10]. If this trend continues, development of the energy sector will remain a major challenge since the private sector is not playing a major role in the activities of this sector. The National Energy Policy being the framework for the development of the energy sector in Malawi sets out policy goals, objectives, strategies and priority actions. The policy also sets out target energy mix for the future as shown in **Table 4** aimed at achieving a shift from an energy sector that heavily relies on biomass to one that better distribution among the various energy sources in the country.

Objectives of Malawi's energy policy are:

- To improve efficiency and effectiveness of commercial energy supply industries;
- To improve the security and reliability of energy supply systems;
- To increase access to affordable and modern energy services;
- To stimulate economic development and rural transformation for poverty reduction;
- To improve energy sector governance;
- To mitigate environmental, safety and health impacts of energy production and utilization.

In fulfilling the above objectives, the energy policy enacted in 2003 aims to achieve three long term goals which are:

- Make the energy sector robust and efficient, to support government's agenda of poverty reduction, sustainable economic development consequently enhancing labor productivity;
- Catalyze the establishment of a more liberalized, private sector-driven energy supply industry;
- Transform the country's energy economy from one that is overly dependent on biomass to one with a high modern energy mix [11].

Looking at **Table 4**, it is clear that the targets are far from being met using the current approaches in developing

Table 4. Energy mix, projections and targets as in national energy policy.

Energy source	2000 (Actual)	2008 (Actual)	2010	2020	2050
Biomass	93.0	88.5	75.0	50.0	30.0
Coal	1.0	2.3	4.0	6.0	6.0
Electricity	2.3	2.8	10.0	30.0	40.0
Liquid fuels	3.5	6.0	5.5	7.0	10.0
Nuclear	0.0	0.0	0.0	0.0	4.0
Renewable	0.2	0.4	5.5	7.0	10.0
TOTAL (%)	100.0	100.0	100.0	100.0	100.0

the energy sector and this is having serious impacts on the economy of the country. The energy sector is still in the infancy stage where energy consumption is relatively low with inefficient energy conversion methods being dominant.

The energy supply sector comprises five key components namely electricity, liquid fuels and gas, coal, biomass and other renewable sources of energy. There have been a number of programs and projects that have been implemented with the intention of achieving the already mentioned goals but there has been little progress so far. A closer look at the components of the energy supply sector brings to light the following.

3.1. Electricity Supply

This sector generates electricity mainly from hydro potential but thermal (mainly diesel and gas based) and small photovoltaic systems are also in use. A significant number of commercial and industrial enterprises have installed their own diesel and petrol operated generators due to unreliable energy supply from the national energy supply utility company, but it is still a challenge to determine the capacity of such due to gaps in regulations. Electricity Supply Corporation of Malawi (ESCOM) Limited is the only electrical power supplier and it is a publicly owned company which was established by an Act of Parliament in 1957 which was revised in 1963 and then 1998. The total installed capacity of ESCOM is about 302 MW, of which 94% is generated by hydropower and the remaining 6% is thermal. Almost all the ESCOM's hydro generation stations are located in the Southern region of Malawi along Shire River (the main outlet of Lake Malawi) except for a capacity of 4.5 MW which is located in the Northern region on Wovwe River. Electrical power is transmitted to all other parts of the country through 132 kV network with 66 kV being used as well in other areas. Distribution network is at 33 kV, 11 kV and 400/230 V. Overall, the electricity network is not in very good condition, resulting in substantial losses on the transmission and distribution networks of about 18% - 22% of the generated electrical energy. One of the important forms of energy is that plays a key role in a country's development is good supply of electrical energy. **Table 5** shows the current installed electricity capacity from various hydro power stations in Malawi being operated by ESCOM; however, available capacity at most of the times is generally about 80% or less of the installed capacity [12].

Malawi has got more potential for generating electricity using hydro potential as evidenced by the outcome of some feasibility studies on various rivers in the country. The following are potential sites and the expected range of power and energy outputs:

Table 5. Hydro electric power generation installed capacity by ESCOM (information from ESCOM).

Year	Site	Capacity (MW)	Cumulative capacity (MW)
1966	Nkula A	24	24
1973	Tedzani I	20	44
1977	Tedzani II	20	64
1980	Nkula B	60	124
1986	Nkula B	20	144
1992	Nkula B	20	164
1995	Wovwe	4.5	168.5
1996	Tedzani III	51.3	219.8
2000	Kapichira I	64	283.8

Manolo with a potential output of 60 to 130 MW, Henga valley with a potential output of 20 to 40 MW, Rumphi with a potential capacity of 3 to 13 MW, Chizuma with a potential capacity of 25 to 50 MW, Chasombo with a potential capacity of 25 to 50 MW, Malenga with a potential capacity of 30 to 60 MW, Mbongozi with a potential capacity of 25 to 50 MW, Kholombizo with a potential capacity of 140 to 280 MW, Mpatamanga with a potential capacity of 135 to 300 MW, Low Fufu with a potential capacity of 75 to 140 MW, Low Fufu and Tran with a potential capacity of 90 to 180 MW, High Fufu with a potential capacity of 90 to 175 MW, Chimgonda with a potential capacity of 20 to 50 MW and Zoa Falls with a potential capacity of 20 to 45 MW [13].

It is a known fact that electrical energy is one of the most preferred forms of energy because of a number of advantages among which are ease of transformation into other forms, it is easy transport, it is easy to change to the required voltage and it is relatively cheap to produce. Looking at the trend portrayed in **Table 5**, it can be observed that the cumulative generation capacity is growing quite slowly and not conducive to national development's requirements. The current suppressed demand for electrical power is currently at about 350 MW and is projected to 600 MW in 2015, 1200 MW in 2025 and 1600 MW in 2030. If the generation capacity growth trend continues the way it has been doing in the past, demand will continue being more than the supply capacity for some years to come. There are plans in what is called the Malawi Electricity Investment Plan to bridge the gap between demand and supply so that demand will be equal to supply by 2016 and thereafter, the demand will be less than supply. This can only be possible if there is action commitment from the government and

development partners and not just paper commitment. The current electricity power supply in Malawi is quite unreliable and according to recent estimates, Malawi loses about USD 16 million annually due to power outages. Investing by the private sector in electricity generation remains a challenge due to the government subsidy provided to ESCOM in electricity generation which gives unfair advantage to ESCOM over any would be investors hence ESCOM has remained the sole electricity generation company to date. Capacity of privately owned generators is difficult to ascertain but according to a survey that was done by National Electricity Council in 2001, it was estimated that the capacity of private generators was 51.3 MW, but there are no later figures. Malawi has got a small electricity supply system as compared to her neighbors which are 2483 MW for Mozambique, 1186 MW for Tanzania and 1737 MW for Zambia (all are installed capacities as of 2008) [14]. In addition to local resources, there are plans to connect Malawi to the Southern African Power Pool in the next few years. The Malawi Rural Electrification Project (MAREP) was also established to increase the number of rural trading centers that have access to the national electricity grid and is now in its fifth phase. Electricity from photovoltaic modules is still insignificant when looking at the overall picture, but it is increasing in utilization, finding applications in telecommunications, lighting, refrigeration and water pumping.

A study was done to determine the correlation between electricity consumption and GDP for Malawi from data between 1970 and 1999 extracted from the statistical bulletins and economic reports published by the National Economic Council and the National Statistical Office. Ganger-causality (GC) and error correction model (ECM) were used to examine causality between kWh and GDP. The results showed that there was a bi-directional causality between kWh and GDP using GC and one-way causality running from GDP to kWh using ECM [15]. In conclusion the author wrote that ECM results reflect better the Malawi economy that of being less dependent on electricity since the economy is heavily dominated by the agricultural sector. However, this might not be an ultimate conclusion since the agricultural sector also depends on electricity although to a small extent in as far as Malawi is concerned because the agriculture sector does not use electricity mainly because of its absence in most of the rural areas where agricultural activities are concentrated. It can be argued that the agricultural activities can be having more electrical energy input if the electricity can be available because this can promote agricultural processing activities which in turn can have an impact in the overall economy of the country. Inadequate electrical power supply has been identified as one of the constraints affecting private sector development in Malawi.

3.2. Liquid Fuels and Gas Supply

In the Energy policy document, the government of Malawi recognized the two main parts of this sector which are the upstream and the downstream. Upstream covers aspects of exploration of oil and gas, production and refinement of crude oil, ethanol and other fuels. Downstream encompasses supply logistics and marketing of liquid fuel products and gas. Some past studies showed that certain thick sedimentary rocks which may have hydrocarbon accumulations are present beneath the northern part of Lake Malawi and in the lower Shire Valley. In July 2011, two companies (Surestream Petroleum Limited and Simkara) were given licenses to carry out oil explorations in Lake Malawi. An extensive Strategic Environmental Assessment will be carried out prior to undertaking any work and the assessment will be presented to Ministry of Natural Resources, Energy and Environment for approval before exploration data acquisition commences. There are also efforts to promote oil and gas exploration ventures in Southern African Development Community (SADC) as a whole and this is through the SADC Energy Activity Plan. Malawi has no refineries for petroleum products because it has been proven to be uneconomical due to the small national market. Malawi therefore imports about 97% of its refined petroleum products; the balance is contributed by locally-produced ethanol which is sold directly to the oil companies for blending with petrol on a maximum 20:80 ratio of ethanol-petrol. In practice, the mixing ratio is usually 12:88 because ethanol production is inadequate and also because there is no legislation to make the blending mandatory resulting some petrol being unblended. Petroleum products are imported into the country mainly through three routes namely Dar Corridor, Nacala Corridor and Beira Corridor. The internal storage capacity of oil for the country to avert supply disruptions by natural or manmade emergencies is supposed to be 30 days but this is not the case on ground due to a number of economical and logistical challenges. The transport sector relies heavily on the oil imported supply; consequently its pricing is heavily influenced by trends on the international oil market.

3.3. Biomass Supply

The availability of biomass energy in Malawi can be made sustainable which currently it is not. Wood resource base is diminishing mainly because woodlands and trees in agricultural areas are being cleared up to start new farming land. Statistics show that between 1991 and 2008 about 669,000 hectares of woodlands were converted to farmland [16]. Diminishing standing stock is leading into gradual reduction of biomass that can be harvested. From the information already given, Malawi is heavily dependent on biomass fuels yet the national energy policy has little information on biomass energy supply. Looking at the biomass side of energy supply, a similar situation exists to that of electricity generation in that very little is being done to address issues concerned with improving the supply and efficiency of biomass, although it is the major energy source for the country. Household sector consumes about 92% of biomass energy and the rest is distributed among the other sectors as shown in **Table 6**. About 76% of firewood is used for cooking, 21.5% for heating water, 2% for space heating and the remainder for other uses.

The national energy policy is focused on shifting energy use away from the current heavy reliance on traditional biomass to modern sources of energy like electricity, liquid fuels and renewable sources but little progress has been achieved so far. High dependence on biomass means that it should be a priority to find means and ways of improving utilization of energy from biomass alongside the issues of fuel switching. There are some policy contradictions that exist concerning this sector that continue to hinder progress for biomass fuel supply. Agricultural activities are diminishing resources for the supply of biomass in that as the population is increasing, more land is converted from forests to farming land. Although there are regulations to safeguard the affected forests, there are no mechanisms to ensure that the regulations

Table 6. Consumption of biomass energy by sector and fuel type in 2008 (adapted from national biomass energy strategy document, 2009).

Sector	Type of biomass energy (TJ)			
	Charcoal	Firewood and sawdust	Residues (crop and wood)	**Total**
Household	8703	115,879	2992	**127,574 (92.2%)**
Industry	31	4092	5562	**9685 (7%)**
Transport	0	0	589	**589 (0.43%)**
Service	102	350	0	**452 (0.33%)**
Total	**8836 (6.4%)**	**120,321 (87%)**	**9143 (6.6%)**	**138,300 (100%)**

are adhered to. Another example is that of charcoal in which its production from indigenous trees is deemed illegal unless it can be proven that it has been produced from a sustainably managed forest for which a production license has been applied for and received (article 81 of the Forestry Act, 1997). Charcoal is therefore continually being confiscated by government authorities yet nearly 40% of the urban households use charcoal for cooking [17]. Existence of such contradictions in various sectors hinders investments or efforts towards modernization in more sustainable alternatives of biomass fuels production. Another challenge is existence of different aspects of the biomass sector under different government departments. The supply side of biomass energy supply is covered under the National Forest Policy (1997), the Forestry Act (1997) and the Land Policy Act (2002) while general issues of energy supply fall under the Energy Policy (2003). While the energy policy envisages an ambitious transition from wood fuels to electricity, liquid fuels, coal and renewable energies, little is written concerning modernization, development or sustainability of the wood fuel sector [18]. To fill this gap a Biomass Energy Strategy was formulated and the document was produced in 2009. This strategy makes a pro-active approach towards managing and developing the biomass energy sector instead of just tolerating biomass fuels as an interim solution while waiting for alternative energy supply sources (which has been the case for some time). The Forest Policy recognizes the importance of wood fuels for livelihood of producers in the rural areas and promotes sustainable wood fuel production [19]. The total consumption of biomass in 2008 was estimated at about 9 million tons of wood equivalent. Rural households accounted for about 80% of total consumption while urban households consumed about 12.2% and the remainder was distributed among the remaining sectors as already shown in **Table 6**.

Malawi has a potential to produce bio-fuels for its local markets which can result in reduction of fuel importation bill. This can be accomplished by blending bio-fuels with petrol or diesel as the case has been of blending ethanol and petrol which has been done for over 25 years. There are two ethanol plants in Malawi, one in Salima and another one in Chikhwawa both plants being connected to sugar factories. Apart from blending with petrol, ethanol has also been converted to gel and liquid fuels for both domestic and industrial use but on a small scale because the products have been deemed not economically viable or not suited to the local cooking practices. The country is capable of producing different non-food feed-stocks by converting unutilized land for bio-fuels crop production. Malawi needs to take the necessary steps to manage risks associated with bio-fuels crop production and increase opportunities for bio-fuel devel-

opment through locally-suited but regionally and internationally aligned good practices in bio-energy production. Malawi's focus should be on emerging trade and investment opportunities for the country, implications for poverty reduction, supply-side constraints to expanding production, use and trade in bio-fuels and promotion of new investment mechanisms.

In the field of biomass and biogas, the country has favorable conditions for the application of a majority of technologies in these fields. So far there have been some initiatives for improved wood cook stoves both at domestic and institutional levels. Biogas digesters have been constructed in a number of areas as pilot projects for the rural communities with variable success levels but uptake of the technology remains a challenge to date. In an attempt to minimize the use of biomass fuels and provide the communities in rural and urban areas with alternative source of energy, the government undertook a number of initiatives. Some of these are: the program for biomass energy conservation (ProBEC) which seeks to use more energy efficient technologies like improved stoves; introducing more efficient firewood management through drying and splitting wood among other ways; improved kitchen management through better ventilation at the cooking place, at domestic and institutional levels and use of alternative renewable energy sources such as solar and gel fuel [20]. Another one is the Promotion of Alternative Energy Sources Project (PAESP) which seeks to promote non-traditional fuels for cooking and heating to reduce effects of environmental degradation.

3.4. Coal Supply

The estimated coal reserves in Malawi are between 80 million tons and 1 billion tons of which 20 million tons are proven reserves of bituminous type. Quality of the coal varies with energy values ranging from 17 to 29 MJ/kg. There are four coal fields in Malawi, three in the northern region and one in the southern region. Three are in Karonga and Rumphi districts while the fourth one is at Ngachira in Chikhwawa district in the Lower Shire Valley. Coal mining started in 1985 and currently two fields are being mined in Rumphi district. The coal produced from these sites is not enough to suffice the country's industrial needs consequently the balance is imported from Mozambique. Coal is used in various industries for heating but is not generally used as a domestic energy supply such that it can be concluded that its use in households is negligible. There are plans by the government of Malawi to start generating electricity from coal and feasibility studies are still being carried out.

3.5. Other Renewable Energy Sources

Malawi is endowed with a number of renewable energy

resources yet utilization of these resources is still in infancy stages. In the field of solar, quite high levels of solar energy in the range of 1200 W/m^2 in the warm months and 900 W/m^2 in the cool months are received in most parts of the country which would enable photovoltaic systems and solar thermal systems to perform well. For wind energy systems, there are quite a good number of areas in the country with mean wind speeds above 5 meters per second almost throughout the year. There have been other programs and projects which promoted the use of renewable energy technologies in Malawi. In terms of bio-fuels, there are some crops that are known to have substantial amounts of oil in their seeds which are still being investigated for use in blending with conventional fuels that are in use today. Malawi lies along the Great Rift Valley and therefore traces of geo-thermal reservoirs have been said to exist. Technologies that use solar energy have a high potential of being implemented successfully since solar energy is a resource that is available in abundance throughout the country. This would be very important for most of the rural areas that are unable to access energy from the national electricity grid which is a common form of energy for various humans needs including water treatment which is a very essential element in healthy livelihood. There have been a number of initiatives in the renewable energy sources sector with the notable ones being the National Sustainable and Renewable Energy Program (NSREP, 1999), Barrier Removal to Renewable Energy in Malawi (BARREM, 1999), Program for Biomass Energy Conservation (ProBEC, 2002) and the Promotion of Alternative Energy Sources Project (PAESP).

3.6. Nuclear Supply

Malawi has about 63,000 tons of proven reserves of uranium at Kayerekera in Karonga district and another deposit which is yet to be quantified at Illomba in Chitipa district. Mining of uranium started in 2008 at Kayerekera but currently all the uranium that is mined is exported. Energy that is currently being used in mining of uranium is not supplied from the national electricity grid due to the site being located in a remote area as well as the supply itself being unreliable. Electrical generators operating on diesel are used for electrical power supply to the mining site. There are plans by the government to build a nuclear power station in the far future.

4. Conclusion

The energy sector in Malawi is still in the early stages of development as evidenced by the heavy utilization of biomass to meet a high proportion of the country's energy demand. The targeted energy mixes as set out in the national energy policy are far from being met and there-fore there is a need for change of approach by formulating new energy development strategies altogether. Energy supply remains a great challenge for Malawi to date especially for people living in rural areas. There has been lack of direction and commitment in addressing issues in the energy sector to such an extent that at certain times when the energy sector is mentioned, it actually means the electricity sector and a proof of this noted through some official government literature in which the words energy and electricity are used interchangeably. The Malawi situation is a sure example that availability of energy resources does not automatically translate into production of energy from the available resources. Lack of financial commitment, lack of favorable conditions for investment in the energy sector, lack of political will and poverty are among the major factors that have contributed to underdevelopment in the energy sector for Malawi. The energy sector will continue being what it is unless deliberate steps are taken to prioritize the sector which will result in better political will, allocation of more funds from the national budget to the sector, mobilization of resources and putting in place some incentives for the private sector to actively participate.

REFERENCES

[1] World Bank, "World Development Indicators," The World Bank, Washington DC, 2011.

[2] Malawi Government, "State of the Environment Report," Ministry of Natural Resources, Energy and Environment, Lilongwe, 2010.

[3] Malawi Government, "Census Main Report," National Statistical Office, Zomba, 2009.

[4] W. W. Liabunya, "Malawi Aquatic Weeds Management at Hydro Power Plants," ESCOM Ltd., Generation Business Unit, Blantyre, 2004.

[5] R. A. Judson, R. Schmelensee and T. M. Stoker, "Economic Development and the Structure of the Demand for Commercial Energy," *The Energy Journal*, Vol. 20, No. 2, 2004, pp. 29-57.

[6] B. R. Mitchell, "International Historical Statistics: The Americas 1750-1993," 4th Edition, Stockton Press, New York, 2004.

[7] A. Maddison, "The World Economy: A Millennial Perspective," Development Centre of the Organization for Economic Co-Operation and Development, Paris, 2004.

[8] IEA Key World Energy Statistics, 2004. http://www.iea.org/publications

[9] A. Grübler, "Transitions in Energy Use," *Encyclopedia of Energy*, Vol. 6, 2004, pp. 163-177.

[10] Publications, Statistical Yearbook, 2010. http://www.nso.malawi.net/data_on_line.html

[11] Malawi Government, "National Energy Policy," Department of Energy Affairs, Lilongwe, 2003.

[12] Escom & Electricity, Power Generation.

http://www.escom.mw

[13] Malawi Government, "Malawi Electricity Investment Plan," Ministry of Natural Resources, Energy and Environment, Lilongwe, 2010.

[14] SAPP Statistics, 2010. http://www.sapp.co.zw/

[15] C. B. L. Jumbe, "Co-Integration and Causality between Electricity Consumption and GDP: Empirical Evidence from Malawi," *Energy Economics*, Vol. 26, No. 1, 2004, pp. 61-68. doi:10.1016/S0140-9883(03)00058-6

[16] Malawi Government, "Malawi Biomass Energy Strategy,"

Ministry of Natural Resources, Energy and Environment, Lilongwe, 2009.

[17] Malawi Government, "Forestry Act," Department of Forestry, Lilongwe, 1997.

[18] Environment, 2010. http://www.sdnp.org.mw

[19] Malawi Government, "National Forest Policy," Department of Forestry, Lilongwe, 1996.

[20] M. Owen and J. Saka, "The Gel Fuel Experience in Malawi," Evaluation for GTZ ProBEC and Malawi Department of Energy Affairs, Lilongwe, 1996.

Promoting Alternative Energy Programs in Developed Countries: A Review

Yasir A. Alturki[1], Noureddine Khelifa[2], Mohamed A. El-Kady[1]
[1]SEC Chair in Power System Reliability and Security, Electrical Engineering Department,
King Saud University, Riyadh, KSA
[2]Ministry of Water and Electricity, Riyadh, KSA
Email: yasir@ksu.edu.sa

ABSTRACT

This paper presents a review of the promoting programs of solar and wind energies in different countries including Germany, Austria, France, Spain, Great Britain, Italy, Japan, USA, Australia, China and India with an emphasis on Germany as a notable example. The success of each program is discussed in detail. The results of the investigation conducted in the paper have shown that some renewable energy laws have proven to be extremely successful. Among such laws, is the Renewable Energy Act (EEG) in Germany, which guaranteed 20 years of stable prices for green power and has already brought many homeowners and businesses to install photovoltaic modules and to feed their electricity into the national grid. By the year 2010, the clean production of electricity by renewable energies in Germany has risen to around 202,000 GWh. In addition, about 70 million tons of CO_2 emissions in 2009 were avoided through the use of renewable energies in electricity generation in Germany.

Keywords: Renewable Energy; Resource Management; Environment; Cost Assessment

1. Introduction

The continuous dependence of many countries on energy imports, growing demand and consequent price increases and climate change and environmental pollution through the use of fossil fuel have increased dramatically. This, along with other factor such as finite sources of energy sources, has led many countries towards development of alternative energies that are effective and environmental friendly by promoting new green technologies through policies and international agreements.

This paper presents promoting programs of solar and wind energies in different countries including Germany, Austria, France, Spain, Great Britain, Italy, Japan, USA, Australia, China and India with an emphasis on Germany as a notable example. The success of each program will be discussed in detail.

2. Introduced Support Programs for Renewable Energies Worldwide

2.1. Germany

To promote the development and use of renewable energies, many supporting programs have been introduced in Germany since 1990 (see **Table 1**). The most important and successful ones are Renewable Energy Act (EEG) in 2000 and its predecessor, the Law on Electricity Feeding into the Grid for Renewable Energy (StrEG) and the Renewable Energy Heat Act (EEWärmeG).

2.1.1. Law on Feeding Electricity to the Grid (StrEG)
The StrEG was decided in December 1990 and since 1991 it has been the governing law on the dispatch of electricity from renewable energy sources in the network grid. The conditions of supply and payment of electricity from renewable energy sources are regulated at a national level. It obliges grid operators to accept and transfer all the offered power produced by renewable energy plants. The renewable energy owners are assured average price of electricity coupled with remuneration.

This Law on Feeding Electricity into the Grid was replaced on April 1, 2000 by Renewable Energies Act (EEG).

2.1.2. Renewable Energies Act (EEG)
The Law on Feeding Electricity into the Grid (StrEG) was replaced by the Renewable Energy Act (EEG) in 2000. It regulates the purchase of electricity produced from renewable energy sources by power suppliers or network operators and determines the remuneration of renewable energy plant operators (owners). Anyone who generates electricity and heat from renewable energy sources in Germany and feeds it into the grid has the right to receive incentive by the Renewable Energies Act (EEG). This

Table 1. German support programs for renewable energies [1].

Year	Programs
1990	1000 Roofs-Photovoltaic-Program
1991	Law on Feeding Electricity into Grid (StrEG)
1999	1000 Roofs-Photovoltaic-Program Market Incentive Program for Promote Renewable Energies (MAP)
2000/2004/2009	Renewable Energy Act (EEG)
2001	CO_2-Building Renovation program
2002	Energy conservation Ordinance (EnEV)
2004	Tax Exemption for Biofuels
2005	Law on the Fuel and Electricity Industries (EnWG) New Edition CO_2-Building Program
2006	Biofuel Quota Law
2008	Building Energy Certification
2009	Renewable Energy Heat Act

Act supports many homeowners, investors and enterprises to install photovoltaic modules and wind turbines and to feed electrical energy into the national power grid.

The owners obtain up to 20 years long a guaranteed fixed payment for their generated power in which the payment depends on the renewable energy sector, size and type of facilities, the amount of generated power, the year of commissioning and location. Based on EEG Act the renewable energy can be produced from hydropower, wind power, solar radiation energy (photovoltaic and solar power plants), geothermal, landfill gas, sewage gas, biomass and methane gas.

The EEG Act was revised, expanded and updated in 2004 and 2009 including modifications of the amount of remuneration. This is justified by the ever decreasing cost and improving the efficiency of solar systems. Further modifications are planned for the year 2012.

After Fukushim crises an energy revolution occurred in Germany with new expected energy laws. Germany plans to close all the 19 nuclear power plants by 2021 in stages. The first power plant was already shut down in two years and the last nuclear power plant will be stopped around 2021.

2.1.3. Advantages of the EEG Act

The EEG has brought many benefits. It is applied to both companies and citizen participants. Every citizen can now supply electricity from renewable energy sources to the grid. Participants are assured of their investments through power purchase guarantee, non-discriminatory network access and equity costs through customized fixed feed-in tariffs for different renewable energy technologies. The payments of renewable energy are recovered by additional costs allocated to all consumers per pay-as-you-go financing.

2.1.4. Renewable Energy Heat Act (EEWärmeG)

The Renewable Energy Heat Act (EEWärmeG) has been introduced to complement the Renewable Energies Act

(EEG) and encourage the continued development of renewable energy for heating and cooling energy in buildings. It was first applied in January 2009 and has obliged owners of new buildings to cover their buildings energy needs in proportion with renewable energy. This may include geothermal, solar thermal or biomass.

Alternatively, similar climate friendly measures have also been implemented such as increased insulation, heat from combined heat and power or waste heat [2]. The obligation applies to the use of heat and cooling for both new and old buildings under certain conditions for up to December 31, 2014.

2.1.5. Feed-in-Tariff for Photovoltaic

The current feed-in-tariff in the respective years is shown in **Tables 2** and **3**. The compensation depends on system type, rated power, and year of commissioning and remains constant for 20 years. As shown in **Table 2**, the feed-in-tariff is gradually reduced as the rated power is increased. For example, for a 40 kW system commissioned in 2009, the first 30 kW is compensated at 43.01 cents/kWh while the remaining 10 kW is paid at rate of 40.9 cents/kWh. The contract lasts until 2029.

The feed-in-tariff has been reduced from year to year. 5% reduction was applied for the years 2004 to 2008. The reduction was increased to 8% and 9% in 2009 and 2010 respectively. Further reduction of 13% and 3% were added on 01/07/2010 and also on 01/10/2010 respectively.

The payment per kilowatt-hour injected from photovoltaic systems to the grid is higher than the price that a consumer pays to the power operator for its electricity consumption. So, the system operator pays consumers incentives for their own use of their photovoltaic generated power as shown in **Table 3**.

The total installed new PV power generation capacity has grown steadily in Germany in 2010 and has reached 17,000 MW. The energy supply has surpassed 12,000 GWh in 2010 (see **Figure 1**). This growth of photovol-

Table 2. Electricity feed-in tariff for photovoltaic systems [3].

Type of Plant		2004	2005	2006	2007	2008	2009	2010	July 2010	Oct. 2010	2011
On a building or a noise barrier	Till 30 kW	57.40	54.53	51.80	49.21	46.75	43.01	39.14	34.06	33.03	28.74
	30 kW till 100 kW	54.6	51.87	49.28	46.82	44.48	40.91	37.23	32.39	31.42	27.33
	Up 100 kW	50.0	51.30	48.74	46.30	43.99	39.58	36.23	30.65	29.73	25.86
Open space system (power independent)	Up to 1000 kW	50.0	51.30	48.74	46.30	43.99	33.00	29.37	25.55	24.79	21.56
	Biased surface	45.7	43.4	40.6	37.96	35.49	31.94	28.43	25.16	25.37	22.07
	Arable land	45.7	43.4	40.6	37.96	35.49	31.94	28.43			

Table 3. Current price for owners of photovoltaic systems [3].

Type of Plant		2004	2005	2006	2007	2008	2009	2010	July 2010	Oct. 2010	2011
Self-Consumption Compensation for Systems on Building	Till 30 kW up 30% self-consumption						25.01	22.76	22.05	21.03	16.74
	Till 100 kW till 30% Self-consumption								16.01	15.04	10.95
	Till 100 kW up 30% self-consumption								20.39	19.42	15.33
	100 to 500 kW till 30% self-consumption								14.27	13.35	9.48
	100 to 500 kW up 30% self-consumption								18.85	17.73	13.86

taics has been created not only by the secured feed-in-tariffs, but also by the minimum generated PV electricity required from new buildings in several communities. With this cumulative installed capacity, Germany is the world's largest producer of PV electricity.

2.1.6. Feed-in-Tariff for Wind Energy

The promotion of renewable energies such as wind energy is also regulated by the Renewable Energies Act (EEG).

2.1.6.1. Feed-in-Tariff for Offshore Plants

The total wind turbines installed in Germany in 2010 reached an output of approximately 27,000 MW (see **Figure 2**) of which one sixth is provided by the onshore facilities while the remaining is by offshores. Both types produce about 42,000 GWh. The relatively large number of installed MW is due to the government support.

Owners of offshore wind plants receive an initial feed-in-tariff of 13 cents/kWh at least 12 years until 2014 after which it decreases by 5% annually until 2015 (see **Table 4**).

The EEG based promotion is applied after the period of 12 years by which the wind plants receive 3.5 cents/kWh till 2014. After 2015, the controlled subsidy rates for electricity from offshore wind turbines also decreases during the entire duration of the remuneration by about 5% on a per year basis.

2.1.6.2. Feed-in-Tariff for Onshore Plants

The StrEG was introduced in January 1991 giving a rapid compensation to onshore wind power facilities for their produced power at initial rate of 8.49 cents/kWh.

After replacing the StrEG in 2000 with the Renewable Energies Act (EEG), the initial rate was increased to 9.2

cents/kWh. From 2009, the incentive was based on a regression of 1% per year as shown in **Table 5**.

In the early years, depending on the efficiency of the wind turbine, an initial rate of 9.2 cents/kWh is paid for a minimum of 5 years. Following this period, the rate is lowered to start from 4.92 cents per kWh where the promotion extends over 20 years.

2.1.7. Market Incentive Program to Promote Renewable Energies

The "market incentive program" is used, among other programs, to promote the solar technology on existing buildings in the following different application areas:

- Space heating;
- Combined water heating and space heating;
- Provide process heat;
- Solar cooling;
- Solar collector systems, which supply the heat predominantly a thermal power network.

This program is funded by the environmental tax levied by the Green Party on gasoline and electricity. For the installation of solar collectors to an area of 40 m² for hot water, the promotion of BAFA (Federal Office of Economics and Export Control) is €60 per m², but not less than €410 [6]. Solar collectors for combined water and space heating, supply process heat or solar cooling are encouraged by a fund of €105 per gross collector area of 40 m² [6]. It is possible to get further support measures, such as low-interest loans, from banks and grants from German states [7].

New support regulations for the market incentive program were introduced on March 15th, 2011. The basic subsidy for solar collectors for combined water and space heating increases to €120 on 30 December 2011, then it was reduced to 90 €/m² [8].

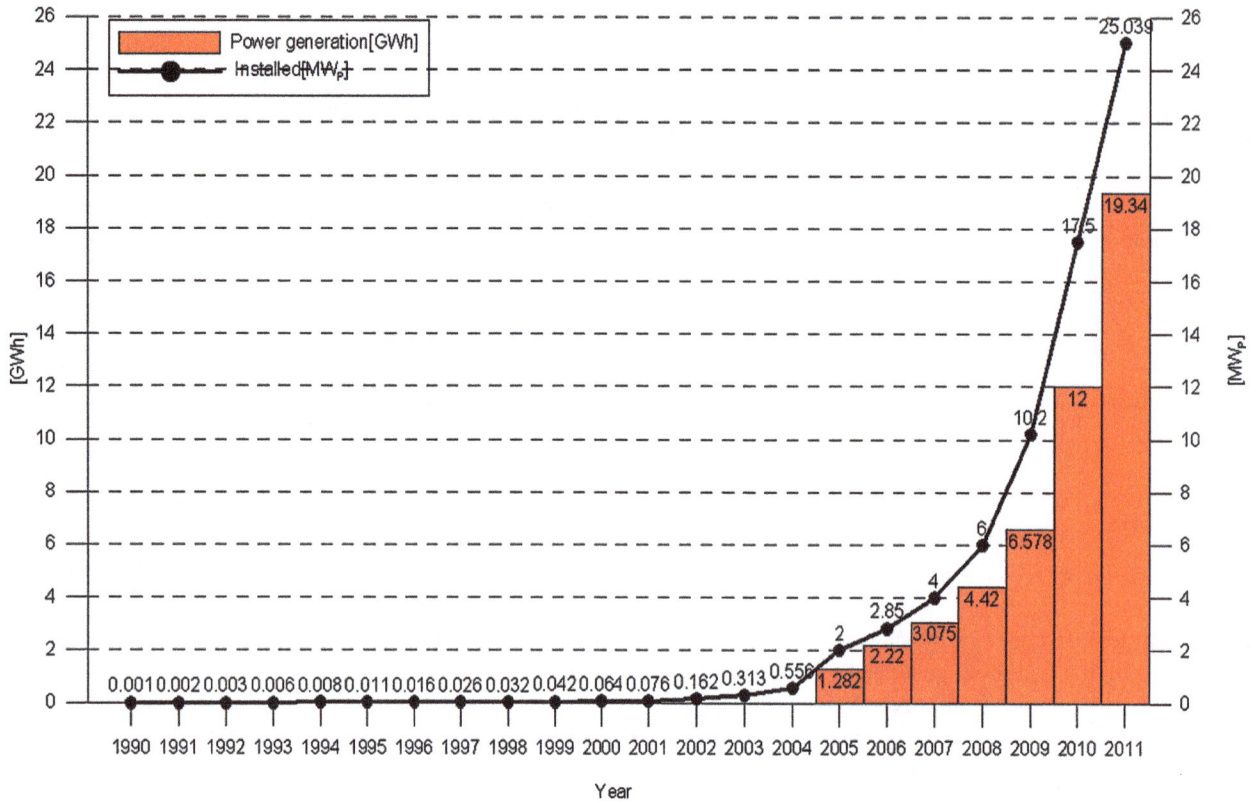

Figure 1. Evolution of electricity generation and installed capacity of photovoltaic systems in Germany [4].

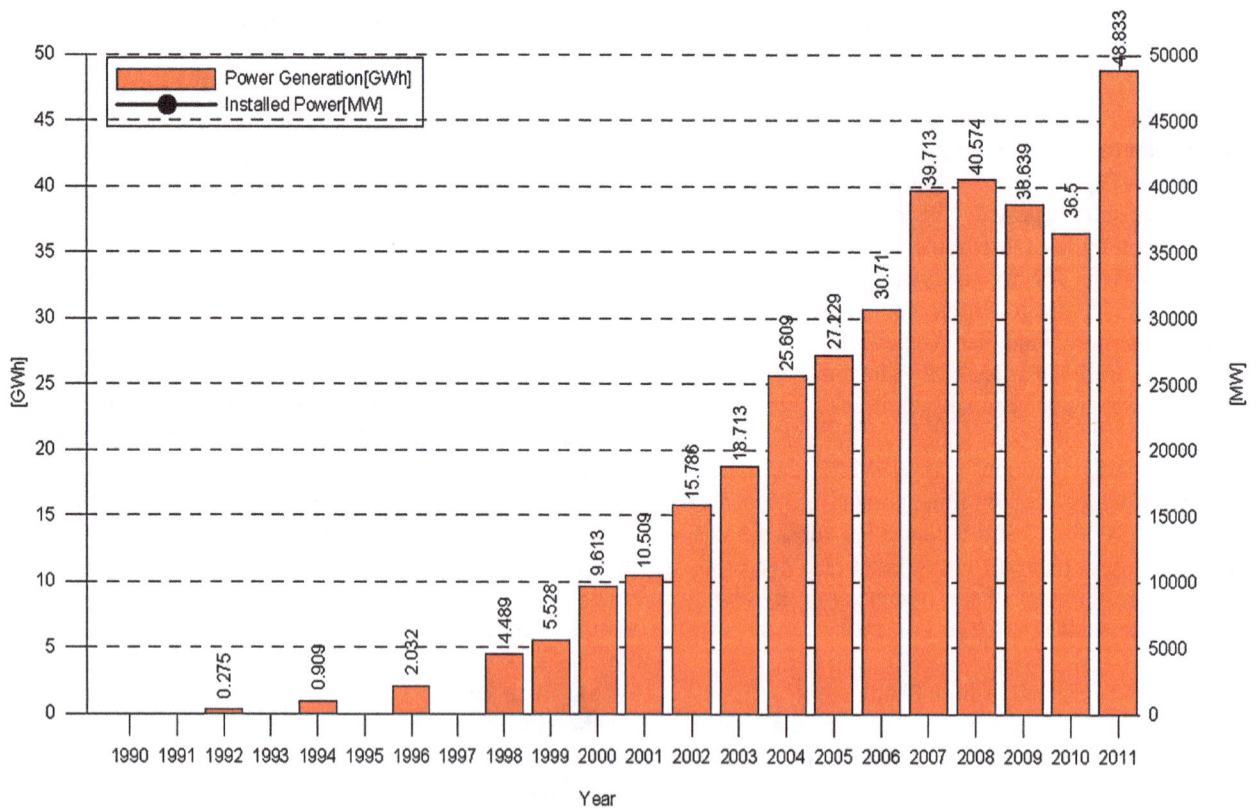

Figure 2. Installed capacity and development of electricity generation from wind turbines in Germany [4].

Table 4. EEG promotion rates for wind turbines (onshore) in Germany [5].

Year in which the wind turbine is put into operation	EEG-subsidy rates Initial feed-in-tariff (cents/kWh)	EEG-subsidy rates Basis feed-in-tariff (cents/kWh)
2009	13	3.5
2010	13	3.5
2011	13	3.5
2012	13	3.5
2013	13	3.5
2014	13	3.5
2015	12.35	3.33
2016	11.73	3.16

Table 5. EEG promotion rates for wind turbines (onshore) in Germany [5].

Year in which the wind turbine is put into operation	EEG-subsidy rates initial feed-in-tariff (cents/kWh)	EEG-subsidy rates basis feed-in-tariff (cents/kWh)
2009	9.20	5.02
2010	9.11	4.97
2011	9.02	4.92
2012	8.93	4.87
2013	8.84	4.82
2014	8.75	4.77
2015	8.66	4.73
2016	8.58	4.68

The bonus for replacing old boiler (without condensing technology) with new condensing boiler is now set at 600 Euros (previously €400) until 30 December 2011, after which it becomes 500 Euros [8].

The combine bonus for solar thermal plus heat pump or solar thermal plus biomass amounts was increased from €500 to €600 until 30th of December 2011 [9], then, it went back to €500. In addition to the basic subsidy granted by BAFA (Federal Office of Economics and Export Control), different special bonuses, some of which are cumulative, have also been incorporated [9].

The installed collector area has reached 15,000,000 m² in 2011 (**Figure 3**). This area was heated by over than 5.6 GWh.

2.1.7.1. Tax Benefits

Large-scale commercial solar systems are exempted from the Value Added Tax (VAT). They also can be depreciated over a period of 20 years [10].

2.1.7.2. Low Interest Rates on Loans

Various loan programs for private and commercial investors offer favorable interest rates and repayment periods over several years.

2.1.8. Results and Achievements of Incentive Programs in Germany

2.1.8.1. Clean Energy

The Renewable Energies Act (EEG) together with its predecessor, the Electricity Feed (StrEG), has initiated a clean power supply. The EEG has resulted in share increase of renewable energies in gross electricity consumption from 6.4% in 2000 to 16.8% in 2010, more than doubling in just ten years period [11].

Table 6 and **Figure 4** [12,13] show the development of electricity supply from renewable sources from 1990 to 2010. The fixed target for 2020 is to reach 30% renewable penetration.

Germany enjoys a mix of various fossil energy sources. The renewable energy accounted for nearly 17% in 2010 while the nuclear power represented nearly 22 percent of the energy mix. The planning of the abolition of nuclear power plants in Germany over the next few years would increase the use of renewable energies.

2.1.8.2. Benefits of EEG is Higher than Cost

The added generation of electricity from renewable energy sources increases the available supply in the electricity market. This has caused the price of electricity on

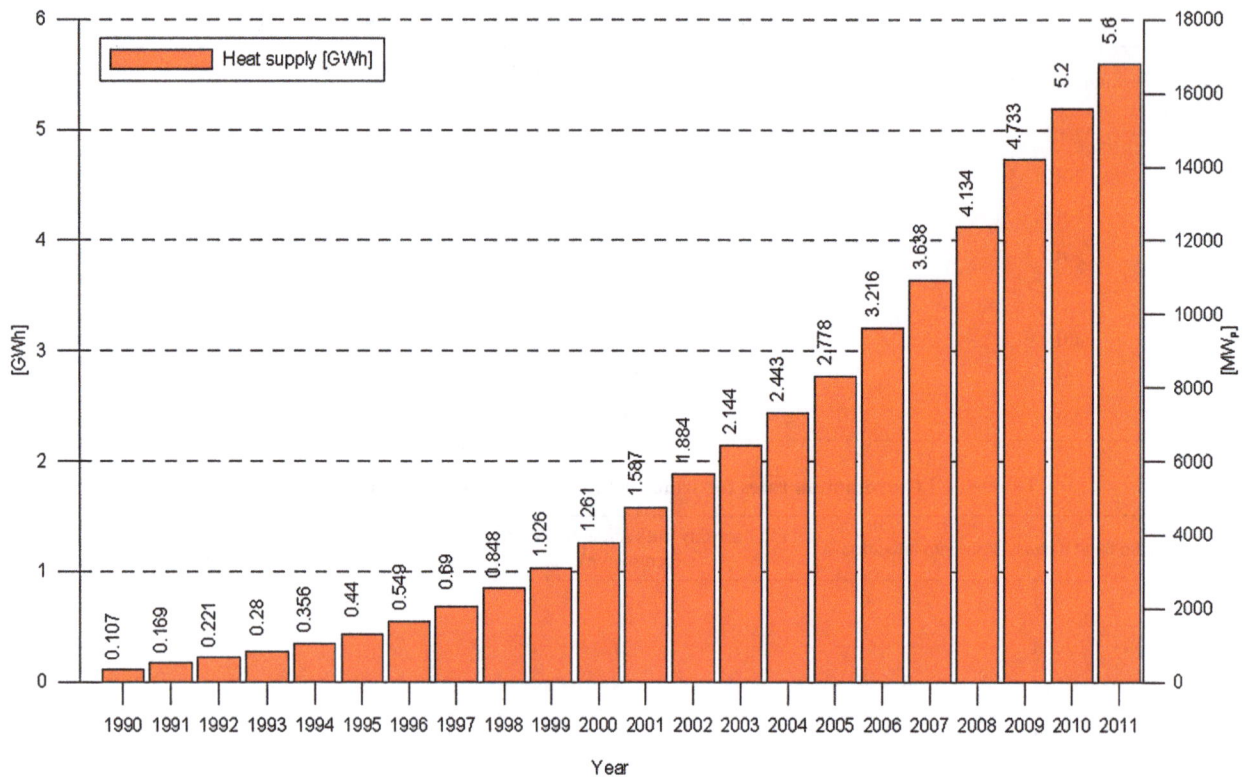

Figure 3. Development of collector and heat from solar thermal plants in Germany [4].

Table 6. Percentage of renewable energies in electric generation in Germany [14].

Renewable Energies	Hydro-Power Plant	Wind Energy	Biomass	Biogen Share of Waste	Photovoltaic	Geothermic	Sun Power Generation	Share of Gross Electricity Consumption
	[GWh]	[GWh]	[GWh]	[GWh]	[GWh]	[GWh]	[GWh]	[%]
1992	15.680	71	222	1.213	1	0	17.086	3.1
1992	18.091	275	296	1.262	3	0	19.927	3.7
1994	19.501	909	569	1.306	8	0	22.293	4.2
1996	18.340	2.032	759	1.343	16	0	22.490	4.1
1998	18.452	4.489	1.642	1.618	32	0	26.233	4.7
2000	24.867	7.550	2.893	1.844	64	0	37.218	6.4
2001	23.241	10.509	3.348	1.859	76	0	39.033	6.7
2002	23.662	15.786	4.089	1.949	162	0	45.648	7.8
2003	17.722	18.713	6.086	2.161	313	0	44.995	7.5
2004	19.910	25.509	7.960	2.117	556	0.2	56.052	9.2
2005	19.576	27.229	10.978	3.047	1.282	0.2	62.112	10.1
2006	20.042	30.710	14.841	3.675	2.220	0.4	71.488	11.8
2007	21.249	39.713	19.750	4.130	3.075	0.4	87.927	14.2
2008	20.446	40.574	22.872	4.559	4.420	17.6	92.989	15.1
2009	19.059	38.639	25.989	4.352	6.578	18.8	94.636	16.3
2010	19.694	36.500	28.710	4.750	12.000	27.2	101.681	16.8
2011	18.074	48.883	31.920	4.950	19.340	18.8	123.186	20.3

Energy mix in Germany in 2010
Renewable energies supploy 16.8% of Gross power consumption

Energy mix in Germany in 2011
Renewable energies supploy 20% of Gross power consumption

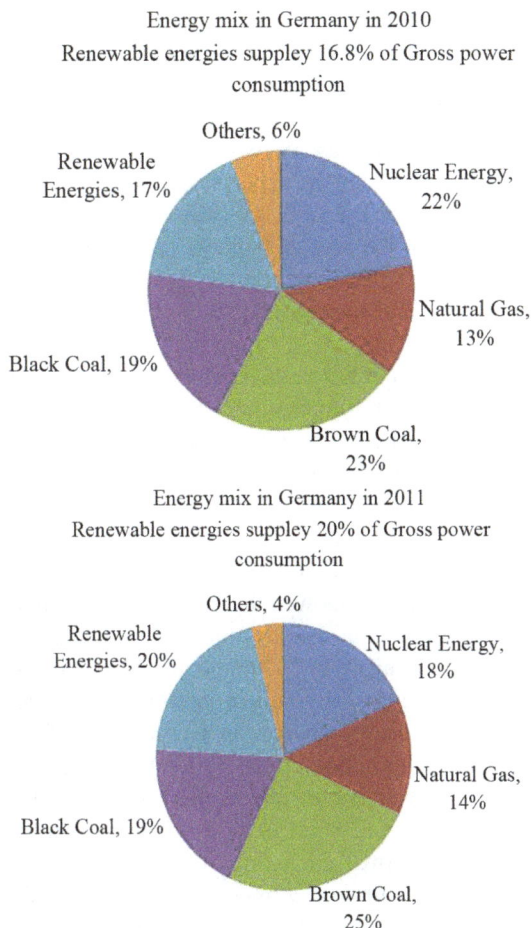

Figure 4. Energy mix in Germany in 2010 and 2011 [12,13].

the wholesale market to fall significantly due to price damping effect ("merit-order effect"). In 2006, this effect led to cost savings of around five billion Euros (**Figure 5**).

In addition, 3.4 billion euros of fuel imports, environmental and climate damage cast was avoided. So, the additional costs of EEG levy of about 3.2 billion euros were recovered from the overall economic savings of more than nine billion Euros.

2.1.8.3. EEG Provides Investment Incentives

The fact that the electricity production has doubled from renewable energies in Germany within a few years is due to the Renewable Energy Act (EEG) and the previous law on feeding electricity into the grid (StrEG). This is a must requirement of investment especially for small and medium-sized enterprises to be able to build production capacities and energy systems. The result was a total saving of more than six billion Euros in 2006.

The share of renewable energy in the electricity mix in 2007 surpassed the target of 12.5 percent, originally intended only for 2010, because the guaranteed feed-in-tariff subsidy rates provided an incentive for the fulfil-

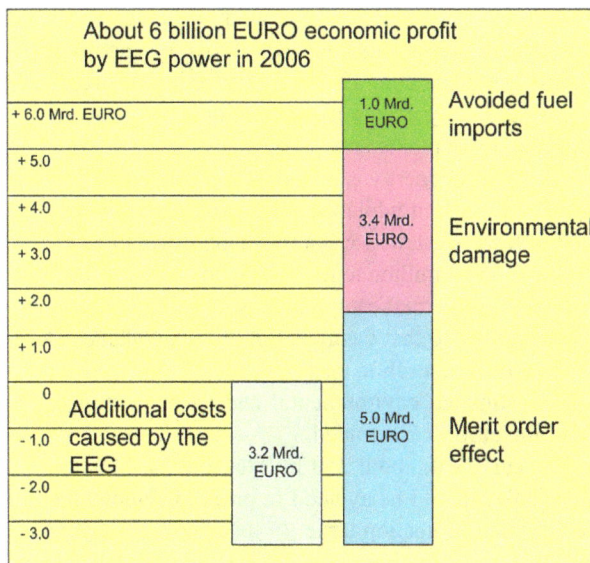

Figure 5. Profit from the EEG power in 2006 [15].

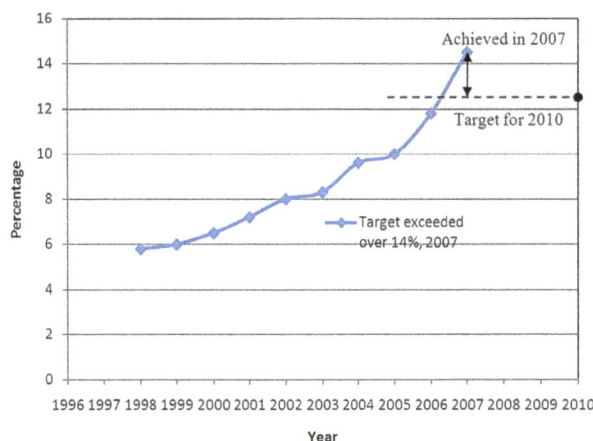

Figure 6. Fixed target for electricity from renewable energy sources in Germany [15].

ment of political goals (**Figure 6**).

In a quota system that would be economically absurd, because the target markets acts at the same time as a lid. Beyond the target rate, the current cannot be profitable marketed [15].

2.1.8.4. Avoidance of Greenhouse Gas Emissions through the Use of Renewable Energies in Electricity Sector

The dependence on fossil fuels, which are still the main fuel source for electricity production, will be reduced by the use of renewable resources. The high level penetration of renewable energy production has a major contribution in greenhouse gases reduction and acidifying air pollutants in Germany. In 2010, electricity sector avoided over 76 million tons of greenhouse gases (**Figure 7**).

The amount of power tempered by the EEG alone led

to a greenhouse gas reduction of around 58 million tones of CO_2 which corresponds to two thirds of the total greenhouse gas reduction. The major reduction of greenhouse gas caused by electricity generation was primarily due to hydropower plants until 1990.

Today, the wind energy is making the largest contribution of 37% reduction while the total biomass eliminates 31%. The newly installed capacity of photovoltaic systems has avoided 7 million tones of CO_2 equivalent greenhouse gases. Geothermal energy has a small share of electricity generation, so that the greenhouse gas avoidance is still well below one million tons of CO_2 [16]. As a result, an annual estimated environmental and climate damage of 3.9 billion euros was avoided.

By the year 2020, about 250 million tons of CO_2, referred to 2007, should be avoided in order to achieve the planned climate protection target (−40% GHE compared to 1990).

The expansion of renewable energies would provide 75 million tons of CO_2, a reduction of about 30% of the total amount [17].

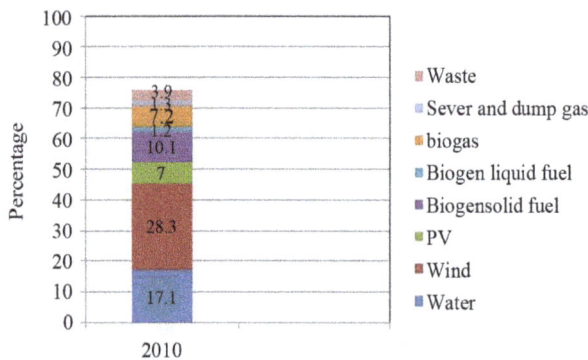

Figure 7. Avoided greenhouse gas emissions through the use of renewable energies in the electricity sector [16].

2.1.8.5. Creation of Jobs
Successful renewable technologies have been developed and installed in the market.

The expansion of renewable energies in Germany creates more jobs than in past years. About 340,000 jobs were created in 2010 in the renewable energy market place (**Figure 8**) [18]. The EEG law secured employment in all industries. In the area of biomass most jobs were created (35.9% of 340,000) followed by solar energy (35.5%), then by wind energy (28.3%).

2.1.8.6. Development of Future Alternative Energy Sector
Through the EEG law, the German economy is a main player in the international market of alternative energies. The German solar companies get benefits from strong market growth of the alternative energy sector. Sales of renewable energies in Germany in 2009 were 33 billion euros. The largest share of the sales of renewable energy was for solar energy followed by biomass and wind power. The investment in facilities for the use of renewable energies in Germany in 2009 amounted to approximately 18 billion Euros. Investment in facilities for the use of renewable energy sources in Germany in 2009 amounted to about 18 billion Euros. By far, the highest in 2009 was the investment in photovoltaic systems followed by Investment in wind turbines, biomass plants, solar thermal, geothermal and hydropower plants [19].

The avoided costs of energy imports were 1.3 and 1.7 billion euros in 2007 and 2010 respectively [17]. The economic benefits of renewable energies in Germany outweigh the costs invested. The total costs incurred by the EEG law are 3.3 billion Euros and the estimated benefits by reducing the wholesale price of electricity, the savings in energy imports, and avoidance of external costs in power generation is 9.4 billion Euros [20].

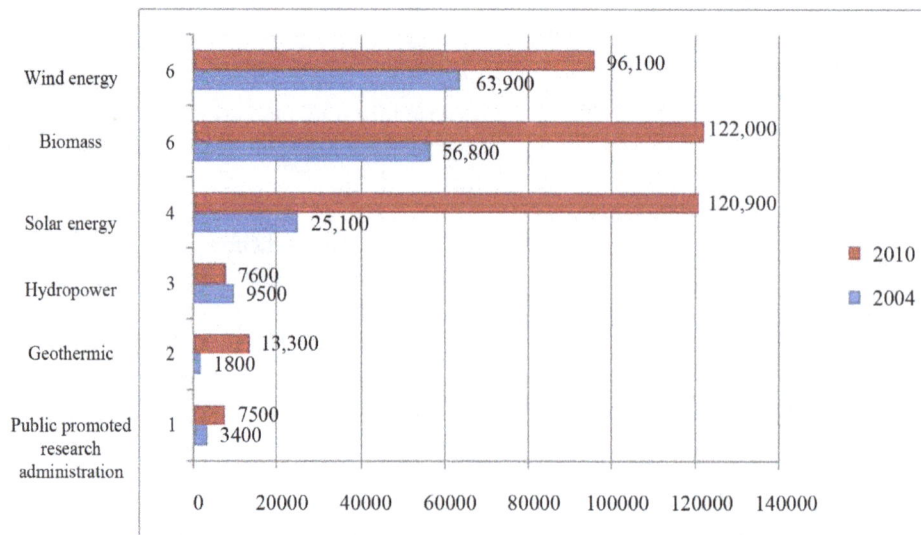

Figure 8. Workers in the field of renewable energies in Germany between 2004 and 2010 [18].

2.1.8.7. EEG Engine for Innovation in Germany

The EEG is a successful technology policy tool. The annual reduction in remuneration for new bioenergy, wind and solar systems exerts a pressure on costs to operators and manufacturers. This makes the systems more efficient, reliable and cheaper. An innovative industry has established itself successfully selling its equipment locally as well as globally. The export quotas for wind and hydroelectric power have already been spent well over 70 percent and a similar development is expected to continue for solar and bio-energy [17].

2.2. Support Programs Worldwide

The German support programs for renewable energies have become international most famous model in recent years. In more than 100 countries around the world the support of renewable energies is already using the same or a similar model introduced and implemented in Germany. In the European Union, the German support method sets as a tool for the development of renewable energies. In addition to Germany, which adopted in 1991 a compensation system (feed-in-tariff) that has now also been used in Estonia (1998), Finland, France (2001), Greece, Latvia, Lithuania, Luxembourg, the Netherlands (2003), Austria (2003), Portugal (1988), Slovenia (2002), Spain (1994), the Czech Republic (2002), Hungary (2003) and Cyprus (2004). **Tables 7-10** show the Feed-in-tariffs for photovoltaic and wind energy with the installed and planned capacities in some countries in the world. Five countries including Japan, India, Australia, USA and China are ready to work out the details, approve and introduce the "Feed-in-Tariff" for renewable energies.

Table 7. Feed-in-tariff for photovoltaic, installed und planned capacities in some European countries [1-24].

Country	Years after Installation	Tariffs [ct/kWh]	PC-Capacity 2010 [MW]	PC-Capacity 2020 [MW]
Germany	20	22.07 - 28.74	17.000	51.750
Spain	25	32 - 44.03	3.808	8.367
Italy	20	29.7 - 40.2	3.479	8.000
France	20	12 - 46	776	4.860
Austria	13	25 - 38	103	1.200
England	25	32.9 - 43.3	52	2.680

Table 8. Feed-in-tariff for wind energy, installed and planned capacities in some European countries [1-24].

Country	Onshore		Offshore		Wind Capacity 2010 [MW]	Wind Capacity 2020 [MW]
	Years after installation	Tariffs [ct/kWh]	Years after installation	Tariffs [ct/kWh]		
Germany	5 20	9.20 5.20	13 7	13 + 3.5 3.5	27.214	45.750
Spain	20 20	7.32 6.12	20	8.43 - 16.4	20.676	35
Italy	15	12 - 30	15	13.5	6.150	12.68
France	10 5	8.2 2.8 - 8.2	10 5	13 3 - 13	6.121	25
Austria	13	9.7	-	-	1.011	3.500
England	20	5.12 - 39.2	20	18	5.692	27.88

Table 9. Feed-in-tariff for photovoltaics, installed and planned capacities in some countries in the world [1-24].

Country	Years after Installation	Tariffs [ct/kWh]	PC-Capacity 2010 [MW]	PC-Capacity 2020 [MW]
Japan	20	Still unknown	6.622	30
China	25	12.2	893	50
USA	20	24.6	2.528	50
India	25	18	0	20
Australia	20	28 - 32	493	20

Table 10. Feed-in-tariff for wind energy installed and planned capacities in some countries in the world [1-24].

Country	Offshore		Wind Capacity 2010 [MW]	Wind Capacity 2020 [MW]
	Years after installation	Tariffs [ct/kWh]		
Japan	15	15.5 - 19.3	2.304	11
China	20	6.2 - 7.5	44.733	200
USA	20	24	40.18	25
India	13	4 - 6	13.065	65
Australia	20	21 - 27	1.88	30

3. Promotion and Status of Renewable Energies Worldwide

Renewable energies are becoming more and more important worldwide. Their use increases steadily counting an important component of global energy supply. Wind, solar, hydropower, geothermal and biomass can provide energy to all continents with more security and promote economic development. Due to the worldwide government supports, the capacity of renewable energy generation has increased rapidly in recent years. The gross electricity generation capacity in the world in 2010 reached 4950 GW with a capacity of 1.313 GW renewable energies representing a share of 26%.

Worldwide, the largest energy capacity (1.005 GW) is provided by hydropower (**Figure 9**) representing 16.1% of the world electricity (**Figure 10**).

The installed capacity of photovoltaic grew from 2004 to 2009 at an annual average increase of 40% as shown in **Figure 11**. In 2010, the solar PV production reached a global level of 40 GW which is almost double the capacity in 2009. This was because of governmental incentives programs and the continued decline in PV modules prices.

The EU dominates the global PV market with 80% share (**Figure 12**). New 17 GW was installed in 2010, enough to power more than 10 million European households.

For the first time ever, Europe added in 2010 more photovoltaic than wind power, led by Germany and followed by Italy. Germany has added more PV (7.4 GW) than the whole world in 2009 and ended in 2010 with 17.3 GW of existing capacity [4].

The installed capacity of wind turbines increased in 2010 by an annual average of 27% reaching198 GW (**Figure 13**).

The top five countries (China, USA, Germany, Spain and India) dominate the total installed wind capacity in the globe with 74% (**Figure 14**). In 2010, China dominated the expansion by 8 GW in just 6 months reaching a total of 52 GW at the end of June.

The majority of European markets of renewable energy in 2011 recorded a stronger growth than previous year. First is Germany with an increase of 766 MW to a

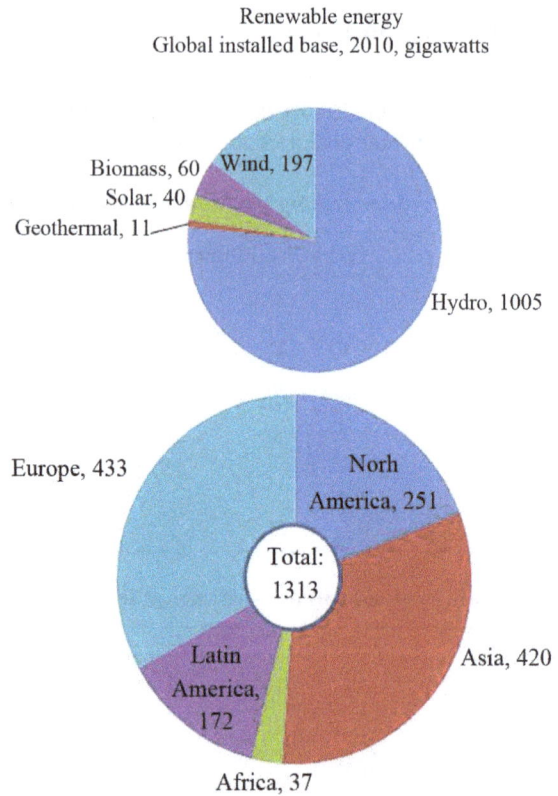

Renewable energy
Global installed base, 2010, gigawatts

Figure 9. Renewable electric power capacity, existing at the end of 2010 [18].

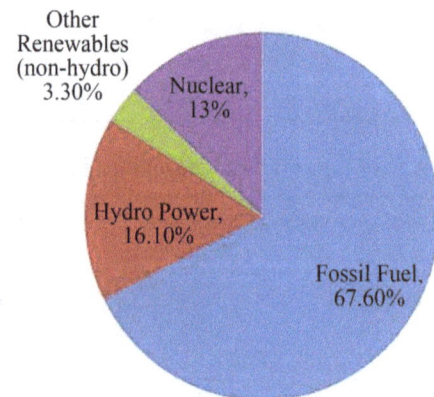

Figure 10. Renewable energy share of global electricity production, 2010 [21].

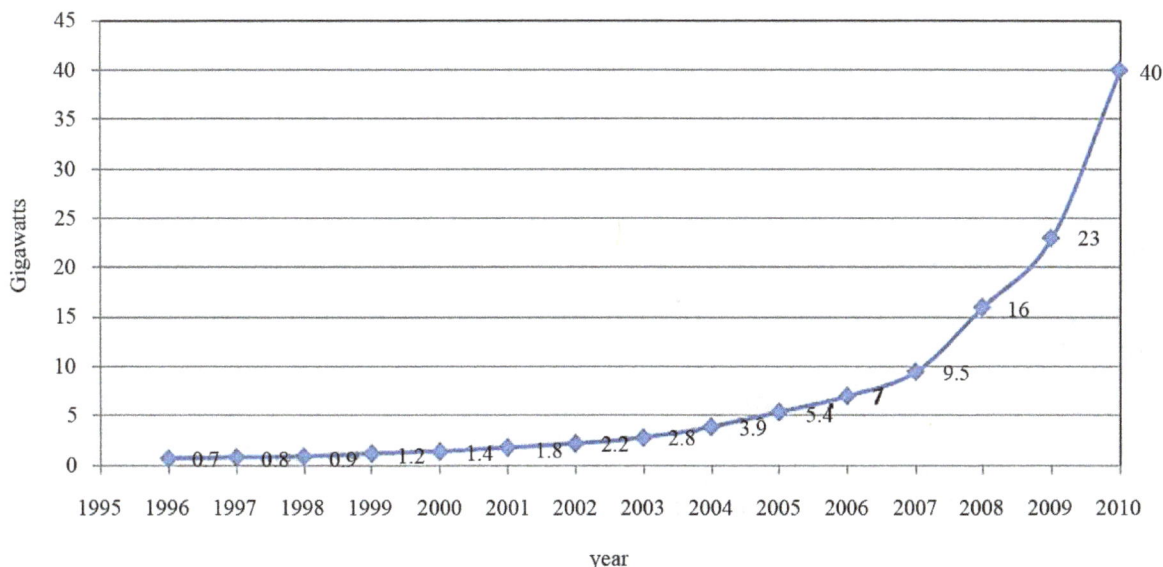

Figure 11. Solar PV, existing world capacity, 1995-2010 [22].

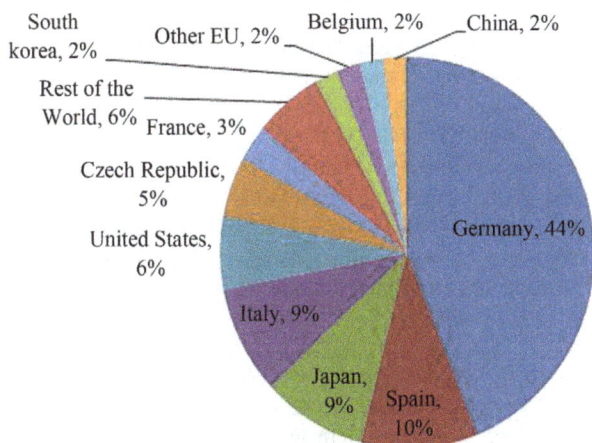

Figure 12. PV-capacity, top 10 countries, 2010 [22].

total power of 27 981 MW (766 MW, 27981 MW), followed by Spain (484 MW, 21150 MW), France (400 MW, 6606 MW), the UK (504 MW, 5707 MW) and Portugal (260 MW, 3960 MW). Only France and Denmark showed a lower expansion than in the first half of 2010. Denmark flew out of the top 10, whereas Portugal is now in 10th Position [23]. On the other side of the Atlantic, US built 2252 MW in the period from January to June 2011, which was 90% higher than the same period of 2010.

A relatively high growth can be observed in Canada where 603 MW were installed in the first half of the year, most of them in Ontario [23]. For the second half of 2011, new installations of 25,500 MW are expected, which would result in an additional building of a total of 43,900 MW. At the end of 2011, the total global wind power capacity is expected to reach 240,500 MW, which is equivalent to 3% of world energy demand [23].

A record of 211 billion dollars investments in renewable energies was reached in 2010, which was roughly a third more than total investments in 2009 (160 billion dollars), and more than five times the investment in the year 2004 (see **Figure 15**). Europe, for the first time, added wind power capacity more than photovoltaic.

Utility companies invested 143 billion dollars in renewable energy and biofuel projects [21]. This increase was mainly due to the incentive funding programs of many countries.

4. Conclusions

The results of this research and development work can be summarized as follows:

1) The EEG law has proved a high successful tool in the broad market of renewable technologies for electricity generation.

2) The Renewable Energy Law (EEG) guarantees 20 years of stable prices for green power and has already brought many homeowners and businesses to install photovoltaic modules and to feed their electricity into the national grid.

3) By the year 2010, the clean production of electricity by renewable energies in Germany has risen to around 202,000 GWh.

4) About 70 million tons of CO_2 emissions in 2009 were avoided through the use of renewable energies in electricity generation in Germany.

5) The economic benefits of the EEG law exceed its cost by several times.

6) The acceptance and payment guarantee by the EEG provides incentives for new investments in renewable power generation.

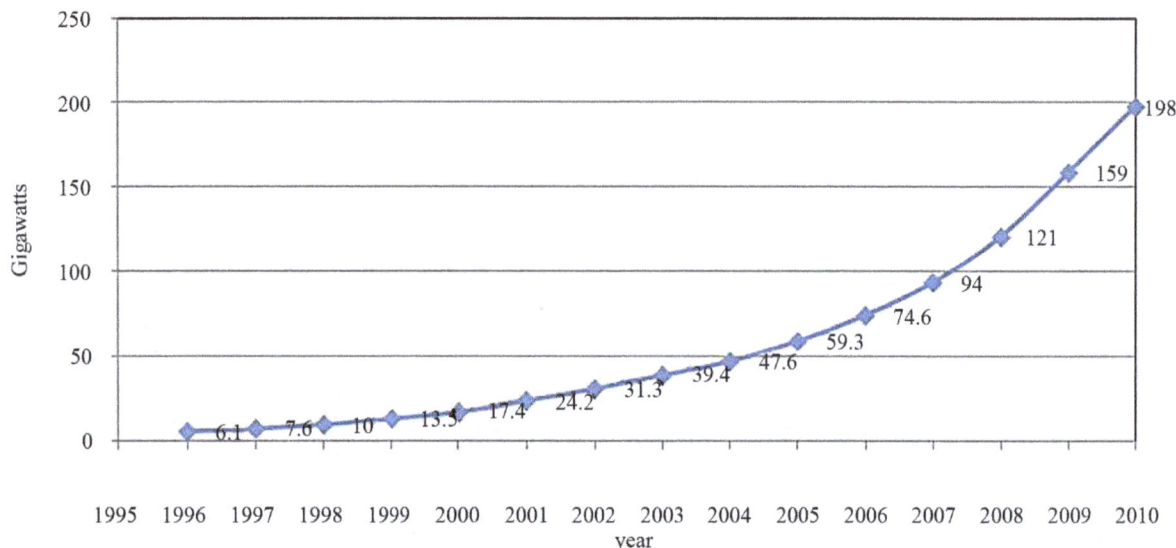

Figure 13. Wind power, existing world capacity, 1996-2010 [22].

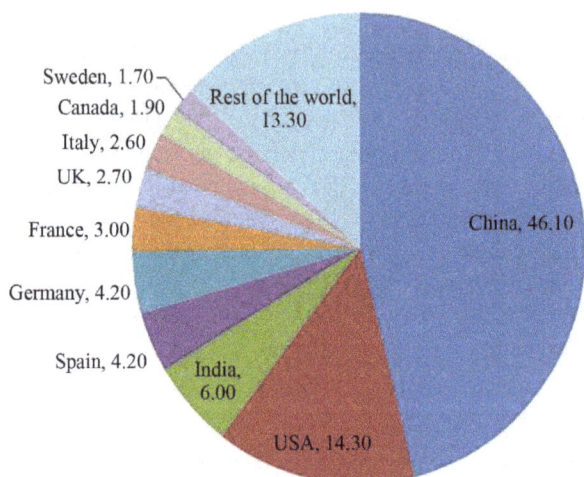

Figure 14. Wind-capacity (%), top 10 countries, 2010 [24].

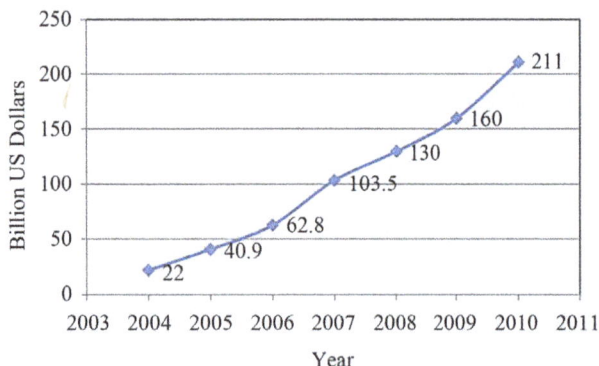

Figure 15. Global new investment in renewable energy, 2004-2010 [21].

7) The introduction of various feed in tariff programs and especially the EEG Laws has created more than 400,000 jobs.

REFERENCES

[1] Eurosolar.org, "25 Years Tschernobyl—Time Bomb Atom Energy—Nuclear Phase-Out Now!" *IPPNW-Congress*, Berlin, Urania, 8-10 April 2011, p. 14. www.tschernobylkongress.de

[2] Domovita-Haustechnik GmbH, "Collection of Information about Energy and Eergy Laws," Energy Laws, pp. 1-3. www. de/zusatz/energiegesetze/Erneuerbare.pdf

[3] Wikipedia, "Renewable-Energies-Laws," Photovoltaic Performance-Based Support Rates in ct/kWh. www.de.wikipedia.org

[4] Federal Ministry for Environment, Nature Conservation and Reactor Safety (BMU), Working Group on Renewable Energies-Statistics (AGEE-Stat), "Development of Renewable Energies in Germany in 2011," 2011. www.erneuerbare-energien.de

[5] H. Herminghaus, "EEG-Support for Wind Power in Germany," Stand, 2011. www.umweltbewusst-heizen.de

[6] Federal Office of Economics and Export control, "Base-, Bonus- und Innovation Support Solar," Stand, 2009. www.markus-energie.de

[7] J. J. Oppermann GmbH & Co. KG, "Renewable Energies, Support and Credits for Solar Collectors." www.alternative-energiequellen.com

[8] Press Office of Consumer Schleswig Flensburg, "Support Renewable Energies." www.schleswig-flensburg.de

[9] Agency for Climate Protection, "E-Heating Laws and Solar Thermal & Wood," 2011, pp. 48-49. www.agentur-fuer-klimaschutz.de

[10] Allianz-Knowledge, "Government Support for Solar Energy Use Worldwide." www.wissen.allianz.de

[11] Federal Ministry for Environment, Nature Conservation and Reactor Security, "Experience Report 2011 for Renewable-Energies-Law (EEG-Experience Report)," Stand,

2011, p. 3. www.bmu.de

[12] German Renewable Energies Agency, "The Electricity Mix in Germany 2010." www.unendlich-viel-energie.de

[13] Germany Information Portal on Renewable Energies, "The Electricity Mix in Germany in 2011," 2011. www.unendlich-viel-energie.de

[14] Federal Ministry for Environment, Nature Conservation and Reactor Security: Working group Renewable Energies-Statistics (AGEE-Stat), "Graphs and Tables with Data of Development of Renewable Energies in Germany in 2011," 2011, p. 12. www.bmu.de

[15] Federal Ministry for Environment, "The Renewable Energies-Law—Success Story." www.unendlich-viel-energie.de

[16] Federal Environment Agency, "Avoided Greenhouse Gas Emissions through the Use of Renewable Energies in the Electricity Sector." www.umweltbundesamt-daten-zur-umwelt.de

[17] Federal Ministry for Environment, "What Will Be the New Renewable Energies-Law EEG?"

http://ebookbrowse.com

[18] Federal Ministry for Environment, Nature Conservation and Reactor Security. Working Group Renewable Energies-Statistics (AGEE-Stat), "Renewable Energies 2010," 2010, p. 16. http://www.renewable-energy-projects.de

[19] S. Handelsblatt, "Data and Facts on the Renewable Energies in Germany." www.de.statista.com

[20] DGS Landesverband Berlin, "Market Economy and Ecology," p. 14. www.dgs-berlin.de

[21] Renewable Energy Policy Network for the 21st Century (REN21), "Renewables 2011, Global Status Report," 2011, pp. 18, 20, 23, 35. www.ren21.net

[22] Globservateur, "Energy > Renewables Exceeds Oil and Coal." www.globservateur.blogs.ouest-france.fr/

[23] Austrian Wind Energy Association (IG Wind Power), "Wind Energy-World Market after a Weak Year 2010 Come back on Track—Semi-Annual Report of WWEA." www.igwindkraft.at

[24] GWEC, "Top 10 New Installed Capacity Jan-Dec 2010." www.sino-report.com

18

Intelligent Decisions Modeling for Energy Saving in Lifts: An Application for Kleemann Hellas Elevators

Vasilios Zarikas[1], Nick Papanikolaou[2], Michalis Loupis[3], Nick Spyropoulos[4]
[1]Department of Electrical Engineering, Theory Division, Academic Institute of Technology of Lamia, ATEI Lamias, Lamia, Greece
[2]Department of Electrical Engineering, Power Electronics Division, ATEI Lamias, Lamia, Greece
[3]Department of Electrical Engineering, Informatics Division, ATEI Lamias, Lamia, Greece
[4]Research and Development Division, Kleeman Hellas, Kilkis, Greece
Email: vzarikas@teilam.gr

ABSTRACT

The present work proposes a methodological approach for modeling decisions regarding energy reduction in an elevator. This is achieved with the integration of existing as well as acquired knowledge, in a decision module implemented in the electronics of an elevator. So far, elevators do not exploit information regarding their recent usage. In the developed system decisions are driven based on information arising from monitoring the use of the elevator. Monitoring provides various records of usage which consequently are used to predict elevator's future usage and to adapt accordingly its functioning. Till now, there are only elevators that encompass in their electronics algorithms with if then rules in order to control elevator's functioning. However, these if then rules are based only on good practice knowledge of similar elevators installed in similar buildings. Even this knowledge which unavoidably is associated with uncertainty is not encoded in a mathematically consisted way in the algorithms. The design, the implementation and a first pilot evaluation study of an elevator's intelligent decision module are presented. The study concludes that the presented application sufficiently reduces energy consumption through properly controlled functioning.

Keywords: Elevators; Energy Consumption Reduction; Energy Engineering; Applied Bayesian Networks; Applied Decision Networks; Applied Influence Diagrams; Applied Intelligent Decisions; Fuzzy Rules

1. Introduction

A large number of techniques in the field of artificial intelligence used to represent knowledge and/or describe decision problems: production rules, semantic nets, Bayesian networks, frameworks, scripts, statements, logic, fuzzy logic and possibility theory, causal networks, among others. They have been shown to be effective especially in domain dependent decision tasks. The choice of a particular technique is based on two main factors: the nature of the application and the skills of the user/designer.

The present work describes a specific technological application of bayesian networks. The bayesian networks that were implemented, incorporated certain and uncertain knowledge as well as fuzzy rules in order to guide the proper functioning of an elevator with respect to energy consumption minimization. bayesian networks are well-known and efficient mathematical models based on subjective probability theory to model knowledge and

making inference. Moreover, probabilistic modeling among these observed parameters of interest are central to science. In order to algorithmically consider causal probabilistic relations, the relations must be placed into a representation that supports manipulation.

Fuzzy rules were used because fuzzy logic is very close to the expert's reasoning. Its utilization became very popular in attempting to resolve the problems of imprecision and uncertainty. It is easier for engineers to express their knowledge using fuzzy If-Then rules instead of using conditional probabilities. Thus, experts can express their knowledge through a number of fuzzy rules to describe the observations and their impact into the problem.

A common problematic feature of these bayesian models is that a very detailed amount of information is used to fill all the involved conditional probability tables (CPTs). Our approach suggests a simple working solution for modeling this particular decision problem of energy saving. This approach determines the CPTs taking probabilities from the measured frequencies of the last week usage of the elevator. In addition the proposed solution suggests a particular novel way for assigning fuzzy linguistic values, extracted from rules, in order to fill conditional probability tables in these BBNs.

Based on searches of the relevant literature, no previous work was found suggesting a bayesian decision system for driving the functioning of an elevator. However, there are some interesting applications of intelligent control units in elevators [1-6].

The paper is organized into the following sections: the second section presents a description on Bayesian networks influence diagrams and Fuzzy rules. The third section provides a description of the designed and developed decision networks. The fourth section describes the first evaluation of the system. Finally, the fifth section discusses the results of the developed tool, outlining is also the main conclusions of the study.

2. Bayesian Networks and Fuzzy Logic

The imprecise and insufficient information that always appears in technology cannot be incorporated in models using the framework of "if-then" rules. However human experts are able to reach decisions with a high level of validity even if the input data are almost always uncertain. This fact led to the development of mathematical models capable to manipulate pieces of information that are associated with an uncertainty value, named Bayesian networks.

Early developed rule based systems were consisted of a knowledge base and an inference system. The knowledge base is a set of rules of the form "If A (with certainty 1) then B (with certainty 1)". Implementing an inference system it is possible to combine these rules and other observations in order to finally reach a decision. Although this framework can tackle a considerable amount of adaptation functioning it cannot be generalized in order to include cases where the condition (A) or/and the act (B) is subject to an uncertainty level.

The construction of a consistent mathematical framework that is allowed to incorporate uncertain pieces of information into a plan of reasoning exists and is the so called belief network or Bayesian network or causal graph [7-11]. It is a graphical representation of a problem domain, which consists of informational nodes (pieces of information) that are known with certainty or with an uncertainty described by a subjective probability. Subjective probabilities express measures of a person's belief, given a certain knowledge background carried by that person. This notion of probability differs from the most used classical probability. The subjective or Bayesian probabilities can describe a value of belief to unique events that are not repeatable. Thus, a subjective prob-

ability $P(O|X,C)$ describes the subjective estimate (belief) of certainty of an event O, given as known that the event X occurred and given a certain background knowledge C.

An important property of this framework is the fact that the direction of probabilistic inference can be reversed. This is allowed due to the Bayes' theorem. Suppose that we know the belief of the influence of a hypothesis H on the observable evidence E, $P(E|H,C)$. Then Bayes' theorem allows us to compute the belief on the influence of E on H (the so called posterior probability) as a function of the prior probability $P(H|C)$.

$$P(H|E,C) = \frac{P(E|H,C)P(H|C)}{P(E|C)} \qquad (1)$$

It is now possible, through this theorem, for domain experts to provide estimates of subjective probabilities in the causal direction and calculates beliefs in the backward the so called diagnostic direction *i.e.* learning the belief in a hypothesis given the relevant evidences. In real life problems we face situations where there are relationships among a large number of variables. A Bayesian network is a representation of such cases.

Formally, a Bayesian network is a graph with the following elements and properties 1) A set of variables (shown as nodes in the diagram). Each variable has a set of mutually exclusive states; 2) A set of directed edges between these variables; 3) The variables and the edges form a directed acyclic graph and 4) To each node-variable there is attached a conditional probability that depends on the parents of the node.

Thus, the design of a Bayesian network is required to draw arcs from cause variables to their immediate effects. In this way causal relationships reveal the conditional dependencies and independencies. After constructing the network based on our prior knowledge and data, appropriate algorithms [7,12-14] exist, in order to determine various probabilities from the network. Probabilistic inferences can be produced with the estimation of a certain probability of interest from our model.

An additional issue is the combination of Utility theory and Bayesian graph theory. This synthesis formulates Decision theory [15-18]. Utility theory provides the axiomatic framework for consistency among preferences and decisions. The axioms introduce the concept of a lottery which is an uncertain situation with various outcomes assigned with a probability of occurrence. Then a set of rules define which outcome is preferred from which lottery. Accepting these axioms we can always define a utility function. A utility function is a scalar that assigns a cardinal scale to each outcome and decision indicating its desirability. The preferred set of decisions is the one that *maximizes the expected utility given, with known uncertainty, the various relevant parameters.*

For a decision problem where there is a set of mutually exclusive decision states D_i with $i = 1, \cdots, n$ and one determining variable V with possible states V_j with $j = 1, \cdots, m$ (an hypothesis that drives the decision). Further, we are interested in cases with non intervening decision states or in other words decisions that their states does not have any correlation with $P(H)$. The determining variable V is part of Bayesian network, and has various parents and child nodes in general.

Constructing an influence network [7] or otherwise called decision network (a Bayesian network with utilities and decision/action nodes) means finally to set the values of a utility table that determine for each action D_i and each state V_j a number that expresses the utility $U(D_i, V_i)$ that we gain. Then the expected utility for taking actions is

$$EU(D_i) = \sum_j U(D_j, V_j) P(V_j) \qquad (2)$$

The preferred decision is associated with the action that gives the maximal expected utility $M(D)$.

$$M(D) = \max_i EU(D_i) \qquad (3)$$

For cases, we have N decision/action variables $D^{(k)}$ with determining variables $V^{(k)}$ where $k = 1, \cdots, N$, and we would like to drive a decision based on all the combinations of actions, then it is straightforward to generalize for a new expected utility as follows

$$EU\left(D^{(1)}, D^{(2)}, \cdots, D^{(N)}\right)$$
$$= \sum_k w_{(k)} \sum_{V_j^{(k)}} U_k\left(D^{(k)}, V^{(k)}\right) P\left(V^{(k)}\right) \qquad (4)$$

where $w_{(k)}$ are the weights that reflect the significance of its action variable with respect to its other. It is important to note that we have treated the various combinations of action states $D_i^{(1)}, D_j^{(2)}, \cdots, D_l^{(N)} \in D^{(1)} \times D^{(2)} \times \cdots \times D^{(N)}$ belonging to different action variables as states of one action variable which is the Cartesian product of the N decision/action variables.

Here we assume that in the previous decision modeling the actions have no impact on the variables in the networks which affect the belief on the hypothesis nodes. In other words we described the framework for sets of non-intervening actions.

Finally, the various nodes in an influence diagram, can be deterministic informational nodes, statistical results or beliefs given as probability distributions, utilities and action (or otherwise called decision) nodes. Decision nodes represent possible actions, informational nodes represent pieces of certain or uncertain relevant knowledge while utility nodes encapsulate designer's preferences, goals etc. Such a diagram then represents a deci-

sion basis.

For the problem of making technological decisions, engineers are usually thinking in order to reach a decision or suggest a solution in a form of if-then rule, *i.e.* Furthermore, a rule-base is more transparent and understandable for engineers. Fuzzy logic is based on fuzzy if-then rules which have the general form "IF X is A THEN Y is B," where A and B are fuzzy sets. A fuzzy set is a set containing elements that have varying degrees of membership in the set. Elements in a fuzzy set can also be members of other fuzzy set on the same universe [19,20], because their membership need not be complete.

Following a knowledge representation viewpoint, a fuzzy if-then rule is a scheme for capturing knowledge that involves imprecision. A key property of reasoning using these rules is its partial matching capability, which enables an inference to be made from a fuzzy rule even when the rule's condition is only partially satisfied.

Randomness describes the uncertainty in the occurrence of an event while fuzziness describes the ambiguity of an event. In classical sets there is no uncertainty, hence they have crisp boundaries, but in the case of a fuzzy set, since uncertainty occurs, the boundaries may be ambiguously specified. The membership function for a set maps each element of the set to a membership value between 0 and 1 and uniquely describes that set. The values 0 and 1 describe "not belonging to" and "belonging to" a conventional set respectively; values in between represent "fuzziness." The determination of the membership function is subjective to varying degrees depending on the situation. It is determined on an individual's perception of the data in question and does not depend on randomness. The latter is a significant point and distinguishes fuzzy set theory from probability theory.

The membership functions that constitute the fuzzy sets which describe the inference of the fuzzy rules are depicted in Equation (5). A numerical value of each fuzzy set is produced after defuzzification with the Center of Area method. The produced numerical value is used to fill the probabilities in CPTs.

3. Designing the Bayesian Network

Some common critics about applied Bayesian networks concern the necessity of filling correctly a lot of conditional probability tables. On the other hand, the appearance of all these probability tables make this decision tool extremely precise, expressive and mathematically consisted. It makes also profound emphatically to the decision builder and/or the interviewed expert how many pieces of information are involved for precise decision making. This set of probabilities by no means can be disregarded unwisely for the sake of simplicity or a fault decision will be driven. On the other hand, experts complained that the human brain does not work in this way and even scientists (not experienced in "Bayesian language") cannot easily report safely all these numbers in order to describe a domain knowledge. Thus, it is vital for the construction of a proper decision network, a correct set of all involved probabilities. This is a task that should be carried from the designer of the bayesian network who usually is a person familiar with the bayesian mathematical framework. Engineers should be allowed to report their knowledge in the form of fuzzy if then rules, thus avoiding misunderstandings with the bayesian reasoning. In the present work engineers reported relevant knowledge in the form of if then rules. A practical solution of this problem is presented in this work for the decision model in study.

The decisions that our decision module has to take are two. First if the elevator after its last use will be on a full activation mode (FM), in a standby mode (SM) or in an off mode (OM). Second the engine should decide in which of the n-th floors the elevator cabinet should rest after its last usage. Both these decisions will be driven according the recorded knowledge of the last week's usage. More specifically both these decision are affected by the following informational deterministic and decision nodes.

- Date.
- Twenty four times seven times $n(24 \times 7 \times n)$ informational nodes that provide the frequency of calls (number of calls of ith floor over total calls of all n floors at the same timezone) from every one of the n-th floors, each for every time zone of each one of the seven days.
- Node expressing the reasoning of common practice fuzzy rule that refers to the preference of the floor chosen to rest, for either a specific season or day or timezone (this helps to make better prediction in cases such as the end of the two weeks Christmas vacation period and the beginning of a full working week).
- Subjective probability expressing the preference of the medium floors as far as the energy consumption is concerned.
- Twenty four times seven (24 × 7) informational nodes that provide the normalized duration of idle use (idle time length over the time duration of the time zone = 1 hour) each for every time zone of each one of the seven days.
- Node expressing the reasoning of common practice fuzzy rule that refers to the preference of the type of functioning mode (FM or SM or OM) for either a specific season or day or timezone (this helps to make better prediction in cases such as the end of the two weeks Christmas vacation period and the beginning of a full working week).
- Subjective probability expressing the preference of

the FM status compared to SM and SM compared to OM mode as far as the user annoyance is concerned (this information makes a difference in case of almost equal balance among two modes).

- One *utility* comprising the overall utility driven by historic data. Utility values are associated on each of the three states the elevator will have to choose: a full activation mode (FM), in a standby mode (SM) or in an off mode (OM).
- One *utility* comprising the utility driven by the preference of each of the three states the elevator: a full activation mode (FM), in a standby mode (SM) or in an off mode (OM).
- One *central multi attribute utility* comprising the overall utility of each of the three states the elevator: a full activation mode (FM), in a standby mode (SM) or in an off mode (OM). This utility is driven be the positive effect of the historic data utility and the negative effect of the preference utility.
- *Decision node* concerning which of the three states the elevator will choose to rest: a full activation mode (FM), in a standby mode (SM) or in an off mode (OM).

Note that in the implemented system we have defined 24 timezones per day and seven days per week. The history records concern last 7 days period data.

In **Figure 1**, a small simplified part (it is not possible to present the full bayesian network due to its size) of the real constructed bayesian network is presented. Note that the preference utility expresses the fact that the FM state is preferable for users contrary to the suggestion of the historic data that choose the most economic mode as far as energy consumption is concerned. The multiattribute utility realizes the trade off of the two utilities according to the designer's tuning.

The fuzzy rules that are integrated into the bayesian network are of the general form:

"If the season is A and the day is B and the time zone is C then the floor N (much less, less, more, much more) preferable".

"If the season is A and the day is B and the time zone is C then the mode (FM or SM or OM) is (much less, less, equally, more, much more) preferable".

In order to translate these fuzzy rules into probabilities through the defuzzification process we have to define first the membership function in use.

$$\mu(B_n) = \begin{cases} \dfrac{x}{x_n - x_{n-1}} + \dfrac{x_{n-1}}{x_{n-1} - x_n}, x_{n-1} \leq x \leq x_n \\ \dfrac{x}{x_n - x_{n+1}} + \dfrac{x_{n+1}}{x_{n+1} - x_n}, x_n \leq x \leq x_{n+1} \end{cases} \quad (5)$$

where $n = 1, 2, \cdots, 5$ with $x_n = n/6$. In Equation (5), x takes values from 0 to 1 and B1 = v.v.weak, B2 = v.weak,

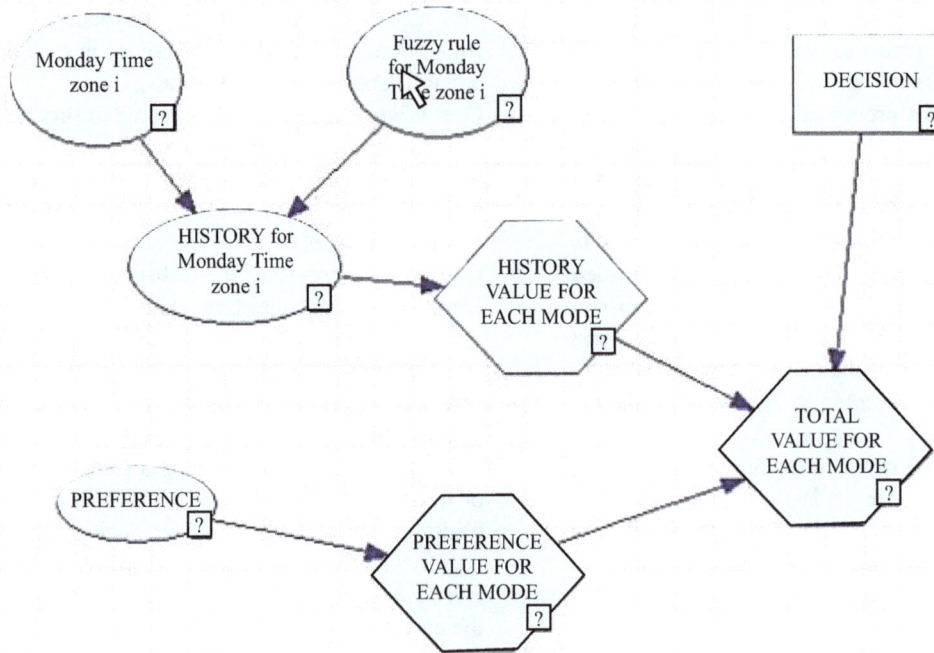

Figure 1. Simplified influence diagram for the decision concerning functioning mode.

B3 = weak, B4 = weak-med, B5 = medium, B6 = med-high, B7 = high, B8 = v.high, B9 = v.v.high. Central values are given in Equation (6).

$$\text{Central}\left[\mu(B_n)\right] = \begin{cases} 0, & - \\ 0.15, & B_1 = \text{much less} \\ 0.35, & B_2 = \text{less} \\ 0.5, & B_3 = \text{equal} \\ 0.65, & B_4 = \text{more} \\ 0.85, & B_5 = \text{much more} \\ 1, & - \end{cases} \quad (6)$$

Based now, on the defined membership functions, linguistic values contained in the rules are transferred to numerical values in order to fill the conditional probability tables, through the defuzzification approach of fuzzy logic. In order to show how the probability tables for BBNs are developed using the above type of if-then rules, a generic approach is provided. Let's consider the following rule for the assessment of preferability to *SM* mode during a timezone: "If the day is *X* and the time zone is *Y* then the mode (*SM*) is (more) preferable". This rule suggests information capable to provide probabilities for the conditional probability table (*CPT*) between the preferability of mode *SM* and the observables *X*, *Y*.

$$P(SM|X,Y) = \begin{cases} \mu(B_4), & \text{timezone} = Y; \\ 1/3, & \text{timezone} \neq Y. \end{cases} \quad (7)$$

Also

$$P(\text{mode}|X,Y) =$$

$$\begin{cases} \dfrac{1-\mu(B_4)}{2}, & \text{timzone} = Y, \text{mode} = OM; \\ \dfrac{1-\mu(B_4)}{2}, & \text{timzone} = Y, \text{mode} = FM; \\ 1/3, & \text{timzone} \neq Y, \text{mode} = FM \text{ or mode} = OM. \end{cases} \quad (8)$$

These probabilities Equations (7) and (8) concern only day X in the simple case that there are no other rules for this day.

Since all informational nodes affect the final decision naturally a first approach is to connect all informational nodes to a central node representing a total utility. However, this leads to a huge and quite confusing probability table. For example, in a case of entering values for a new rule, it would be necessary to change the values for the already entered rules. Furthermore, a table containing values gained from more than two or three nodes is highly inflexible. It's not possible to distinguish the values between the different nodes. Therefore intermediate deterministic nodes are necessary for a solvable network topology, see **Figure 1**.

4. Evaluation

Genie tool has provided the C code which consequently

has been integrated for research purposes in the electronics of the control unit see **Figure 2**.

4.1. The VDI Directive

The reduction of energy consumption in modern lifts has become a critical parameter for their commercial competitiveness, especially in the EU market. For this reason, all important lift developing companies are making Research and Development efforts through the introduction of improved driving techniques (in terms of the electromechanical part) as well as of "smart" energy management methods (in cases of multiple installed lifts at the same building).

However, the determination of the exact energy consumption for a lift is a rather complicated issue, taking into account their stochastic use and the variety of the load and of the total trip—even on an annual basis. Hence, for the ensuring of the healthy competitiveness in the lift market specific and common energy consumption measuring rules have been set. To this direction VDI 4707 is a commonly accepted guideline [21] which describes a transparent method for the assessment and classification of the power requirement and consumption of lifts. According to VDI 4707 the lift consumption can be distinguished to the following categories: A. Travel Demand, is the total energy consumption of the lift during trips at specified trip cycles with a defined load. Travel demand is expressed as energy consumption per traveled distance multiplied by load weight, Wh/(m·kg). B. Standby Demand, is the total energy consumption of the lift in standby mode. Relevant studies in this issue [22-25], have highlighted that the standby mode consumption is a very important part of the annual energy consumption, which may reach (in some cases of rarely used lifts) even the 80 percent. Standby demand is expressed as power consumption, (W).

Today commercial lifts are adopting energy saving methods in order to be included in the higher possible energy category (according to VDI 4707). These energy saving techniques are aiming to the reduction of energy consumption in terms of travel demand as well as of standby demand. In more details, travel demand is reduced by using regenerative breaking through AC/DC/AC power electronic drives as **Figure 3**. Presents [26,27]. For the standby mode energy saving is achieved by switching off lighting and inverter electronic circuits. Additionally, some modern lifts use more sophisticated software in order to cut down standby consumption; these lifts are programmed in such a way, so as to power off during intervals that are not used. The main difficulty in this type of lift control is that after powering off a critical time interval is necessary until the lift comes again in full operation (reset of electronic circuits, self test, etc). Moreover, this time interval protects the inverter from frequent restarts which would jeopardize its health status (mainly due to the start up overcurrents). Nevertheless, this sophisticated operation is not adaptive, but it is standard for each building category. In the present work the option of implementing an adaptive controller for lifts is investigated. The proposed control algorithm modifies the standby and power off modes time intervals according to the lift use—as daily as seasonal—achieving so lower annual energy consumption. Furthermore, the proposed algorithm aims also at the travel

Figure 2. Control unit.

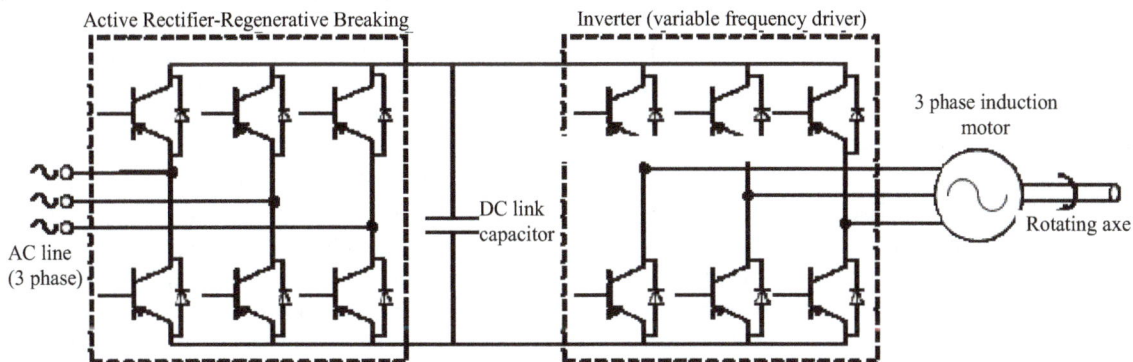

Figure 3. AC/DC/AC inverter drive schematic with regenerative breaking.

demand reduction through the decrease of the annual lift traveled distance.

The proposed algorithm has been tested to the following elevator type—derived by VDI 4707 sample calculation:

- Type of building: Residential block/ doctor's practice.
- Nominal load: 320 kg.
- Speed: 0.63 m/s.
- Stops: 5.
- Vertical rise: 11.2 m.
- Trips per day: approximately 100.
- Travel distance: 1,134 m per day.
- Travel demand: 8.93 mWh/(m·kg).
- Standby demand: 200 W.
- Usage category (according to VDI 4707): category 1.
- Travel energy demand per day: 8.93 mWh/(m·kg) 1134 m 320 kg = 3.24 kWh.
- Standby energy demand per day: 110 W 23 h = 2.59 kWh.
- Total energy demand per day: 3.24 kWh + 2.59 kWh = 5.82 kWh.
- • Specific energy demand: 5.82 kWh/(1134 m 320 kg) = 16.05 mWh/(m·kg).
- Energy efficiency class (according to VDI 4707 calculation): class F.

4.2. Test Procedure/Results

Kleemann Hellas was founded in 1983, based on the know-how and licence of the german company Kleemann HUBTECHNIK GmbH. The head office of the company is based in the Industrial Area of Kilkis in Northern Greece. Kleemann company's activities concern both the manufacturing and trading of Complete Lift Systems. Kleemann is enlisted among the largest companies of the lift industry in the European and international market (more than 12,000 new systems or three percent of the world's new lift units annually).

Tests were performed for a Kleeman elevator carrying the developed research decision support system on its electronic control unit. The evaluation took place in the Tower for Experiments based on the headquarters of the company. This tower belongs to the main industrial campus in Kilkis city, Macedonia, Greece, see **Figure 4**. The test procedure comprised two main scenarios: 1) A heavy load scenario (meaning 100 trips/day) with two peaks morning (08:00 - 10:00) and afternoon (16:00 - 18:00) and 2) An average use scenario (meaning 50 trips/day and less dense peaks). Each scenario was followed for one week. The final outcome of this pilot study evaluation was: A fourteen percent energy saving, for the heavy load scenario, was achieved compared to the energy consumption of the same elevator with a conventional control unit. In addition, a five percent energy saving, for

Figure 4. Kleemann tower for elevators testing.

the average load scenario, was achieved compared to the energy consumption of the same elevator with a conventional control unit.

These first results together with an analytic list of all decisions that have been performed by the decision system will be the input for improving the whole system during the second evaluation run.

5. Discussion

Engineers as experts have reported certain and uncertain scientific knowledge in the form of fuzzy rules. The decision system suggests actions based on their reported fuzzy rules and the frequencies that the control unit evaluates from the last week usage of the elevator. The decision support system encapsulates two specific topology bayesian networks which were designed for the two main decisions that ensure energy saving.

After construction of Bayesian network using the Genie tool, a number of test cases have been examined in order to set evidences to the network and illustrate its decision making capabilities. Specifically, one hundred fifty (150) trips each associated with the relevant decision making actions were performed. The decision making capabilities of the system were tested and they will guide the reforming of the whole system in order to eliminate false responses. However even in this first pilot study a non negligible energy saving was achieved.

The developed system gives a front-end decision system about the functioning of elevators with the aim to save energy. Of course, more trials and real tests are needed for a large number of cases in order to confirm or improve the system. Future work will be directed towards this direction. The aim of this research paper was two fold. First, a new techniques for integrating into BNs fuzzy if-then rules was proposed and second an efficient modeling and reasoning concerning the integration of all informational nodes into the network with a specific workable and solvable topology was presented.

This approach for modeling decisions has several advantages: 1) a graphical way of encoding the information is used; 2) certain and uncertain pieces of knowledge can be consistently incorporated and interweaved; 3) nodes in these diagrams are able to represent information coming from engineers/experts and historic data; 4) the system can be adjusted to different decision policies and strategies; 5) beliefs associated with some of the informational nodes can be altered dynamically and a new decision based on the new evidences can be driven; 6) decisions are not hard-coded into the system, this means that the influence diagram comprises a high level description of the decision reasoning that could be easily modified, customized and re-used; and, 7) if during the intermediate evaluation of the application, new knowledge (rules) has to be built in, the designer either simply corrects the various conditional probabilities or more alters the topology of the network and the probabilities. Some difficulties and disadvantages are: 1) the designer or developer needs to understand the Bayesian reasoning; 2) the designer or developer needs to test the network for sensitivity in order to check if the changes of the various probabilities have the correct and desired impact on the utilities driven decision. However, the latter is not a difficult task since for most of the cases only a small part of the network is activated.

Novel ideas that have been materialized in the present work are: 1) Experts/engineers have not been involved for the probability assignments but only for reporting fuzzy rules; 2) Fuzzy rules have been translated to probabilities according to a specific technique; 3) Historic data concerning the last week usage of elevator alter non trivially the functioning of the elevator; 4) There is an intermediate layer of utilities that transfer their values to a central utility node

A particular set of cases were studied as a pilot preliminary study. In upcoming work, more tests and trials have to be made for model validation. Future work will be focused to analyze and implement this approach in other type of elevators too as well as on a different context, *i.e.* large buildings with more than one elevator.

6. Acknowledgements

This work has been supported by the research program "Less energy consumption in elevators (LESS)", Project Code 09SYN-32-829, within the Greek Research Activity "COOPERATION". This is co-financed by the European Union (European Social Fund) and Greek national funds.

REFERENCES

[1] L. Cui, "An Elevator Intelligent Scheduling Method Using Neural Network Control," *International Journal of Digital Content Technology and Its Applications*, Vol. 7, No. 3, 2013, pp. 174-181.

[2] T. Chen, Y.-Y. Hsu and Y.-J. Huang, "Optimizing the Intelligent Elevator Group Control System by Using Genetic Algorithm," *Advanced Science Letters*, Vol. 9, No. 1, 2012, pp. 957-962. doi:10.1166/asl.2012.2654

[3] P. E. Utgoff and M. E. Connell, "Real-Time Combinatorial Optimization for Elevator Group Dispatching," *IEEE Transactions on Systems, Man, and Cybernetics Part A: Systems and Humans*, Vol. 42, No. 1, 2012, pp. 130-146.

[4] N. A. Rahim, H. W. Ping and J. Jamaludin, "A Novel Self-Tuning Scheme for Fuzzy Logic Elevator Group Controller," *IEICE Electronics Express*, Vol. 7, No. 13, 2010, pp. 892-898.

[5] Y. Cheng, X. Wang and Y. Zhang, "A Bayesian Reinforcement Learning Algorithm Based on Abstract States for Elevator Group Scheduling Systems," *Chinese Journal of Electronics*, Vol. 19, No. 3, 2010, pp. 394-398.

[6] X.-C. Wang and D.-M. Yang, "Intelligent Algorithm of Elevator Group Control by Statistic Approximation," *Xitong Fangzhen Xuebao/Journal of System Simulation*, Vol. 13, 2001, pp. 100-101.

[7] F. V. Jensen, "An Introduction to Bayesian Networks," UCL Press Limited, London, 2000.

[8] J. Pearl, "Probabilistic Reasoning in Intelligent Systems: Networks of Plausible Inference," Morgan Kaufmann, San Mateo, 1988.

[9] J. Stutz and P. Cheeseman, "A Short Exposition on Bayesian Inference and Probability," National Aeronautic and Space Administration Ames Research Centre: Computational Sciences Division, Data Learning Group, 1994.

[10] N. Friedman and M. Goldszmidt, "Learning Bayesian Network from Data," SRI International, Menlo Park, 1998.

[11] D. Heckerman and D. Geiger, "Learning Bayesian Networks," Microsoft Research, Redmond, 1994, p. 3.

[12] J. Pearl, "Fusion, Propagation and Structuring in Belief Networks," *Artificial Intelligence*, Vol. 29, No. 3, 1986, pp. 241-288. doi:10.1016/0004-3702(86)90072-X

[13] J. Pearl, "Evidential Reasoning Using Stochastic Simulation of Causal Models," *Artificial Intelligence*, Vol. 32, No. 2, 1987, pp. 245-258. doi:10.1016/0004-3702(87)90012-9

[14] J. Pearl and T. Verma, "The Logic of Representing De-

pendencies by Directed Graphs," *Proceedings*, *AAAI Conference*, Seattle, 13-17 July 1987, pp. 374-379.

[15] R. L. Winkler, "An Introduction to Bayesian Inference and Decision," Holt, Rinehart and Winston, Toronto, 1972.

[16] E. J. Horvitz, J. S. Breese and M. Henrion, "Decision Theory in Expert Systems and Artificial Intelligence," *International Journal of Approximate Reasoning*, Vol. 2, No. 3, 1988, pp. 247-302.

[17] B. W. Morgan, "An Introduction to Bayesian Statistical Decision Processes," Prentice-Hall Inc., Englewood Cliffs, 1968, p. 15.

[18] J. Pearl, "Influence Diagrams–Historical and Personal Perspectives," *Decision Analysis*, Vol. 2, No. 4, 2005, pp. 232-234. doi:10.1287/deca.1050.0055

[19] L. A. Zadeh, "The Concept of a Linguistic Variable and Its Application to Approximate Reasoning," *Information Science*, Vol. 8, No. 3, 1975, pp. 199-249. doi:10.1016/0020-0255(75)90036-5

[20] H. M. Saraoglu and S. Sanli, "A Fuzzy Logic-Based Decision Support System on Anesthetic Depth Control for Helping Anesthetists in Surgeries," *Journal of Medical Systems*, Vol. 31, No. 6, 2007, pp. 511-519.

[21] VDI 4707 Guideline, "Lifts Energy Efficiency," 2008.

[22] G. Barney, "Vertical Transportation in Tall Buildings," *Elevator World*, Vol. LI, No. 5, 2003, pp. 66-75.

[23] CIBSE, "Guide D Transportation Systems in Buildings," 2005.

[24] J. Nipkow, "Electricity Consumption and Efficiency Potentials of Lifts," Report of Swiss Agency for Efficient Energy Use SAFE, HTW Chur University of Applied Sciences, Zurich, 2005.

[25] N. Spyropoulos and L. Asvestopoulos, "Hydraulic vs. Traction Lifts: Environment Friendliness and Quality of Service to the User," *The 17th International Congress on Vertical Transportation Technologies*, Thessaloniki, 11-13 June 2008, pp. 247-251.

[26] N. Mutoh, Y. Hayano, H. Yahagi and K. Takita, "Electric Braking Control Methods for Electric Vehicles with Independently Driven Front and Rear Wheels," *IEEE Transactions on Industrial Electronics*, Vol. 54, No. 2, 2007, pp. 1168-1176. doi:10.1109/TIE.2007.892731

[27] M.-J. Yang, H.-L. Jhou, B.-Y. Ma and K.-K. Shyu, "A Cost-Effective Method of Electric Brake with Energy-Regeneration for Electric Vehicles," *IEEE Transactions on Industrial Electronics*, Vol. 56, No. 6, 2009, pp. 2203-2212. doi:10.1109/TIE.2009.2015356

19

Research on New Compressed Air Energy Storage Technology[*]

Xian Ma[1], Jingtian Bi[1], Weili Chen[1], Zhisen Li[2], Tong Jiang[1]
[1]State Key Laboratory of Alternate Electrical Power, System with Renewable Energy Sources,
North China Electric Power University, Beijing, China
[2]Tai'an Power Supply Company, Shandong Electric Power Corporation
Email: maxiansdu1989@163.com

ABSTRACT

In recent years, wind power generation and photovoltaic power generation have been developing rapidly, and the installed capacity of the new resources generation has been keeping a fast growth every year. But with the incorporation into the grid, the new resources generation that has the properties such as randomness and volatility causes certain risks to the power grid, which results in the falling of the incorporation proportion instead of rising. This paper describes the current status and development problems of the new energy in China, and gives a brief introduction of characteristics of various energy storage technologies. This paper focuses on the analysis of the compressed air energy storage technology in recent years and new developments and the latest technology at home and abroad, additionally, the paper introduces a new concept of the compressed air energy storage system.

Keywords: New Energy; Wind Power; Power Storage Technology; Compressed Air Energy Storage

1. Introduction

UN Secretary-General Ban Ki-moon, the "Sustainable Energy for All" initiative was launched in late 2011, specified the year of 2012 as "Sustainable Energy for All" International Year. In the period of fast development of new energy, China, the United States, Germany, Japan and many other countries are all striving to fight for the exploitations of new energy to stimulate the future economy growth, so the development and utilization of new energy, energy-saving technologies and energy-saving products have become the key points of the national energy strategy for every country.

Coal, oil, natural gas and other fossil fuels will eventually dried up, coupled with the needs of environmental protection, so that large-scale application of renewable energy is imperative. The renewable energy such as wind power and solar energy have the properties as randomness and volatility, and frequent wind power off-gird pose a threat to the safe operation of the power grid, the above situations result in increasing proportion of abandoned wind yearly[1]. The combination of Energy storage technologies and renewable energy generation technology not only can improve the stability of the system and improve power quality, but also improve resource utilization. Therefore, the large-scale use of renewable energy need the support of large-scale and distributed energy storage technology.

2. Analysis of Energy Storage Technology

2.1. Premise of Energy Storage Technology

According to statistics released by the National Grid, at the end of June 2012, China's grid-connected wind power has reached 52.58 million kilowatts and overtaken the U.S. to become the world's biggest wind power country. But at the same time the year of 2011 as a turning point in the development of wind power industry in China entered a period of steady development instead of rapid growth. In the year of 2011, the china's wind turbines in operation were suspended for 19 hours every day, the amount of 100 billion kWh of wind power was abandoned and the percentage of abandoned wind power was over 12%, which is the equivalent of 330 million tons of standard coal consumption, or 10 million tons of carbon dioxide emissions to the atmosphere. The loss was more than 50 billion that is accounting for nearly 50% of the wind power industry profitability.

If the new energy generation plants such as wind power and solar photovoltaic power plants were equipped with energy storage devices, firstly, the energy storage components would adjust the output of the unit and solve the

[*]Supported by the National High Technology Research and Development of China 863 Program (2012AA050208).

randomness and uncontrollable problems of the new energy power generation itself to reduce the impact of the output variation of the new energy on the grid, secondly, the devices would store the electrical energy during the period of abundant electricity and release the electricity when it is necessary. The combination of energy storage with large capacity wind power generation system is an important part of the renewable energy, especially for wind power plants.

Figure 1 represents the development of wind power in China from 2008 to 2012. It is not difficult to find that the cumulative installed capacity of wind power grows fast, but the wind turbines grid ratio shows a downward trend.

2.2. Introduction of Energy Storage Technology

Existing electrical energy storage technologies include pumped storage, compressed air energy storage, flywheel energy storage, battery energy storage, superconducting magnetic energy, super capacitor energy storage and so on [2].

- The flywheel energy storage system absorbs electrical power from grid and transfers it into mechanical energy during a low load period and the fast rotating flywheel functions as prime motor to drag the generators run when the load need is tedious [3]. Flywheel system has the advantage that no friction loss, wind resistance, long life and no environmental impact. The disadvantage is that the relatively low the energy density is relatively low and the cost of system security is high, so this technology doesn't show the priority in small system and it now mainly works as assistant for the battery system.

- The superconducting magnetic energy storage system transfers the grid power to magnetic field energy directly, which doesn't need the energy exchange [4]. This storage system is characterized by simple techniques, timeproof equipment, high density, fast response speed, low loss and costliness. This system is not adapted by distributed power systems until now.

- Super capacitor is mainly used for load smoothing and power quality high peak power occasions for a short time in the power system and have not been widely used because of the expensive price [5].

- Battery energy storage system is a chemical power storage system and the batteries need to maintain a certain temperature in the work process. Proportion of the cost for insulation takes a big part in existing battery energy storage power station and when the temperature reaches a certain limit, there will be a certain security risk [6].

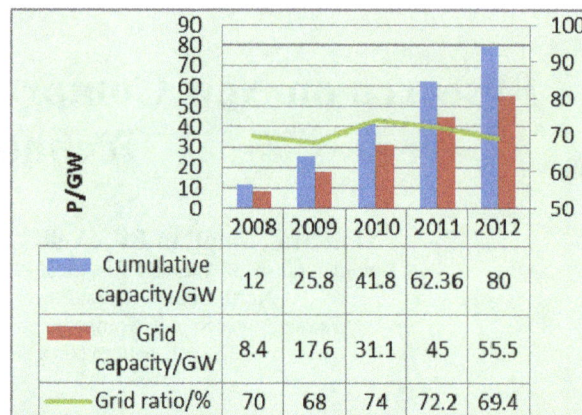

Figure 1. China's wind power development in the 2008-2012.

- The pumped storage power plant is flexible to start and stop and able to improve and stabilize the system voltage. But it requires special geographical conditions for the construction of reservoirs and dams, and the construction period is long, the initial investment is huge, and the large area of vegetation will be submerged and even cities, which will lead to ecological and immigration issues.

3. Compressed Air Energy Storage, CAES

Compressed air energy storage is second to pumped storage in the large-capacity storage technology. Although pumped storage technology has been developed widely, but because of its own constraints, there is a growing hope is attached to compressed air energy storage technology in large-capacity storage technology [7].

Compressed air energy storage technology made many breakthroughs in the decade's years from the traditional hot generation technology using gas turbine to the cool generation technology transferring the gas potential energy of the compressed air to other forms of energy, and the concept of compressed air energy storage is not limited to using gas turbine. The following content will analyze both characteristics and latest technical development of hot generation and cool generation of compressed air energy storage.

3.1. Hot Generation Compressed Air Energy Storage

Hot generation compressed air energy storage is referred to traditional technology, which is actually gas turbine power plant for peaking regulation. It uses power energy to press the air into the underground gas chamber and store the energy, and releases the high-pressure air that burning with combustible gas for generation. Compared with other energy storage technologies, hot generation

CAES has advantages such as large capacity, long working hours, good economic performance and long life of charge-discharge cycle and so on. But it has disadvantages, for example, hot generation CAES must still rely on burning fossil fuels to provide heat with the usage of gas turbine, which is unable to reduce carbon emissions and doesn't meet the requirements of the development of green renewable energy.

3.2. Cool Generation Compressed Air Energy Storage

The innovative concept of cool generation CAES was proposed in the past one or two years, which achieve the exchange between the gas potential energy and other forms of energy. Such a storage system does not need to burn the combustible gas, so it will reduce carbon emissions to achieve the goal of the green energy storage.

March 2012, CAES Technology Company Sustain X obtained the patent of constant-temperature CAES system. The patent technology is like: During the high-pressure air expansion, its temperature tends to fall according to the ideal gas law, and then the spray mechanism will release the spray at a suitable high temperature to make the temperature in the vessel remain at a constant level, so the spray transfers thermal energy to the gas. During air compression, its temperature tends to rise, and then the spray mechanism will release the spray at a suitable low temperature to make the temperature in the vessel remain at a constant level, so the gas will transfer thermal energy to the spray and the hot water can also be used for other forms of acting [8]. **Figure 2** is patent schematic diagram of Sustain X Company. Although in theory, the efficiency of the constant temperature CAES system is better than that of conventional systems, the director Mark Johnson of Advanced Research Projects Agency of the U.S. Department of Energy project thought that it may need five years or longer to prove its economy and find a wide range of applications.

In early 2013, Sustain X Company also announced a U.S. patent, which will collect waste heat for generation to save the leak energy that was taken away by steam. Dax Kepshire, Sustain X company vice president, said that if the system was equipped to a conventional power plant, which will act as a peak-regulation plant and will be cheaper than gas peak-regulation plant.

Sustain X Company uses ground air tank to replace the cave, which diminish the problem of geographical restrictions, and it uses the piston instead of the turbines for generation in order to reduce the size and cost of the gas tank.

In addition to Sustain X Company, Danielle A • Fong, the co-founder of Light Sail Energy Company, founded a compressed air energy storage technology, which is still based on CAES, but the difference is that the technique uses the piston divides the cylinders into two parts. The piston will move during the high-pressure gas expansion or the gas compression and drives piston rod, which will achieve the exchange between the gas potential energy and mechanical energy [9]. **Figure 3** is patent schematic diagram of Light Sail Energy Company. While the existing difficulty is that the temperature will reach $1000\,°C$ when the air is compressed, this means a large part of loss of energy by the way of heat. She invents a technique that the hot water is separated away in the process of the air compression compressed air and uses the thermal energy through the circulation loop to minimize the loss [10]. Danielle believes that her approach would cost only 1/10 compared with the same power of the battery storage.

Although China's CAES technology researches and projects are still focused to the use of the gas turbine power generation, the concept of combination between compressed air energy storage and pumped storage is proposed, and the patent of water-gas encompassing vessel energy storage system is one example. During air expansion, the compressed air stored in the vessel is not used to combust in a gas turbine, but rather the air push the water flows from the vessel to the low-pressure pool

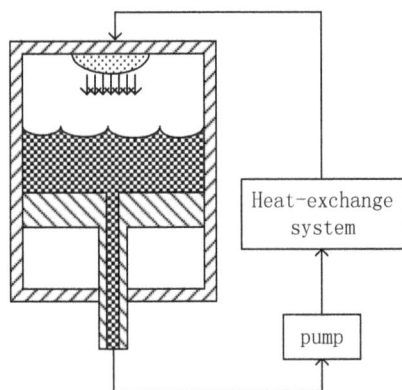

Figure 2. Diagram of thermostat compressed air system of Sustain X.

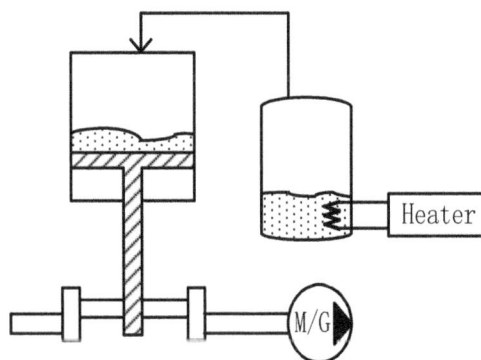

Figure 3. Diagram of Danielle A • Fong's patent technology.

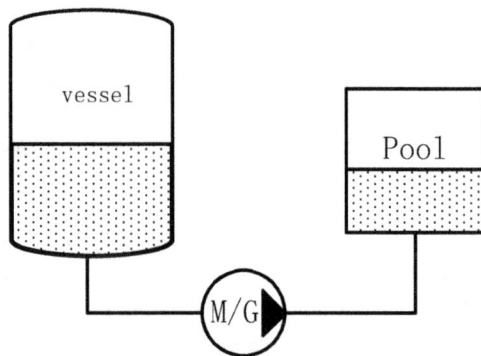

Figure 4. Diagram of patent technology of water-gas encompassing vessel energy storage system.

and drive the turbine into generation. During air compression, the air pump uses the power energy squeeze the air into the vessel, and the water pump extracts the water from low-pressure pool into the vessel [11]. **Figure 4** is patent schematic diagram of water-gas encompassing vessel energy storage system. The system uses the hydroturbine for generation, which will obtain higher efficiency of energy exchange than the traditional CAES technology. But the technique has shortcoming either, for example, during the water flow promoted by the high-pressure air, the gas volume increases, and the pressure gradually reduces, so the pressure forced on the hydroturbine is not constant, which will lead to low-efficiency operation of hydroturbine and off-generation beyond a certain pressure range.

The innovation of CAES technology has raised the possibility of a new green and low consumption energy storage technology; however we also need to constantly improve and perfect these technologies in order to achieve the goal of the green and economic energy storage.

4. Conclusions

The large-scale usage of renewable energy and grid operation needs the support of large-scale and distributed energy storage technology, and now the rapid develop-

ment of various energy storage technologies also make a new turn for the new energy grid operation. The cool generation compressed air energy system technology brings a possibility for green and economic energy storage technology. The future energy storage system needs a variety of storage patterns to match up with each other.

REFERENCES

[1] U.S., "Energy Information Administration," *Renewable Energy Consumption and Electricity, Preliminary Statistics*, 2010,Washington, DC: U.S., Energy Information Administration, 2011.

[2] G. Rachel, "2009 Renewable Energy Data Book,Washington," DC:US Department of Energy, 2010.

[3] C. S. Sun, Y. N. Wang and X. R. Li, "Synthesized Power and Frequency Control of Wind Power Generation," *Proceedings of the CSEE*, Vol. 28, No. 29, 2008, pp. 111-116.

[4] S. J. Cheng, J. Y. Wen and H. S. Sun, "The Storage Technology and Its Application in Modern Power System," *Electrotechnical Journal*, Vol. 24, No. 4, 2005, pp. 1-8.

[5] J. M. Geng, T. M. Li and W. Wang, "Research on the Characteristics of Energy Storage for Supercapacitor," *Low Voltage Apparatus*, Vol. 5, 2008, pp. 16-18.

[6] M. Ding, Y. Y. Zhang, M. Q. Mao, X. P. Liu and N .Z. Xu, "Economic Operation Optimization for Microgrids Including Na/S Battery Storage, *Proceedings of the CSEE*, Vol. 31, No. 4, 2011, pp. 7-14.

[7] B. Zhou, "Development of Large Compressed Air Storage Power Generation System," *China Electrical Equipment Industry*, Vol. 3, 2004, pp. 43-45.

[8] X. Sustain, Inc.High-Efficiency Liquid Heat Exchange in Compressed-gas Energy Storage Systems: US, Vol. 8, No. 171, 728B2[P/OL].2012-05-08[201-01-23].

[9] "Light Sail Energy," Inc., Compressed Gas Storage: US, 2011/ 0204064A1[P/OL].2011-08-25[2013-01-23].

[10] "Light Sail Energy," Inc., Compressed Air Energy Storage System Utilizing Two-phase Flow to Facilitate Exchange: US,2012/0269651A1[P/OL].2012-10-25[2013-01-23].

[11] H. R. Wang, "Water-gas Encompassing Electric Power," E*nergy Storage System*: CN, 102434362A [P/OL]. 2012-05-02[2013-01-23]

20

A Model for Regional Energy Utilization by Offline Heat Transport System and Distributed Energy Systems —Case Study in a Smart Community, Japan

Liyang Fan[1,2], Weijun Gao[2], Zhu Wang[1]

[1]Department of Civil Engineering and Architecture, Zhejiang University, Hangzhou, China
[2]Department of Architecture, The University of Kitakyushu, Kitakyushu, Japan
Email: happylamb68@hotmail.com

ABSTRACT

Under the Kyoto Protocol, Japan was supposed to reduce six percent of the green house gas (GHG) emission in 2012. However, until the year 2010, the statistics suggested that the GHG emission increased 4.2%. What is more challenge is, after Fukushima crisis, without the nuclear energy, Japan may produce about 15 percent more GHG emissions than 1990 in this fiscal year. It still has to struggle to meet the target set by Kyoto Protocol. The demonstration area of "smart community" suggests Japanese exploration for new low carbon strategies. The study proposed a demand side response energy system, a dynamic tree-like hierarchical model for smart community. The model not only conveyed the concept of smart grid, but also built up a smart heat energy supply chain by offline heat transport system. Further, this model promoted a collaborative energy utilization mode between the industrial sector and the civil sector. In addition, the research chose the smart community in Kitakyushu as case study and executed the model. The simulation and the analysis of the model not only evaluate the environmental effect of different technologies but also suggest that the smart community in Japan has the potential but not easy to achieve the target, cut down 50% of the CO_2 emission.

Keywords: Smart Community; Demand Side Response; Distributed Energy System; Reutilize Factory Exhaust Heat; Offline Heat Transport System

1. Introduction

Distributed energy systems (DES) have been drawing increasing attention as a substitute for grid in the low-carbon society development [1,2]. Compared with the traditional centralized energy supply system, the distributed energy generations are easy for renewable energy using and can avoid the loss in energy delivery as well. However, as the integration of distributed energy generation become major concerns, one problem occurred that the conventional energy supply model, the unidirectional top-down grid could hardly be multipurpose to it [3]. It can only support the energy flow from the energy station to static users. A much smarter energy supply system will be desirable to support multi-direction energy flows that can dynamically switch between the user and local energy providers. It needs for more observable, accessible, and controllable network infrastructures. The future energy system, termed as smart grid, is the system emerging as these requires. It can in-

telligently and automatically control and optimize operations.

The Japanese motivation toward "smart grid" can be suggested in its new energy strategies (decided on Dec. 30th, 2009). Four areas are conducted to be the demonstration trial sites, known as "smart community". In its concept, smart community is the basic unit in the smart evolution for the country [4]. The three main aspects of the smart community were distributed energy generation (DEG), distributed energy storage (DES), and demand side response (DSR) [5].

The DEG referred to the energy generation in the smart community, which is distributed into the power grid, including *PV* systems, micro-turbines and combined heat and power (*CHP*) plants. The prevailing of the distributed renewable resource was the focus for the Europe smart grid development. In some sense, the DEG can be the fundamental element that the energy generation and consumption can be carried out in an islanded manner. In another words, it is expected that the smart community only import a small amount of electricity from the outside, or even export their energy surplus to the neighborhood. Mathiesen *et al.* presented the analysis and results of a 100% renewable energy system by the year 2050 including transport. It revealed that 100% renewable energy systems would be technically possible in the future, and may be economically beneficial compared to the business-as-usual energy system [6].

The DES under the DSR control is a key underpinning of smart grid [5]. The intelligently controlled DES can serve to shift the electricity demand away from the peak periods, making the energy supply system more efficiently. Elma *et al.* developed model for stand-alone house that only supported by the renewable resource such as *PV* and wind system [7].

The energy system model for the smart community (grid) should convey the concept of DEG, DES and DSR. M. Welsch *et al.* demonstrated the flexibility and ease-of-use of open source energy modeling System with regard to modifications of its code. It may therefore serve as a useful test-bed for new functionality in tools with widespread use and larger applications, such as MESSAGE, TIMES, MARKAL, or LEAP [8]. B. B. Alagoz *et al.* draw a framework for the future digital power grid concept and assess its viability in relation to volatile, diverse generation and consumption possibilities [5].

It can be found that the models mentioned above are mostly focused on electricity supply chain, among which the heat supply system hardly be mentioned. The heat supply should also convey the concept of DEG, DES and DSR. It can make use of onsite exhaust heat, such as recovery heat of *CHP* plant and nearby factory exhaust heat (FEH) [9]. It should be a dynamic controllable as well, which can smooth out the heat fluctuation.

In this paper, the research introduced a smarter heat supply infrastructure into the smart community, paralleled with the electricity supply system. The proposed district energy system includes the use of diverse renewable and untapped energy resource, demand-responsive intelligent management, and efficient energy delivery. It not constructed an intelligent distributed electricity supply chain system with *PV* and *CHP* plants, but also promoted DEG, DES and DSR concept on heat supply chain system.

An intelligent heat supply chain system should have efficient heat storage, delivering system for the heat sharing between buildings. Not like the electricity, the heat delivering always faced to two mean problems. With the traditional pipe system, it will be limited by the delivering distance because of the high infrastructure fee and heat lost during the way. Further, the traditional pipe system can hardly use the temperature lower than 90°C. To come over these two problems, in this research, the model introduced the offline heat supply system (PCM)

to realize the DEG, DES and DSR concept of heat.

Offline heat transport system is a truck with a container that full of phase change material. It was firstly developed by the German National Aerospace Laboratory in 1980, and put into practical use in the year 2001 by a chemical company in Frankfurt [10]. There were many researches that use the PCM material as a heat storage component in the buildings or other area [11-13]. Japan introduced this technology in 2003, and creatively used it as heat supply system [10]. In Aomori, the PCM heat transport system was firstly put into trial in the year 2008. It collected the exhaust heat from a sewage factory, transported and supplied to a fishing center. Now it becomes a business for the SANKI Company, termed as "Heat home delivery". H. Kiyoto proposed the PCM system for exhaust thermal energy utilization [14]. As the smart system expect, this system can intelligently response to the demand side, switching between the energy provider and energy storage. It can collect the heat from the buildings where have surplus heat, store and then transport to the buildings where have heat demand.

This paper presented a model of a controllable, demand-responsive and balanced distributed energy systems network under the concept of smart community in Japan. Various technologies such as *PV*, *CHP* plant and PCM system were considered in the model. As a case study, the model was conducted into one demonstration smart community in Kitakyushu, Japan. Through the execution of the model, the research evaluated the environmental effect of every technology and estimated the potential of the smart community that whether the place can finally cut off 50% CO_2 emission as it set.

2. Concept, Definition and Modeling

2.1. The Demand-Response Network Model

The demand-response network (DRN) model for smart community is a tree-like hierarchical model that comprises the community energy management system (CEMS), energy station (ES) and building energy management system (BEMS).

Figure 1 demonstrated the hierarchical model. The end users (managed by BEMS) reside at the bottom of the hierarchy. They will be prioritized and organized into groups. Every group is managed by one ES. The ES is at the lowest rank unit for the energy strategies decision that controls the introduction of DEG, DES and DSR. The ES collect information of the energy generation and consumption in the group and send signals to the CEMS. The CEMS connected with each other and formed city energy net work, which organized in a topological structure. ES is assumed to have two modes, the energy surplus mode (SUR) and the energy insufficient mode (INS). The ES can dynamically switching its mode depending

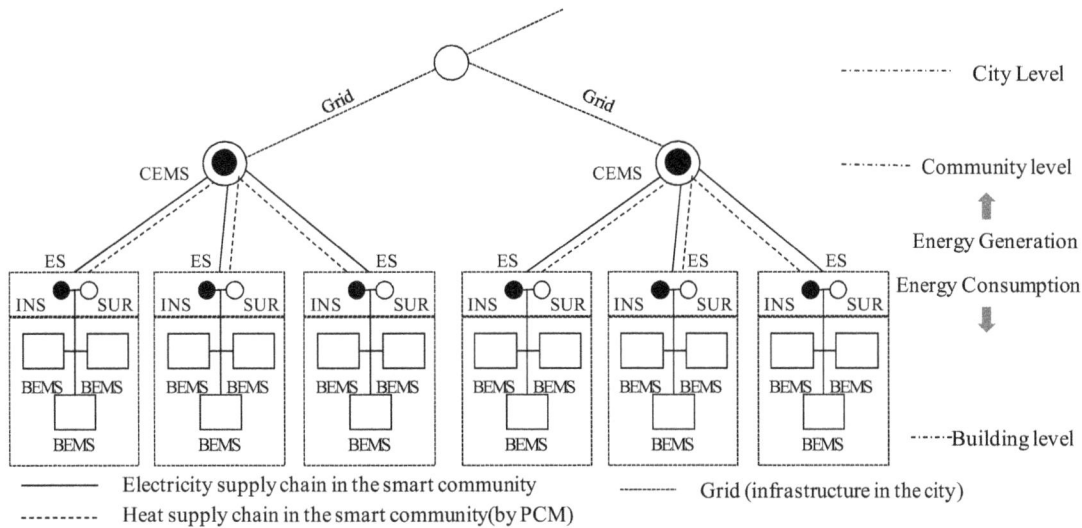

Figure 1. The tree-like hierarchy of DRS.

on the energy generation and consumption in the group. There mode signal will send to CEMS who collected and distribute the energy.

The proposed DRN system in this research is different from the smart grid that it not only has electricity supply chain but also has heat supply chain as well. **Figure 2(a)** illustrated the electricity supply chain. The DES in DRN system only acts as a back up and the buffer unit. The energy produced by DEG is supplied to the end users from the buffer unit. When the energy generated by DEG is more than the energy consumption in the group, the ES will in SUR mode and become an energy supplier to other ES. Oppositely, the ES will in INS mode when the energy generated by DEG is less than the energy consumption, and become an energy consumer. **Figure 2(a)** illustrated the heat supply chain. Similar with the electricity supply chain, it has a buffering unit that comprises the PCM tank and the heat exchanger. Under the INS mode, the CEMS will transport heat to the ES by trucks that with PCM tank. Under the SUR mode, the tank in the buffer unit will collect the surplus heat and be transported to other ES when it received the order from the CEMS. The mode signals in the heat supply and that in the electricity supply are self-governed.

2.2. District Energy Using Concept and Operation Hypothesis

As the DRN system illustrated before, the building in the community will be divided into Groups. Every group is managed by ES, the basic unit to make energy strategies. **Figure 3** described the district energy using concept.

- Introduction of the renewable energy: all the buildings will be introduced with *PV* system. The electricity generated by *PV* system will be preferentially used by the building themselves and the left electricity will

be sent back to the grid.

- Introduction of the *CHP* system: The CEMS will characterize buildings by their demand types. The buildings have both high electricity consumption and heat consumption (such like commercial buildings and public buildings) will be introduced with the *CHP* system, named as *CHP* group. The capacity of the *CHP* system is set as electricity peak load of the group. The buildings without *CHP* system is considered as Non-*CHP* (NCHP) group. The electricity produced by *CHP* plant will satisfy them first and then send the left electricity to NCHP system. The *CHP* group will generate all their own demand beside *PV*. Therefore, as the DRN described before, the electricity of these groups are only in SUR mode. The NCHP group will be in INS mode if *PV* cannot afford their electricity consumption.

The CEMS will manage the model signal, control and dispatch the electricity. It will preferentially use the DEG, thus maximum the output of *CHP* plant. The electricity produced by *PV* can sell back to grid but the electricity produced by *CHP* plant cannot. In that case, when the electricity generated by *CHP* more than the district electricity demand, the CEMS will restrict the *CHP* output. It will preferentially chose the *CHP* plant with higher efficiency and lower down the *CHP* plant with low efficiency. If the efficiencies of the *CHP* plants are the same, CEMS will cut down the *CHP* plants in same rate.

- Reutilization of the onsite exhausts heat and the FEH: the recovery heat of the *CHP* system will be preferentially used by the group first. However, if the recovery heat is more than the heat demand, the heat supply mode of the ES will turn to SUR. This part of heat surplus will be collected by PCM.

Further, the CEMS will select out the FEH resource

(a)

(b)

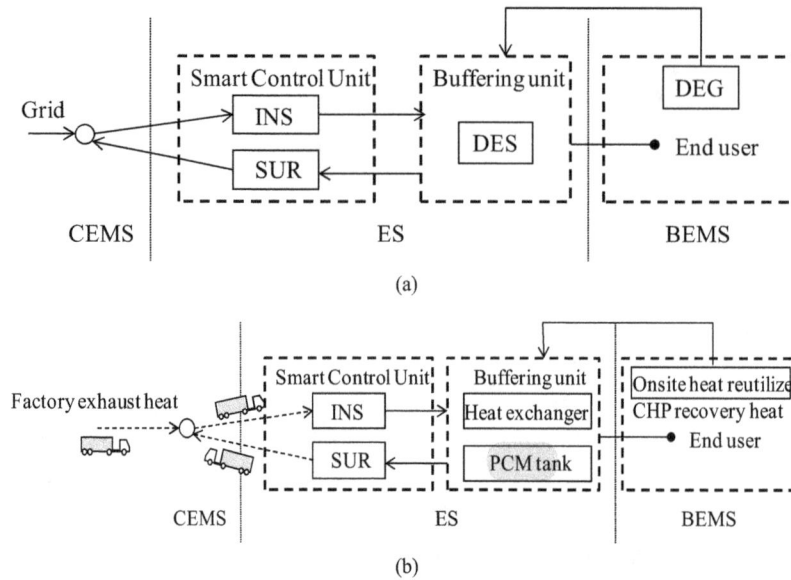

Figure 2. DRN system energy supply chain; (a) Electricity supply chain; (b) Heat supply chain.

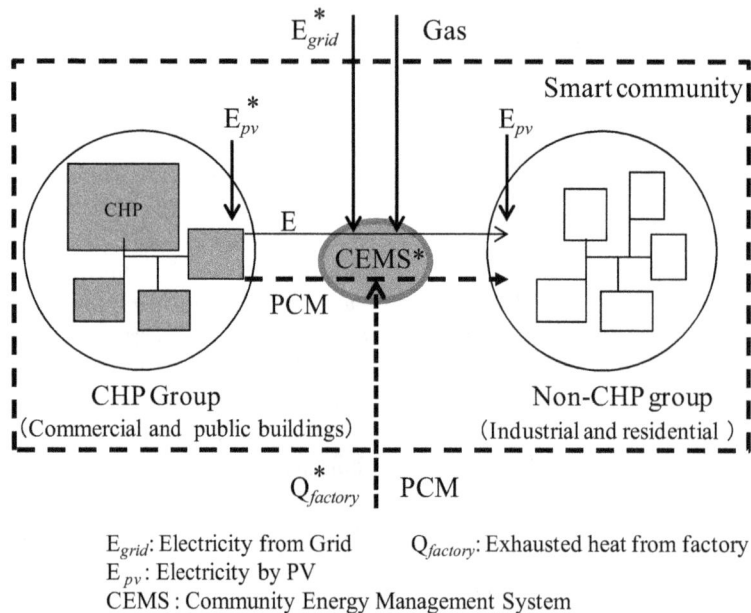

E_{grid}: Electricity from Grid $Q_{factory}$: Exhausted heat from factory
E_{pv}: Electricity by PV
CEMS : Community Energy Management System

Figure 3. District energy using concept.

based on the characteristic of the PCM system, which collects the FEH and utilizes it in the community.

The onsite *CHP* exhaust heat and the FEH stored in the PCM system will be preferentially used. The CEMS will distribute the heat according to the SUR signal from the ES. It will be sent to the ES which have the higher heat insufficient amount.

2.3. Energy Balance Management and Simulation Modeling

The energy balance management and the simulation flow

are conducted as **Figure 4**. The simulation of the DRN system is also a bottom-up model. Firstly, based on the district zoning, the research will estimate the building energy consumption and described profiles by groups. As the tree-like hierarchy described in the second part, buildings in one group will be managed by one ES. Secondly, the CEMS will characterize the groups by its energy consumption character and introduce proper DEG in every ES, some are with *CHP* system but some are not. The simulation separated them into *CHP* group and the NCHP group. Thirdly, the research executed the simulation.

During the simulation, ES will dynamically switch between the INS mode and the SUR mode by estimating the energy consumption and the generation. Finally, the research will calculate the primary energy consumption and evaluate the environmental effect of the every technology as well as the whole community.

1) Estimation of district energy consumption

The energy consumption of the whole community $\left(E_{\text{demand}}^{\text{community}}\right)$ is calculated as Equation (1)

$$E_{\text{demand}}^{\text{community}} = E^1 + E^2 + \cdots + E^n$$
$$= \sum_n \sum_m \sum_d \sum_i \left(ELEC_{mdh}^n + HEAT_{mdh}^n \right) \quad (1)$$

$ELEC_{mdh}^n$ is hourly electricity load, calculated as Equation (2)

$$ELEC_{mdh}^n = \sum_k e_{mdh}^n \times s_k \quad (2)$$

$HEAT_{mdh}^n$ is hourly heat load, calculated as Equation (3)

$$HEAT_{mdh}^n = \sum_k h_{mdh}^n \times s_k \quad (3)$$

n is the group number;

m is month; d is date, h is hour;

$E^1 \cdots E^n$ is the energy consumption of every group;

e_{mdh}^n and h_{mdh}^h is the energy consumption unit in

Kyushu area, Japan [15];

k is the building function;

s_k is the building area for one function (k).

2) The electricity balanced management

Figure 5 illustrated the simulation model for the electricity balance. The buildings will preferentially use the electricity produced by PV. The electricity produced by PV system in one group $\left(PV_{mdh}^n\right)$ is calculated as Equation (4):

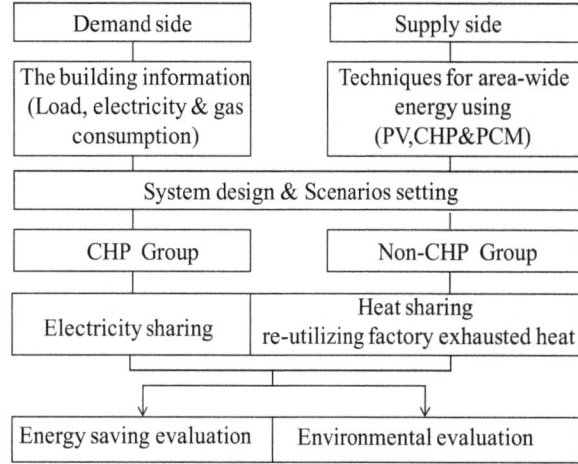

Figure 4. The simulation flow.

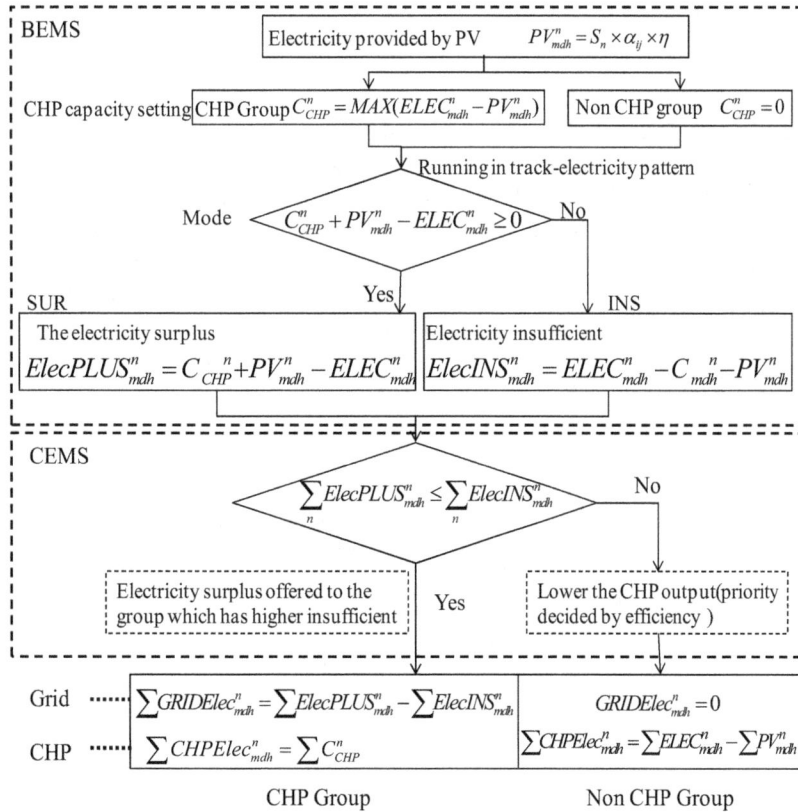

Figure 5. The electricity balanced model.

$$PV_{mdh}^n = S_n \times \alpha_{mdh} \times \eta \qquad (4)$$

s_n is the area for PV penal in a group (n);

α_{mdh} is the hourly sun radiation rate [16];

η is the efficiency of the PV penal [16].

The CHP capacity $\left(C_{CHP}^n \right)$ is decided as Equation (5):

$$C_{CHP}^n = \begin{cases} \text{MAX}\left(ELEC_{mdh}^n - PV_{mdh}^n \right)(CHP\text{group}) \\ 0 \qquad\qquad\qquad\qquad (NCHP\text{group}) \end{cases} \qquad (5)$$

The ES will decide the mode by the prediction of electricity load profile of the CHP system, PV system and electricity demand.

When $C_{CHP}^n + PV_{mdh}^n - E_{mdh}^n \geq 0$, the group is in SUR mode. The expected surplus electricity $\left(ElecPLUS_{mdh}^n \right)$ is calculated as Equation (6)

$$ElecPLUS_{mdh}^n = C_{mdh}^n + PV_{mdh}^n - E_{mdh}^n \qquad (6)$$

On the contrary, when the group is in INS model, the expected electricity insufficiency $\left(ElecINS_{mdh}^n \right)$ is calculated as Equation (7)

$$ElecINS_{mdh}^n = E_{mdh}^n - PV_{mdh}^n - C_{mdh}^n \qquad (7)$$

If $\sum^n ElecPLUS_{mdh}^n \leq \sum^n ElecINS_{mdh}^n$, CEMS would lower down the total CHP output (prior use the equipment with higher efficiency). Under this situation, there was no electricity supplement from the grid. The electricity generated by CHP plant $\left(CHPElec_{mdh}^n \right)$ is calculated as Equation (8):

$$\sum CHPElec_{mdh}^n = \sum ELEC_{mdh}^n - \sum PV_{mdh}^n \qquad (8)$$

If $\sum^n ElecPLUS_{mdh}^n > \sum^n ElecINS_{mdh}^n$, the surplus electricity from CHP group will be offered to the NCHP group. Under this situation, the electricity from the grid $\left(GRIDElec_{mdh}^n \right)$ is calculated as Equation (9):

$$\sum GRIDElec_{mdh}^n = \sum ElecPLUS_{mdh}^n - \sum ElecINS_{mdh}^n \qquad (9)$$

Electricity offered by CHP is calculated as Equation (10):

$$\sum CHPElec_{mdh}^n = \sum C_{CHP}^n \qquad (10)$$

3) The design and modeling for the PCM system

• PCM for collecting the FEH

According to the system parameter, economically the system can utilize heat with in 135 km and 20 km round trip [14]. The CEMS will economically select out the possible utilized rescource, and make a plan for the PCM system. The collecting schedule of the PCM trucks should match with the factory working hour. It will become more complecate as the factory heat resource increase. Considering the various factors for making the plan, the research assumed that the FEH collected would be transported to the demand side and used in the following day.

The number of the tanks for collecting FEH used in one day (x) is desided by the capacity (listed in **Table 1** [17]). It should satisfy equation (11):

$$Q_{pcm} \cdot x \geq \sum Q_{fac}, x \in (1,p) \qquad (11)$$

Q_{pcm} is the capacity for the PCM tank;

Q_{fac} is the daily factory exhaust of the selected resources;

The exhaust heat that can be used in the demand side is limited by the energy lost during the heat storage, transport and heat exchange. CEMS will estimate it and select out the proper resource. The amount of the heat $\left(HEAT_{recFAC} \right)$ that can use in the demand side is as Equation (12)

$$HEAT_{recFAC} = \mu \cdot Q_{fac} \qquad (12)$$

μ is the overall efficiency of the PCM system, set as 0.9 in this research [16].

• PCM for the heat delivery between the groups

ES will use the estimated consumption pattern for the consistent prediction and send the mode signal to CEMS.

For every group, CHP recovery heat $\left(CHPREC_{mdh}^n \right)$ is as Equation (13):

$$CHPREC_{mdh}^n = \begin{cases} \eta_h / \eta_e \cdot ElecCHP_{mdh}^n \ (CHP\text{group}) \\ 0 \qquad\qquad\qquad\quad (NCHP\text{group}) \end{cases} \qquad (13)$$

Table 1. The type and parameters of the PCM tank.

Type	Melting Point/°C	Heat temperature/°C	Tank capacity/MWh	Usage Hot Water	Usage Heating	Usage Cooling
Type 1	58	85(70)	0.8 ~ 1.1	○	○	—
Type 2	78	100(90)	—※	○	○	—
Type 3	116	150(130)	—※	○	○	○
Type 4	118	150(130)	1.1 ~ 1.4	○	○	○

※Types 2 and 3 are used outside Japanese; ○The function it has; —The function it doesn't has.

η_e is the electricity generating efficiency of *CHP* plant;

η_h is the heat recovery efficiency of *CHP* plant;

If $CHPREC_{mdh}^n - HEAT_{mdh}^n \geq 0$, the ES is in SUR mode and the expected value of heat surplus $\left(\text{Heat}_0 SUR_{mdh}^n \right)$ is as Equation (14):

$$\text{Heat}_0 SUR_{mdh}^n = CHPREC_{mdh}^n - HEAT_{mdh}^n \tag{14}$$

If $CHPREC_{mdh}^n - HEAT_{mdh}^n < 0$, the ES is in INS mode and the expected value heat insufficient $\left(\text{Heat}_0 INS_{mdh}^n \right)$ is as Equation (15):

$$\text{Heat}_0 INS_{mdh}^n = HEAT_{mdh}^n - CHPREC_{mdh}^n \tag{15}$$

Every day, the PCM system will carry the FEH and input into the community from the first peak time in the morning, set as h_0. During the day, the system will preferentially use the heat stored in the PCM and release it before the next day. Therefore, every day at the time h_0, the heat amount stored in the PCM system is reset.

The amount of stored heat energy in the PCM that can be supplied to the ES in SUR mode at h time in one group $\left(PCMREC_{mdh}^n \right)$ is as Equation (16)

$$\sum^n PCMREC_{mdh}^n = \begin{cases} \sum^n PCMREC_{md(h-1)}^n + \sum^n \text{Heat}_0 SUR_{mdh}^n - \sum^n \text{Heat}_0 INS_{mdh}^n & \left(h \neq h_0 \right) \\ HEAT_{recFAC} & \left(h = h_0 \right) \end{cases} \tag{16}$$

The total amount of PCM truck (p) should satisfy Equation (17)

$$\text{MAX}\left(\sum^n PCMREC_{mdh}^n \right) \leq Q_{pcm} \cdot p \tag{17}$$

$\text{MAX}(\cdot)$ is an function to determine the maximum value of the stored heat in PCM system by the expected value.

4) The heat balanced management

Figure 6 illustrated the heat balanced management.

The collected heat in the PCM system including the recovery heat of *CHP* system and FEH are used for heating, cooling and hot water in the community. It is also managed by CEMS following total quantity priority that supplied to the group, which had lager amount of heat insufficient, $\text{MAX}\left(\text{Heat}_0 INS_{mdh}^n \right)$.

With the use of the waste heat that collected by the PCM system, the heat insufficient $\left(\text{Heat}_R INS_{mdh}^n \right)$ is as Equation (18):

$$\sum_n \text{Heat}_R INS_{mdh}^n = \sum^n \left(HEAT_{mdh}^n - CHPREC_{mdh}^n - PCMREC_{md(h-1)}^n \right) \tag{18}$$

When $\text{Heat}_R INS_{mdh}^n \leq 0$, the heat demand can be satisfied with the onsite exhaust heat reutilization that the heat-source equipment $\left(AUSHEAT_{mdh}^n \right)$ is not required as Equation (19):

$$AUSHEAT_{mdh}^n = 0 \tag{19}$$

When $\text{Heat}_R INS_{mdh}^n > 0$, the heat-source equipment is used as supplement. The heat offered by the heat-source equipment is as Equation (20):

$$AUSHEAT_{mdh}^n = \text{Heat}_R INS_{mdh}^n / \eta^n \tag{20}$$

η^n is the efficiency of heat source equipment.

2.4. Assessment Index Setting

1) Energy saving ratio

ESR is energy saving ratio, defined as Equation (21):

$$ESR = \left(Q_{input}^{Conv} - Q_{input}^{CHP} \right) / Q_{input}^{Conv} \tag{21}$$

For *CHP* system, the primary energy input is as Equation (22):

$$Q_{input}^{CHP} = E_{Utility}^{CHP} \times \varepsilon_{Grid} + \left(V^{CHP} + V^{Boiler} \right) \times \varepsilon_{gas} \tag{22}$$

$E_{Utility}^{CHP}$ is the electricity input in *CHP* system; V^{CHP}, V^{Boiler} is the gas input to the *CHP* plant and boiler.

For conventional system, the primary energy input is as Equation (23):

$$Q_{input}^{Conv} = E_{Utility}^{Conv} \times \varepsilon_{Grid} + V^{Conv} \times \varepsilon_{gas} \tag{23}$$

$E_{Utility}^{Conv}$ is the electricity input in conventional system; V^{Conv} is the gas input to conventional system for hot water;

ε_{Grid} is primary energy consumption unit of grid in Japan (11.4 MJ/kWh); ε_{gas} is the primary energy consumption unit of gas (45 MJ).

2) CO_2 reduction ratio

$\eta_{\Delta CO_2}$ is CO_2 reduction ratio, defined as Equation (24):

$$\eta_{\Delta CO_2} = \left(EX_{CO_2}^{Conv} - EX_{CO_2}^{CHP} \right) / EX_{CO_2}^{Conv} \tag{24}$$

$EX_{CO_2}^{CHP}$ is CO_2 emission for *CHP* system, calculated as Equation (25);

$$EX_{CO_2}^{CHP} = ex_{CO_2}^{gas} \times \left(V^{CHP} + V^{Boiler} \right) \times \varepsilon_{gas} + ex_{CO_2}^{Pow} \times E_{Utility}^{CHP} \times \varepsilon_{Grid} \tag{25}$$

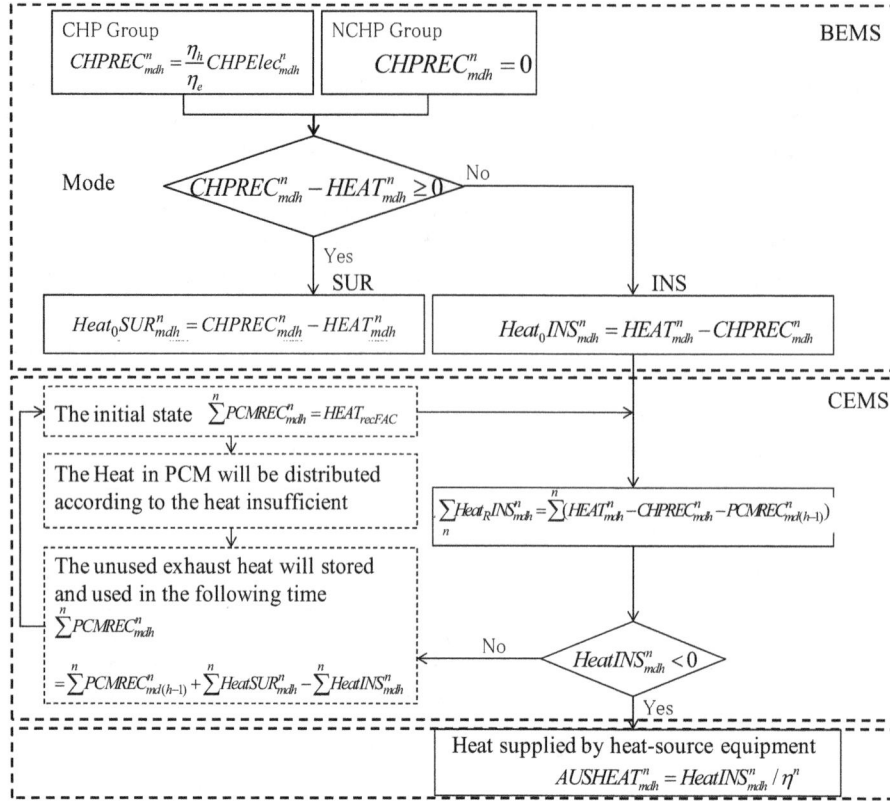

Figure 6. Heat supply calculation flow.

$EX_{CO_2}^{Conv}$ is CO_2 emission for conventional system, calculated as Equation (26).

$$EX_{CO_2}^{conv} = ex_{CO_2}^{gas} \times V^{conv} \times \varepsilon_{gas} + ex_{CO_2}^{Pow} \times E_{Utility}^{Conv} \times \varepsilon_{Grid} \quad (26)$$

$ex_{CO_2}^{gas}$ is the CO_2 emission unit for gas in Japan (13.8 g-C/MJ);

$ex_{CO_2}^{Grid}$ is the CO_2 emission unit for grid in Japan (153 g-C/kwh).

3. Numerical Study

3.1. Research Site

Kitakyushu lied in the northern part of Kyushu, the westernmost of the four main islands in the Japanese archipelago. It used to be one of Japan's four leading Industrial regions and contributed greatly to the rapid economic growth of Japan.

The smart community creation project is newly launched in Yahata Higashisa district, where used to be the factory district of the steel company. The government invested 16.3 billion yen over the five-year period from 2010 to 2014. It has already cut 30% of the CO_2 emission compared with the other place in the city. However, the target for the smart community was to cut 50% of the existing emission, still 20% need to get [4].

3.2. Energy Load

Detailed knowledge about energy end-use loads is important for the energy system design and optimization. In this study, the hourly load demand for electricity, cooling, heating and hot water have been calculated according to the energy consumption unit (the system in Japan that displays energy consumption intensities) of various buildings in Kyushu, Japan [15]. As the method descried in part 2, the whole community is divided into 4 groups. **Figure 7** displayed the image of community and district zoning.

Figure 8 described the detailed hourly load profiles for every group in summer (Aug.), winter (Jan.), spring and autumn time (May). The energy consumption profile firmly related with the building function.

1) The group4 is the residential area, thus the peak of the energy consumption comes during the night. The group1 also has considerably higher energy consumption compared with groups2 and 3, because there is a hotel in the group.

2) The commercial group (group2) has a higher energy consumption during the day, but almost no energy consumption during the night.

3) The hot water load is higher in residential group (group4), but lower in groups2 and 3.

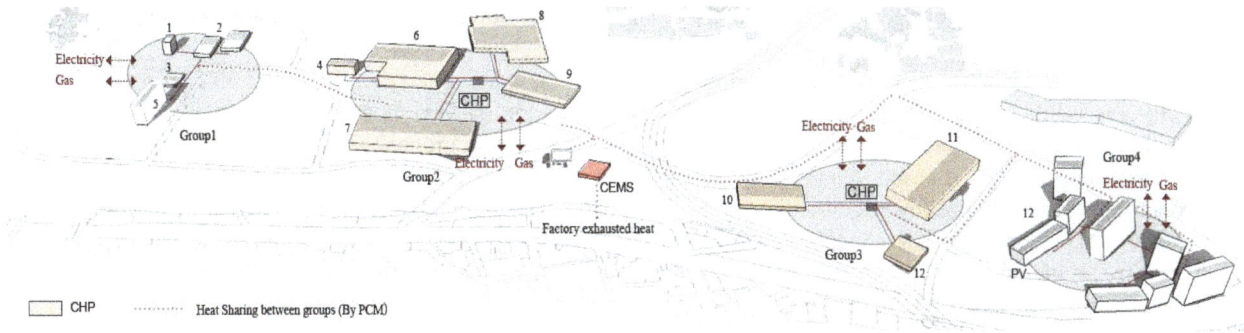

Figure 7. The community model and energy system.

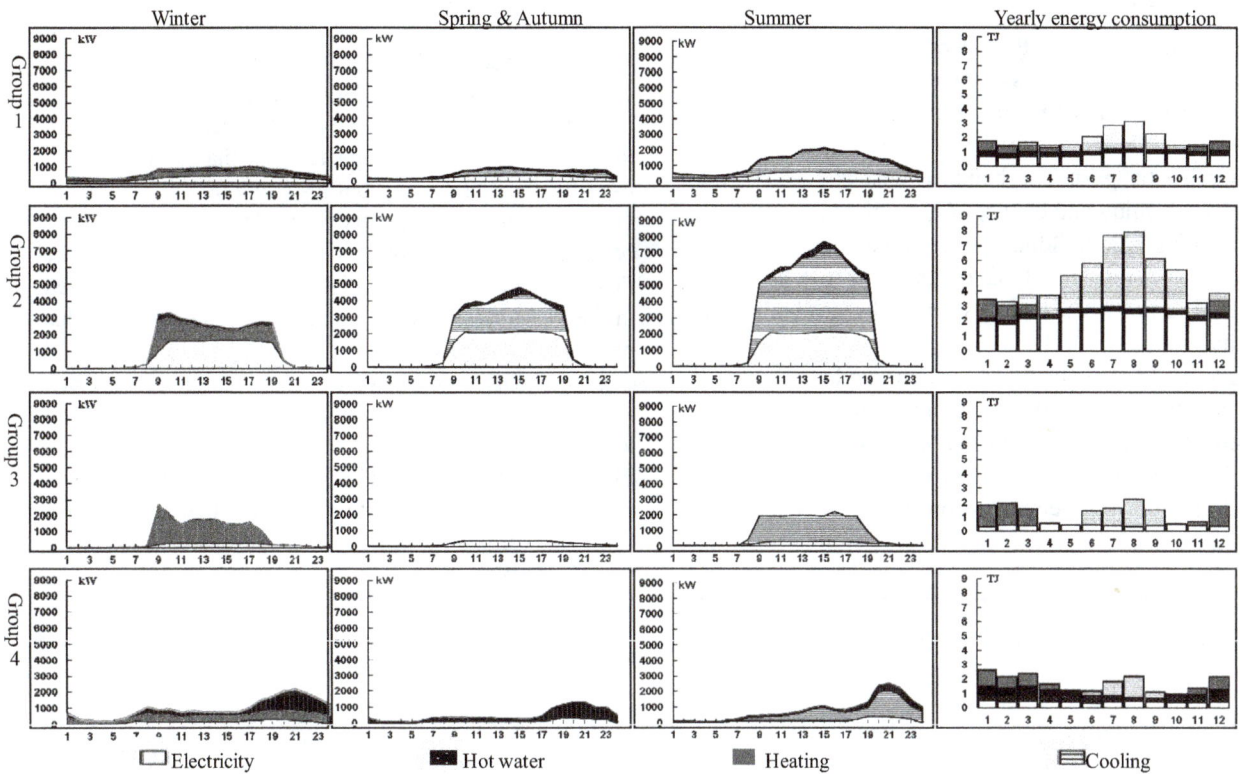

Figure 8. The district energy consumption.

3.3. FEH Load

It is reported that the factory exhaust heat in Japan can satisfy the heat consumption of all the residential buildings for five years [17]. In that case, there is a great potential to make use of the factory exhaust heat. It can cut the energy consumption on the civil side as well as the CO_2 emission on the factory side.

Another important input to the energy system is the reutilization of the FEH. It collected by PCM system and used in the community for heating, cooling and hot water. The study adopted the database of the FEH based on GIS built in the research before and selected out the four potential factory resources (within 20 km) [18]. Usually, the temperature for FEH is higher than 300°C and daily

exhaust heat is around 38.9 GJ. The tank type with the capacity of 1.4 MWh will be introduced into the system. As this research only discussed the environmental effect of the PCM system, thus it is supposed that there are enough tanks for collecting all the exhaust heat (the heat of the factory and the unused *CHP* recovery heat).

3.4. DEG Technologies and District Energy System

This district is the demonstration area that the latest technologies are expected to be introduced into the area. The smart community is also undertaken the Kitakyushu Hydrogen Town project. The project marks as the world-first attempt to use a pipeline recycling the hydrogen generated in the iron manufacturing and operating the fuel cells as an energy supply to the district. The demonstration testing is processed jointly by Fukuoka Prefectural and city gas utilities [19]. The pipeline connected with the hydrogen station and hydrogen fuel cells that installed in buildings in this district. These fuel cells generate electricity by combining hydrogen and oxygen. **Table 2** showed DEG technologies assumed in this study and their properties, including gas engine (GE), fuel cell (FC), Hydrogen fuel cell (HFC) and *PV*. All equipments are city gas fired.

3.5. Setting of Cases

In order to investigate the effect of technologies in the DRN, the following cases are assumed for analysis.

Base case: conventional system. Base case indicated conventional energy supply system. The electricity load is satisfied by grid. The buildings also used air conditioner for heating and cooling. The commercial buildings, office and public buildings use multiple air-conditioning systems and residential buildings use room air-conditioner. The hot water load is satisfied by gas boiler fired by city gas.

Case 1: The conventional system combined with *PV* systems. In this case, the community still keeps the conventional system, but facilitated with *PV*. The electricity will be supplied by *PV* system, or by grid, or by combined of both. The electricity from the *PV* system will be

used by the buildings themselves, and left electricity will be send back to the grid.

Case 2: Individually introduced DEG systems, displayed in **Figure 9(a)**. In case 2, the *CHP* plants with GE are introduced in groups2 and 3. The *CHP* plants and *PV* systems can satisfy the electricity load of these two groups. The thermal load can also be supplied by the recovery heat of the *CHP* plants and the deficiency supplemented by gas boiler. In this case, the electricity and recovery heat of the *CHP* plants cannot supply to other groups or return back to the grid. Therefore, the NCHP groups still get electricity from the *PV* and grid, keep as the conventional system.

Case 3: DRN system without using factory exhaust heat, described in **Figure 9(b)**. In the DRN system, the community uses the same DEG technologies with case 2, but controlled and managed by CEMS. Under the CEMS, the electricity produced by the *CHP* plants not only be used for the *CHP* group but also supplied to the NCHP group as well. The recovery heat of the *CHP* group will be used in the *CHP* group first and then recycled by the PCM system. The CEMS distributed the heat that stored in the PCM system with thermal insufficient and surplus profile of every group.

Case 4: DRN system with the utilization of the FEH, as **Figure 9(b)**. Besides the technologies and DRN system that assumed in cases 3 and 4 also make use of the FEH by PCM system. The PCM system collected the exhaust heat from the factory resource that set in part 3

Table 2. Technical parameters of system.

Facility		Parameter	COP
Grid		η	0.35
Gas engine (GE)	Electricity generation	η_e	0.3
	Heat recovery	η_{rec}	0.45
Fuel cell (FC)	Electricity generation	η_e	0.4
	Heat recovery	η_{rec}	0.3
Hydrogen fuel cell	Electricity generation	η_e	0.48
	Heat recovery	η_{rec}	0.42
Boiler		η_b	0.8
Absorption chiller		COP_{ac}	1.1
Heat exchanger (H-EX)		COP_{he}	1
Multiple air-conditioning system	Cooling	COP_1	4
	Heating	COP_2	3.9
Room air conditioner	Cooling	COP_1	3.22
	Heating	COP_2	2.83
PCM system		η_{rec2}	0.9

(a)

(b)

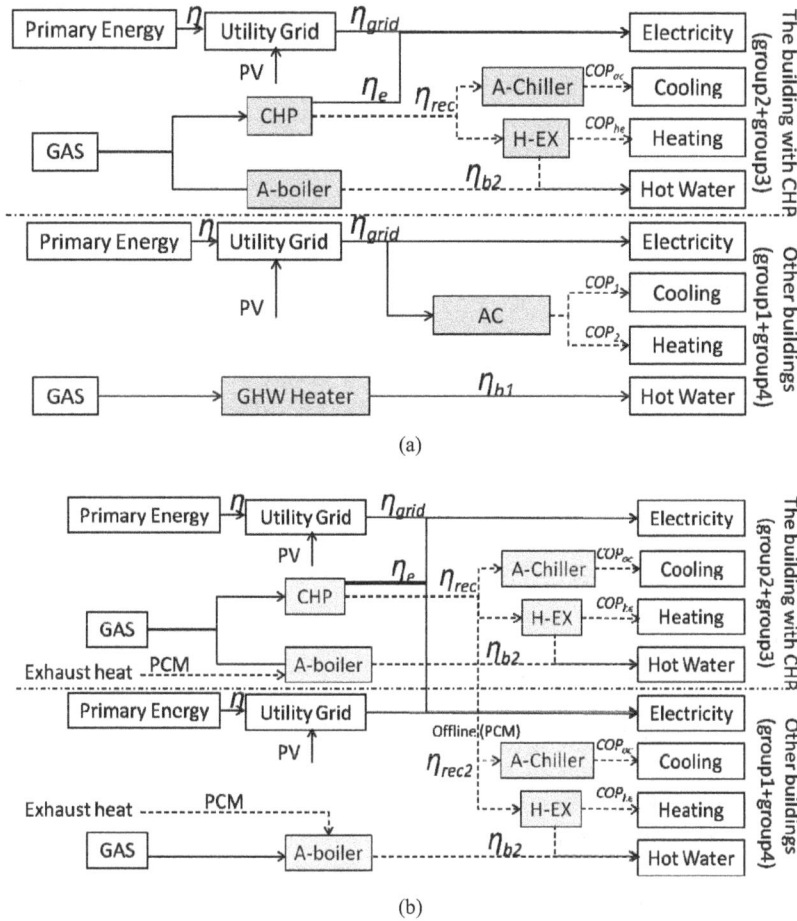

Figure 9. Case setting.

and transport it to the CEMS in the community. Besides the surplus *CHP* recovery heat, this part of heat will also be distributed by CEMS.

Cases 5 and 6: the DRN system with the *CHP* plants of FC and HFC. Beyond the DRN systems that build in cases 4 and 5 introduced the *CHP* plant of FC and case 6 introduced HFC.

4. Results and Discussions

4.1. The Effect of Electricity Sharing in DRN System

Figure 10 is the electricity balance in the community with the individually introduced DEG systems (case 2) and the DRN system (case 3). Both of the cases use *CHP* plant with GE and *PV*. The comparison between the two cases can show the effect of the electricity sharing between them. It can suggest that *PV* system can provide 35% of the community electricity consumption and the individual *CHP* plant can produce 41% electricity. By electricity sharing, the *CHP* group can offer 2 GWh electricity to the Non *CHP* Group, which occupied 52% of their electricity consumption. As a whole, the community

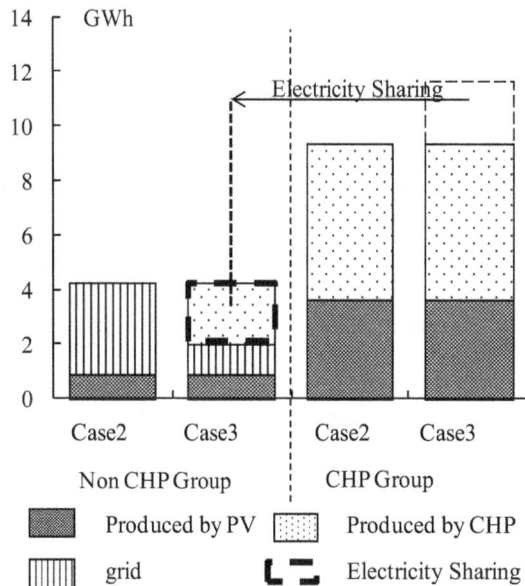

Figure 10. Yearly electricity balances of cases 2 and 3.

can produce 58% of the electricity by *CHP*, and only 7% from the grid, while the individual system need 24%.

As we know, the electricity produced by DEG has less energy loss during the electricity delivery. Therefore, the system can save more energy as it gets less electricity from the grid. In the DRN system, the CEMS can operate the *CHP* plants and distribute the electricity to the whole community. Therefore, it will increase the output and working hours of the *CHP* plant reduce the electricity from the grid.

The electricity sharing used in DRN system can shift the electricity demand from the peak. Just as **Figure 11** suggested, without *CHP* plant, the peak hour should come during the noontime, but now it shift to 8 o'clock in the morning and 18 o'clock in the afternoon. Further, from the city level, the less relay on the grid will alleviate electricity shortage especially during the peak hours. That means with the DRN, the city can smooth out the electricity flucuate.

4.2. The Effect of Heat Sharing in DRN System

The DEG with *CHP* plants not only reduce the energy loss, but can make use of the recovery heat as well. In case 2, the individually *CHP* system can only use the recovery heat by the *CHP* group itself. However, under the CEMS, in case 3, the DRN system can distribute the recovery heat to other group with the PCM system. In that case, it improved the utilization rate of the recovery heat. As **Figure 12** illustrated, the individual *CHP* has 37.9 GWh recovery heat every year and 31.1 GWh is used for thermal consumption in *CHP* group. In DRN system, the yearly *CHP* recovery heat is 47.3 GWh, among which 6.4 GWh heat is offered to the NCHP group. This part of heat occupied 33.8% of heat consumption in NCHP. Under this condition, 85% of the *CHP* recovery heat can

be reused which possessed 68.8% of the community heat demand.

Figure 13 illustrates the daily heat balance in the community, taking the wintertime as example. The plus value means the heat surplus of each group. Groups2 and 3 are the *CHP* groups and their heat surplus means the left heat after their own utilization. PCM system can collect this part of heat and used for heat supply in other groups. The minus part means the heat insufficient. For Groups2 and 3, it means the heat deficient after utilizing the *CHP* recovery heat. **Figure 13** can suggest that the first peak of the heat insufficient comes on 9 o'clock in the morning and the peak of the heat surplus comes on 19 o'clock. Groups2 and 3 have no heat demand from 19 o'clock to the next 9 o'clock, thus during this time all the *CHP* recovery heat will be supplied to NCHP group. From the 9 o'clock to 19 o'clock, group3 has the largest heat insufficient, thus the stored heat in the PCM system will be preferentially supplied to group3. That means the heat sharing not only between the *CHP* group and NCHP group, but also between the *CHP* groups. After the CEMS collects the heat and stores in the PCM system, it only distributes the heat according to the heat insufficient volume.

4.3. The Effect of Using Factory Exhausted Heat

Until now, the city of Kitakyushu still has 1412 factories and industries, which have exhaust heat. The existing research put forwards questionnaire to all the factories, estimating and setting up a database by GIS for the yearly exhaust heat. As the result, the yearly exhausted thermal energy is about 18,000 TJ.

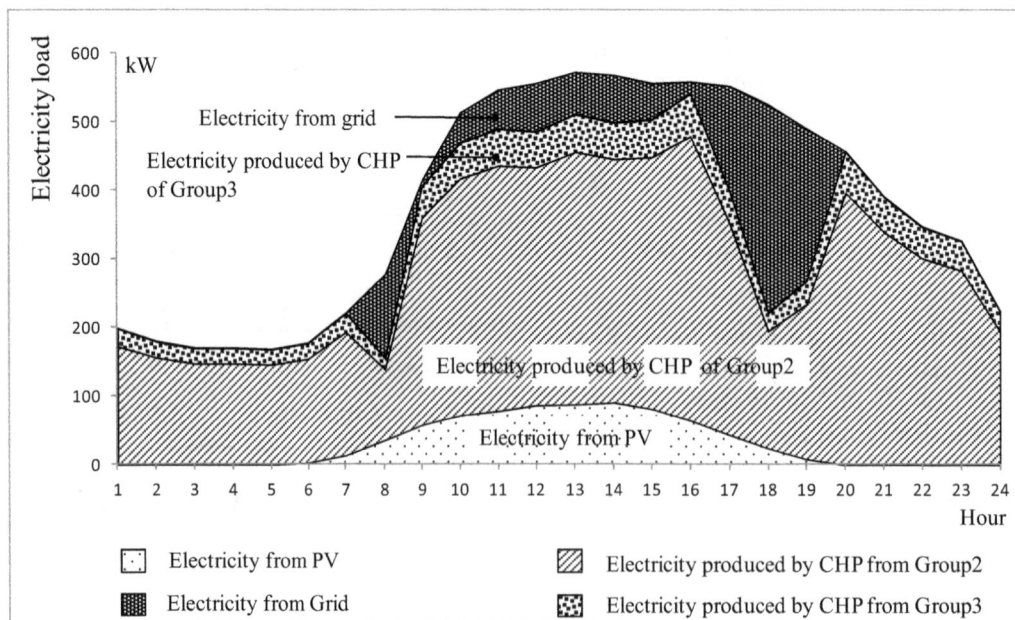

Figure. 11. Daily electricity balance of group1 (summer).

For this community, four factories were set as the resources and the total yearly heat amount that can offer to the community was 14.2 TJ (38.9 GJ per day) [18].

In this research, it is set that exhaust heat will be averagely supplied to the community from the first peak hour in the set time range. **Figure 14** is the relationship between time range and the heat volume, as well as the energy saving result. It can suggest that in this case, 6 hours is the optimal time range and it can cut 42.8% of primary energy beyond the *PV* and *CHP* system.

4.4. The Effect of Introduction of Different *CHP* Plant

As the techniques of *CHP* plant improved, the environmental performance of the system changed as well. The gas engine and the fuel cell have already been widely used in Japan. As a trial project, the community introduced hydrogen fuel cell. **Figure 15** is the energy saving ratio of these three kinds of *CHP* plant. The fuel cell and gas engine had similar effect when the capacity is low, but after 1000 kW, the fuel cell improved obviously. The hydrogen fuel cell had a higher efficiency on both electricity generating (48%) and heat recovery (42%), thus the system can reach an optimal energy saving ratio around 53%.

4.5. The Assessment of Environmental Effect from the Community Side

Figure 16 is the low carbon ratio for every technology. By introduced the *PV* system and the *CHP* plant (gas engine), it can cut off 29.4% of the carbon emission. The networking *CHP* system can reduce energy consumption

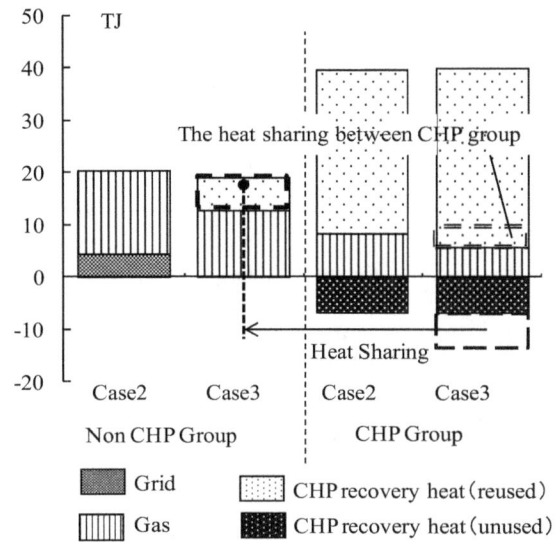

Figure 12. Yearly heat balances of cases 2 and 3.

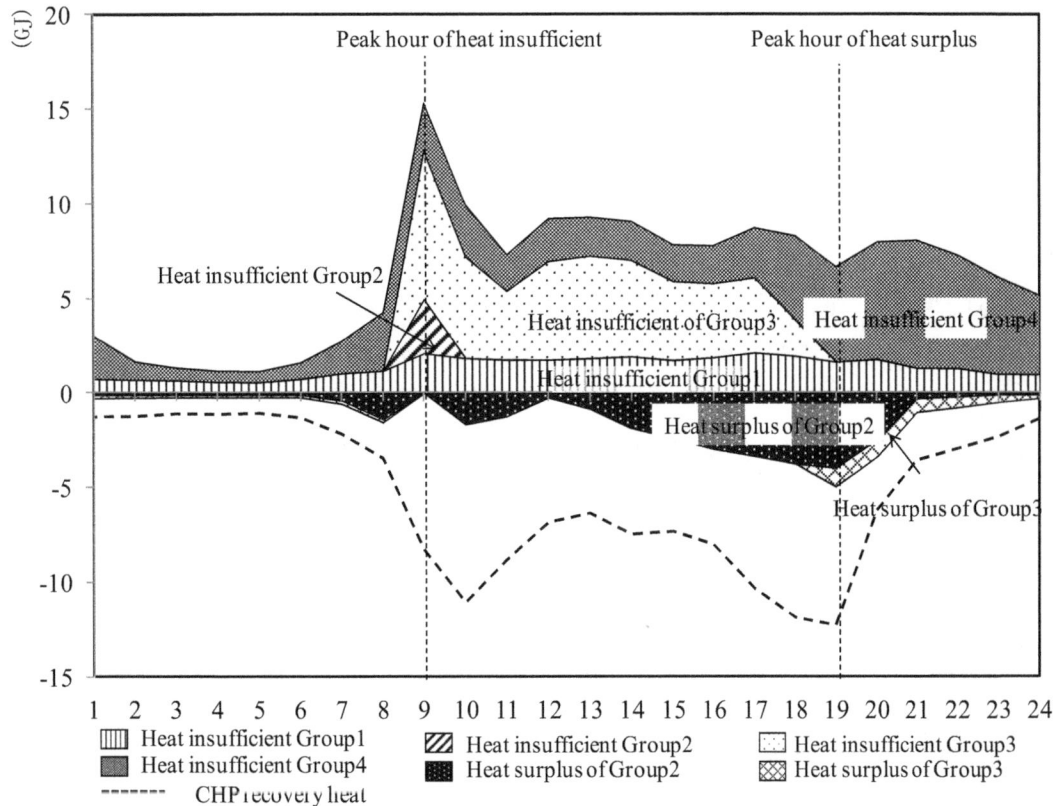

Figure 13. Daily heat balances of the community.

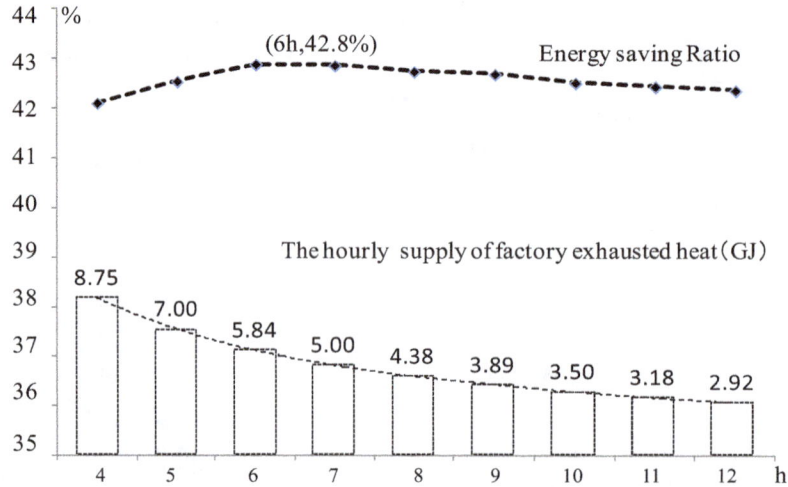

Figure 14. Energy saving ratio and the supply span of factory exhaust heat.

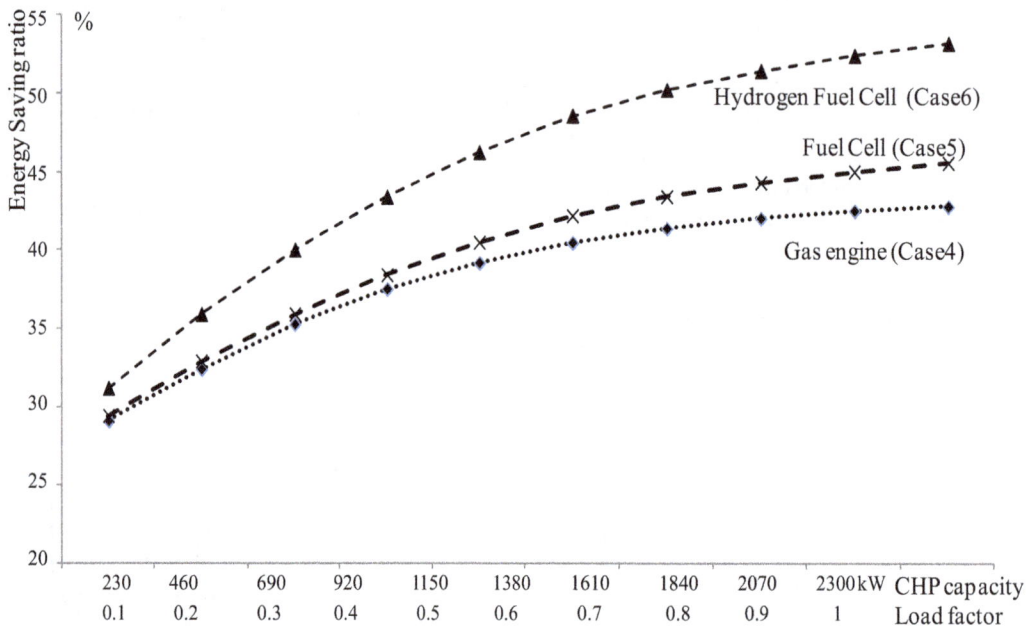

Figure 15. Energy saving ratio with different kind of *CHP* plant.

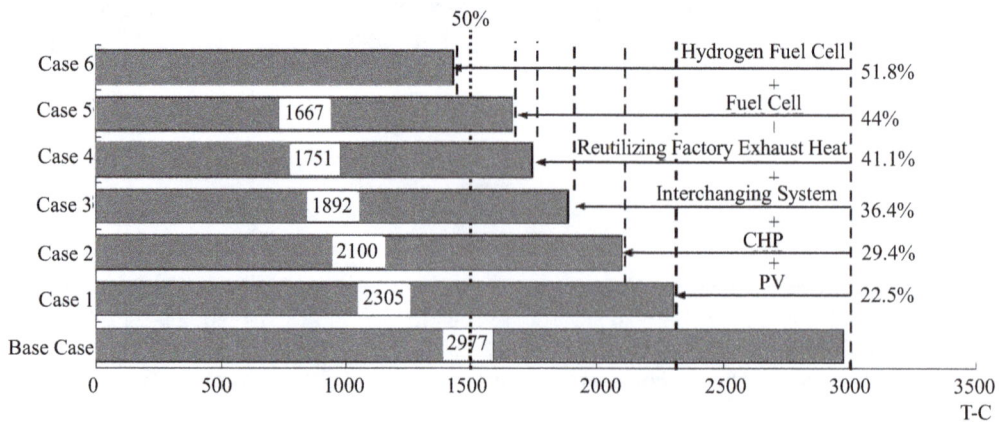

Figure 16. Yearly CO$_2$ emissions for cases.

and cut off another 7% carbon emission. Beside these, the reusing of factory rejected heat energy can cut off 41.1% CO_2 emission. With the introduction of fuel cell and hydrogen fuel cell, it is proved that the community can get 51.8% CO_2 emission reduction ratio.

5. Conclusions

The study proposed a DRN energy system model for smart community in Japan. One innovation is that the model not only has a smart grid but also has a smart heat energy supply chain by PCM system. The PCM system controlled by CEMS conducted the heat sharing between buildings. In that way, it can maximize onsite use of *CHP* recovery heat. Further, this model promoted a collaborative energy utilization mode between the industrial sector and the civil sector. The introduced PCM system will also collected the exhaust heat from the nearby factory.

1) The CEMS can dispatch the energy, including heat and electricity in the district, by the information received from the ES. The electricity sharing between the groups can improve the working hour and output of the *CHP* system. In that case, the distributed energy system can satisfy 93% of electricity consumption by itself. It enhances the reliability and independence of the energy system, shift the energy consumption away from the peak hour as well. The heat sharing can also enhances the independence of the energy system and satisfy the 68.8% of the thermal demand by *CHP* recovery heat.

2) There are different kinds of *CHP* plants, as gas engine and fuel cells. They have different characteristics, different electricity generation efficiency and heat recovery efficiency. The latest HFC, firstly under trail in this district, is the new kind *CHP* plant that can obviously improved the environmental effect of the system.

3) In general, the introduction of nature energy resource (*PV*) can cut 22% of the CO_2 emission. The introduction of *CHP* systems can cut around 30% CO_2 emission. Beyond that by DRN control, the district energy sharing can cut 36.4% CO_2 emission. The using of factory exhausted heat and the development of the *CHP* plant can help the district finally reach the target, that cut more than 50% of the primary energy consumption and the CO_2 emission.

REFERENCES

[1] C. Sollia, R. Anantharaman, A. H. Strømmana, X. Zhanga and E. G. Hertwicha, "Evaluation of Different *CHP* Options for Refinery Integration in the Context of a Low Carbon Future," *International Journal of Greenhouse Gas Control*, Vol. 3, No. 2, 2009, pp. 152-160. doi:10.1016/j.ijggc.2008.07.008

[2] R. Evans, "Environmental and Economic Implications of Small-Scale *CHP*," *Energy Policy*, Vol. 21, No. 1, 1993, pp. 79-91. doi:10.1016/0301-4215(93)90211-W

[3] M. Wissner, "The Smart Grid—A Saucerful of Secrets?" *Applied Energy*, Vol. 88, No. 7, 2011, pp. 2509-2518. doi:10.1016/j.apenergy.2011.01.042

[4] Government's New Growth Strategy. http://www.kantei.go.jp/jp/sinseichousenryaku/sinseichou01.pdf

[5] B. B. Alagoz, A. Kaygusuz and A. Karabiber, "A User-Mode Distributed Energy Management Architecture for Smart Grid Applications," *Energy*, Vol. 44, No. 1, 2012, pp. 167-177. doi:10.1016/j.energy.2012.06.051

[6] B. V. Mathiesen, H. Lund and K. Karlsson, "100% Renewable Energy Systems, Climate Mitigation and Economic Growth," *Applied Energy*, Vol. 88, No. 2, 2011, pp. 488-501. doi:10.1016/j.apenergy.2010.03.001

[7] O. Elma and U. S. Selamogullari, "A Comparative Sizing Analysis of a Renewable Energy Supplied Stand-Alone House Considering Both Demand Side and Source Side Dynamics," *Applied Energy*, Vol. 96, 2012, pp. 400-408. doi:10.1016/j.apenergy.2012.02.080

[8] M. Welsch, M. Howells, M. Bazilian, J. F. DeCarolis, S. Hermann and H. H. Rogner, "Modelling Elements of Smart Grids—Enhancing the OSeMOSYS (Open Source Energy Modelling System) Code," *Energy*, Vol. 46, No. 1, 2012, pp. 337-350. doi:10.1016/j.energy.2012.08.017

[9] R. S. Adhikari, N. Aste and M. Manfren, "Multi-Commodity Network Flow Models for Dynamic Energy Management—Smart Grid Applications," *Energy Procedia*, Vol. 14, 2012, pp. 1374-1379. doi:10.1016/j.egypro.2011.12.1104

[10] Introduction of Offline Heat Transport System, SANKI. http://www.sanki.co.jp/product/thc/thc/point.html

[11] R. Niemi, J. Mikkola and P. D. Lund, "Urban Energy Systems with Smart Multi-Carrier Energy Networks and Renewable Energy Generation," *Renewable Energy*, Vol. 48, 2012, pp. 524-536.

[12] Heat Supply System Offline Using the Latent Heat Storage Material-Heat Transformer Container System, 2009.3, JEFMA No. 57, pp. 53-55.

[13] M. Medrano, M. O. Yilmaz, M. Nogués, I. Martorell, J. Roca and L. F. Cabeza, "Experimental Evaluation of Commercial Heat Exchangers for Use as PCM Thermal Storage Systems," *Applied Energy*, Vol. 86, No. 10, 2009, pp. 2047-2055. doi:10.1016/j.apenergy.2009.01.014

[14] H. Kiyoto, "Study on the Exhaust Thermal Energy Utilization by Using the PCM Transportation System," *Summaries of Technical Papers of Annual Meeting Architectural Institute of Japan*, 20 July 2008, pp. 747-748.

[15] O. Lab, "Consumption Unit of Electricity, Heating, Cooling and Hot Water," Waseda University, Tokyo, 2002.

[16] H. Ren, "Effect of Carbon Tax and Electricity Buy-Back on the Optimal Economic Adoption of *PV* System for Residential Buildings," *Journal of Environmental Engineering*, Vol. 622, 2007, pp. 49-55.

[17] Introduction of Offline Heat Transport System, SANKI. http://www.sanki.co.jp/product/thc/thc/point.html

[18] L. Y. Fan, "Potential Analysis on the Area-Wide Factory Exhaust Thermal Energy Utilization by PCM Transportation System in a Recycling-Oriented Community," *Summaries of Technical Papers of Annual Meeting, Architectural Institute of Japan, Kyushu Chapter*, Vol. 51, 1 March 2012, pp. 341-344.

[19] Highlighting JAPAN July 2011, "Stepping Stones to 'Smart'ness," 2011.
dl.gov-online.go.jp/public_html/gov/pdf/hlj/.../12-13.pdf

21

Energy and Exergy Analysis of a Vegetable Oil Refinery

Musediq Adedoyin Sulaiman[1], Abayomi Olufemi Oni[2], David Abimbola Fadare[3]

[1]Mechanical Engineering Department, Olabisi Onabanjo University, Ago-Iwoye, Nigeria
[2]Mechanical Engineering Department, University of Agriculture Abeokuta, Abeokuta, Nigeria
[3]Mechanical Engineering Department, University of Ibadan, Ibadan, Nigeria
Email: fadareda@gmail.com

ABSTRACT

Energy and exergy analysis was conducted for a vegetable oil refinery in the Southwest of Nigeria. The plant, powered by two boilers and a 500 kVA generator, refines 100 tonnes of crude palm kernel oil (CPKO) into edible vegetable oil per day. The production system consists of four main group operations: neutralizer, bleacher, filter, and deodorizer. The performance of the plant was evaluated by considering energy and exergy losses of each unit operation of the production process. The energy intensity for processing 100 tonnes of palm kennel oil into edible oil was estimated as 487.04 MJ/tonne with electrical energy accounting for 4.65%, thermal energy, 95.23% and manual energy, 0.12%. The most energy intensive group operation was the deodorizer accounting for 56.26% of the net energy input. The calculated exergy efficiency of the plant is 38.6% with a total exergy loss of 29,919 MJ. Consequently, the exergy analysis revealed that the deodorizer is the most inefficient group operation accounting for 52.41% of the losses in the production processes. Furthermore, a critical look at the different component of the plant revealed that the boilers are the most inefficient units accounting for 69.7% of the overall losses. Other critical points of exergy losses of the plant were also identified. The increase in the total capacity of the plant was suggested in order to reduce the heating load of the boilers. Furthermore, the implementation of appropriate process heat integration can also help to improve the energy efficiency of the system. The suggestion may help the company to reduce its high expenditure on energy and thus improve the profit margin.

Keywords: Vegetable Oil Refining; Crude Palm Kernel Oil; Energy; Exergy; Irreversibility

1. Introduction

The production of edible oil has been in practice in Nigeria for centuries, and it remains an essential ingredient in much of the southwest of the country cuisine. It uses are also found in animal feed, for medicinal purposes, and for certain technical applications. Its enormous demand for various purposes has made the industry very important. According to the vegetable oil producers, Nigeria is self-suficient in edible oil requirement. Statistics show that the country produces altogether 500,000 metric tonnes of edible oil annually made up of 320,000 metric tonnes from the organized sector while the remaining 180,000 metric tonnes come from small-scale producers in the unorganized sector [1]. Total demand of refined edible oil in Nigeria is put at 250,000 metric tonnes annually [1]. By this statistics, the country is supposed to export about 250,000 metric tonnes of edible oil to neighboring West African countries annually. The vegetable oil industry has played a significant role in the national economy generating business worth more than 10 billion naira for transportation and allied sectors. It has provided over 25,000 direct employments and indirect employment for more than one million farming families. Altogether, the industry is believed to have direct effect on the lives of more than six million Nigerians [1]. However, the Nigerian edible oil refineries are facing more challenges today than ever before as a result of increased competitiveness and varying energy demand. In addition, high cost of production, the epileptic supply of electrical energy supply from the national grid, inadequate production and distribution of petroleum products and the growing concern over global warming have created complex and sometimes conflicting challenges for industrial operations. The industries must operate a system that runs effectively in order to survive in today's highly competitive world market. Hence, the Nigerian industries are seeking cost effective energy saving technology and practices that will minimize energy use while maintaining or increasing product quality and quantity.

The possibilities of minimizing energy wastage in production processes cannot be overemphasized. This necessitate the need to investigate the useful part of energy, in any particular section on the production line, which economically helps save cost of production and automatically improves the reduction in the energy wastage. This fact therefore sets in the effective and efficient use of energy which is very important to industries. As much as energy is highly important for the process industries, minimum amount of energy should be wasted under normal situations.

The traditional method of assessing the energy disposition of an operation involves the application of first law of thermodynamic. The advent of exergy method which is based on the second law of thermodynamic exposes the inadequacies of the first law. Exergy analysis provides information about the irreversibility state of thermodynamic processes. It thus indicates means of assessing the locations, types and magnitudes of wastes and losses and to identify meaningful efficiencies of the system. The wide spread of the use of exergy method by several researcher has brought about steps towards cutting down on energy cost, conservation of scarce energy resources and reduction of environmental damage. Exergy analysis methodologies have been applied to many industrial systems such as: sugarcane bagasse gasification [2], malt drink production [3], flavored yogurt [4], and fruit juice [5]. Although a considerable volume of energy and exergy-related analyses of industrial processes exists in literature, limited work has been reported on energy and exergy analyses of vegetable oil from palm kennel oil processing operations. Energy and exergy analysis for production of vegetable oil from soybean oil, sunflower and olive oil has been reported for Turkey. Only the work of Fadare *et al.* [2] and Waheed *et al.* [5] has respectively reported the energy and exergy analyses of malt drink and fruit juice processing operations in Nigeria. To the best of the authors' knowledge, no work has been conducted on the energy and exergy analyses of edible vegetable oil production in Nigeria. Therefore, the aim of this study is to analyze the energy consumption pattern and exergy inefficiency of palm kennel oil refining operations in Nigeria, in view of improving the efficiency of the system, reduce the production costs and hence, increase the profitability of edible

vegetable oil production.

2. Materials and Method

2.1. Plant Description

Energy and exergy studies of refining crude palm kennel oil into edible vegetable oil were conducted in a factory located in the southwestern Nigeria. The energy requirement and exergy inefficiency for processing 100 tonnes per day of edible vegetable oil from palm kernel oil was estimated. The plant operates on a 3-shift of 8-working hours per day with a total of 55 workers per shift out of which about 27 workers are involved in the production process due to the automation level of the factory. The main sources of energy utility for the plant are electrical, thermal and manual. The primary source of electrical energy used is either from the national grid or from the company's power generating set. Steam generated from diesel-fueled boilers is used for heating purpose, while cooling is effected through the use of condensers.

2.2. Process Description

The production steps consist of four group operations: neutralizer, bleacher, filter, and deodorizer, which are powered with two boilers for steam generation and a 500 kVA generator for electricity generation. The process flow line for processing of edible vegetable oil from palm kernel oil is shown in **Figure 1**. The crude oil (palm kernel oil) is kept under vacuum in a buffer feed tank. The oil is first pumped through a heat exchanger, where it is heated by the outgoing hot deodorized oil. Phosphorus acid solution is pumped with the aid of dosing pump into the static acid mixer and is mixed with the crude oil. The mixture is then routed to the neutralizer. The purpose of neutralization is to change the structure of gums or phosphatides present

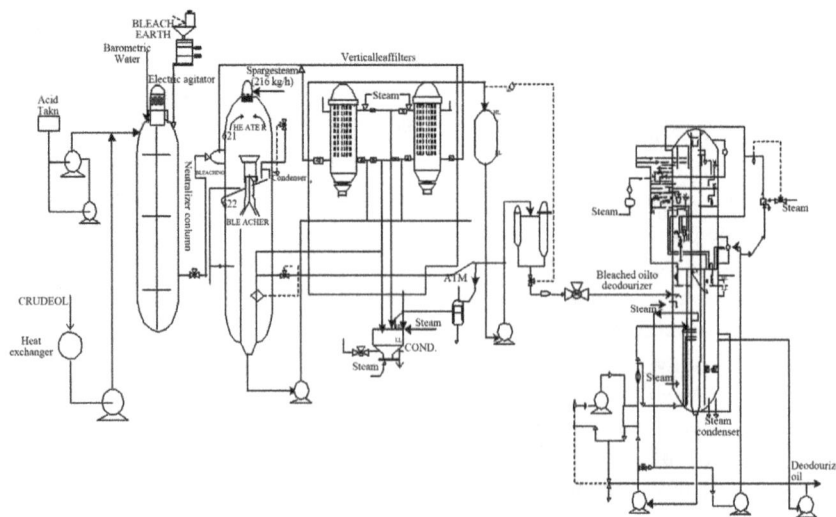

Figure 1. Schematic diagram of the production of edible vegetable oil.

in the crude oil in such a way that they can be removed during the bleaching process. The required amount of bleaching earth is added in the tank by the bleaching earth dosing device. Under atmospheric pressure, the oil is pushed into the bleacher. The neutralized oil is treated with bleaching earth/activated carbon for the removal of colouring pigments. The mixture is then heated by steam while the contents are maintained under vacuum by the barometric condenser and vacuum pump. When the required temperature is reached, the moisture present in the oil is completely removed upon completion of bleaching. This oil-bleached suspension is routed to the hermetic leaf filters where bleaching earth and precipitated matters are removed. Deodorization is the last stage in the vegetable oil refining process. The operation is carried out at high temperature by injecting open stream and maintaining high vacuum at which time all the odiferous matter is distilled off and carried away to barometric condensers through vacuum system. Under reduced pressure, volatile free fatty acids and other substances are also removed in order to produce bland final product. The result is odourless product with an acceptable, colour and taste. The bland oil is then sent to final storage with the addition of oxidants to prolong the product shelf life.

2.3. Data Collection

The plant utilized electrical, thermal and manual energies for the production process. The required parameters for evaluating energy consumption and exergy efficiency in each unit operation were measured directly or obtained from the factory's energy department. An inventory of the power rating of electric motors, properties of steam, coolant and product streams, boiler and chiller operating conditions, number of man-power required for manual labour and time taken for each operation were determined. The data were collected from the plant over a period of two months. The measuring quantities used in the course of the data acquisition include: 1) A stopwatch for measuring the time spent in each operation; 2) A measuring cylinder for measuring the amount of fuel consumed and 3) A weight balance for measuring the quantity of crude and processed oil.

2.3.1. Evaluation of Electrical Energy
The electrical energy input, E_p, in kW h was obtained by multiplying the rated power of the electric motor, P, in kW with the corresponding hours of operation, t. The motor efficiency, η, was assumed to be, 80% [2]:

$$E_P = \eta P t \tag{1}$$

2.3.2. Evaluation of Thermal Energy
Thermal energy input, E_F, was calculated based on quantity of fuel (diesel or oil-cake) used to generate steam in the boiler. The quantity of fuel, W, in kg used was converted to energy (MJ) by multiplying the quantity consumed by the corresponding calorific value, C_f, of fuel (J/kg) [2]:

$$E_F = C_f W \tag{2}$$

Note that the calorific value for diesel and oil-cake are respectively 42 and 37 MJ/kg [2].

2.3.3. Evaluation of Manual Energy
Manual energy, E_m, in kW was estimated based on the value recommended by Odigboh [5]. According to him, at maximum continuous energy consumption rate of 0.30 kW and conversion efficiency of 25%, the physical power output of a normal human labour in tropical climates is approximately 0.075 kW sustained for an 8 - 10 hour workday:

$$E_m = 0.075 N t \ (\text{kW} \cdot \text{h}) \tag{3}$$

N is the number of persons involved in the operation and t is the useful time spent to accomplish a given task in hours.

2.3.4. Evaluation of Energy Intensity
Energy intensity is the amount of the energy required per unit output of the production. Production volumes in this case are expressed in tonnes. The energy intensity was evaluated as the ratio of total energy input, E_T, in MJ and the volume of edible vegetable oil produced, V_t in tonnes:

$$E_i = \frac{E_T}{V_t} \tag{4}$$

The required parameters for evaluating energy and exergy in the four unit operations are represented in **Table 1**.

2.4. Exergy Change of the Process Stream

The exergy of process stream (E_x) can be expressed as the sum of the physical (E_{PH}), chemical (E_{CH}), kinetic (E_K) and potential (E_{PT}) exergy. Mathematically,

$$E_x = E_{PH} + E_{CH} + E_K + E_{PT} \tag{5}$$

where

$$E_{PH} = (h - h_0) - T_0(s - s_0) \tag{6}$$

$$E_{PT} = m g h \tag{7}$$

$$E_K = \frac{m v^2}{2} \tag{8}$$

$$E_{CII} = \sum_i \mu_{0,i} N_i \tag{9}$$

where $\mu_{0,i} = h_{0,i} - T_{0,i} S_{0,i}$ and $N_{0,i}$ = number of moles.

Table 1. Required parameters for evaluating energy and exergy values in the refining of palm kennel oil.

Unit operation	Required parameters	Value
Neutralization	Number of persons	4
	Time taken (h)	5
	Electrical power (kW)	37
	Crude oil inlet temperature (K)	303
	Crude oil outlet temperature (K)	358
	Weight fraction of water in oil	0.03
Bleaching	Number of persons	7
	Time taken (h)	6
	Electrical power (kW)	5.5
	Neutralized oil inlet temperature (K)	358
	Neutralized oil outlet temperature (K)	373
	Steam mass requirement (kg/h)	215
	Density of neutralized oil (kg/l)	0.9
	Weight fraction of water in oil	0.04
Filtration	Number of persons	2
	Time taken (h)	5
	Electrical power (kW)	5.5
	Bleached oil inlet temperature (K)	373
	Bleached oil outlet temperature (K)	353
	Weight fraction of water in oil	0.04
Deodorizing	Number of persons	7
	Time Taken (h)	7
	Electrical power (kW)	26.2
	Steam mass requirement (kg/h)	513
	Filtered oil inlet temperature (K)	353
	Deodorized oil outlet Temperature (K)	473
	Density of oil (kg/l)	0.85
	Weight fraction of water in oil	0.06

In Equations (6)-(9) h is the specific enthalpy (kJ/kg), s is the specific entropy (kJ/kg·K), both evaluated at T and P of each process stream; h_0 and s_0 are, respectively, the specific enthalpy and specific entropy evaluated at the reference state $(T_0 = 298.15 \text{ K and } P_0 = 100 \text{ kP})$.

For a typical control volume with steady flow and accumulation of exergy occurring in the system, the exergy balance of the system can be represented as [6]:

$$\sum_j \left(1 - \frac{T_0}{T_j}\right)\dot{Q}_j - \dot{W}_{cv} + \sum_i \dot{m}_i\, e_i - \sum_0 \dot{m}_i e_i - \dot{I}_{cv} = 0 \quad (10)$$

The term $\dot{Q}$ represents the time rate of heat transfer across the boundary, T_j is the instantaneous temperature of the boundary, $\dot{W}_{CV}$ is the time rate of exergy transfer

by work, $\dot{I}$ is the time rate of exergy loss due to irreversibility within the system, the term $\dot{m}_i e_i$ accounts for the time rate of exergy transfer accompanying mass flow and flow work, while subscripts i and o represents the inlet and outlet, respectively.

The specific flow exergy (e) of the system can be expressed as:

$$e = h - h_0 - T_0\left(S - S_0\right) + \frac{V^2}{2} + gz \quad (11)$$

where h and s denote the enthalpy and entropy of the system respectively, h_0, s_0 and T_0 are the enthalpy, entropy, and temperature, respectively at the dead state (environment).

Neglecting the kinetic and potential energy, the net change of exergy of the system can be expressed as:

$$e_2 - e_1 = h_2 - h_1 - T_0\left(S_2 - S_1\right) \quad (12)$$

The net exergy change of the process stream in and out of each unit operation in the edible vegetable oil production system was evaluated using the predictive model proposed by Singh [7]:

$$e_2 - e_1 = c_p\left(T_2 - T_1\right)\left[1 - \frac{T_0}{\left(T_2 - T_1\right)ml}\right] \quad (13)$$

where

$$\left(T_2 - T_1\right)ml = \frac{T_2 - T_1}{\ln\left(T_2/T_1\right)} \quad (14)$$

The specific heat capacity of the edible vegetable oil can be determined using the expression

$$c_p = 4.1868\left(0.3823 + 0.6183x\right) \quad (15)$$

where x is the weight fraction of water in the oil.

2.5. Exergy Inefficiency and Useful Work of the System

The exergy inefficiency can be evaluated with the expression

$$I_{ff} = \frac{I}{\sum I_{all}} \quad (16)$$

where I_{ff} is the inefficiency of the system and is defined as the ratio of the irreversibility in each unit operation to the irreversibility in the overall operations.

The useful work input into the system can be expressed as [7]:

$$W_u = \left(e_2 - e_1\right) - T_0 R_s \quad (17)$$

where W_u is the useful work, R_s the production of entropy and T_0 the ambient temperature. The exergy difference $e_2 - e_1$ is defined in terms of each component exergy e_x per unit mass and the mass flow rate m. From Equation

(14), it is obvious that the exergy change is a balance of useful work and the entropy production term, which can be regarded as work loss because of irreversibilities.

2.6. Exergy Efficiency

In order to determine how well the desired effect of the system is accomplished, the efficiency is calculated as the ratio of the net produced exergy to the net supplied exergy or as a fraction of the net supplied exergy used by the system to perform its function [8]:

$$\eta = 1 - \left(\frac{I_{loss}}{e_{in}} \right) \qquad (18)$$

3. Results and Discussion

3.1. Energy Expenditure of the Plant

A total of 23 h was required to process a batch of 100 tonnes of palm kennel oil in edible vegetable oil. For this case, there was a total outage of electricity supply from national grid, hence the power generating set was used for the entire production process. The average rate of fuel consumption by the power generating set (diesel), boiler 1 (oil cake), and boiler 2 (diesel) are respectively 21.0, 45.7 and 56.3 kg/h. The total energy consumption for the entire production was therefore, estimated to be 48,703 MJ while the average energy intensity was 487.0 MJ/tonne.

The energy consumption pattern for the main group operations is shown in **Table 2**. The net energy input into the production units was estimated as 23333.64 MJ with thermal (95.23%), electrical (4.65%) and manual (0.12%) of the energy input. The deodorizing operation consumed the highest energy with 13127.96 MJ (52.26%), followed

by the bleaching process with 9224.21 MJ (39.53%), neutralization unit with 899.18 MJ (3.85%) while the filtration operation accounted for the least energy with 81.9 MJ (0.35%).

3.2. Exergy Expenditure of the Plant

The exergy analysis of the system gave insight to the inefficiencies and the opportunities for exergy loss minimization of each of the unit operations in the four main group operations involved in the production of palm kennel oil. Conceptually the exergy calculation of the system was divided into process stream exergy and utility exergy. Exergy accounts for individual process were presented in order to identify major losses and evaluate the potential for further technical improvements in the production of the palm kennel oil processes.

The exergy expenditure of the plant was divided into two main categories. The first evaluation looks into the four main group operations while the second category examines the exergy losses in all the plant components including the utility sections (boiler 1 and 2). In the first category, the exergy change in process stream, useful work, steam exergy (utilities), entropy generated, effluent losses and the inefficiency associated with the different unit operations in the overall production system was evaluated. The change in the oil exergy was only associated with operations where there was change in the inlet and outlet temperatures. This can be seen to occur in all the unit operations considered. The negative value of exergy change in the filtration unit was due to the drop in temperature of the oil during the process. Furthermore, the useful work comprised both electrical and manual energy (**Table 3**).

Table 2. Time and energy use data in refining of palm kernel oil.

Unit operation	Time taken (h)	Electrical energy, E_l (MJ)	Thermal energy, E_t (MJ)	Manual energy, E_m (MJ)	Total energy, E_o (MJ)	E_o/E_{tt} (%)
Neutralization	5	532.8	361	5.4	899.18	3.85
Bleaching	6	95.04	9118.23	11.34	9224.61	39.53
Filtration	5	79.2	-	2.7	81.90	0.35
Deodorizing	7	377.28	12741.23	9.45	13127.96	56.26
Total	23	1084.32	22220.43	28.89	23333.64	100

Table 3. Exergy balance in the edible vegetable processing operation.

Components	Exergy change (MJ)	Useful work (MJ)	Utilities/process stream exergy change (MJ)	Irreversibility (MJ)	Effluent losses (MJ)	Total exergy losses (MJ)	Inefficiency (%)
Neutralization	193	538.2	360.98	706		706	7.79
Bleaching	124	106.38	2916	2898	475	3373	37.21
Filtration	−157	81.9	0	239		239	2.63
Deodorizer	1687	386.73	2915	1615	3132	4747	52.37
Total	1847	4715	6191	5457	3607	9065	100

The electrical energy is an energy source that consists of pure exergy while the inability to account for the entropy generated by a human labourer justified its inclusion in the useful work [3]. The results obtained for the different unit operations are presented in **Table 4**. An overall evaluation of the production processes from an exergy perspective reveals the ranking of order of the energy killers in the plant. From this stand point, the highest entropy was generated in the deodorizer (accounting for more than half of the losses) followed by bleaching, neutralizer and filtration. The irreversibilities within the systems are as a result of high temperature difference between the inlet and the outlet stream of both the oil streams, heating and cooling utilities. From the technical analysis of the different component, a considerable part of the losses in the deodorizer can be attributed to wasteful energy losses during heating and cooling respectively. Of the total exergy losses in the deodorizer, the deodorizer column accounted for 34 percent while deodorizer steam condensers accounted for 66 percent. This is an indication that the heating and cooling resulted into wasteful losses. This is always the case for exergy calculations and is due to the fact that the exergy value of heat is often much lower than its energy value, particularly at temperatures close to reference temperature Fadare *et al.* [3]. The irreversible and effluent losses are minimal in other units.

In the second category, a holistic overview of the entire component (boilers included) was considered. The exergy efficiencies along with the percentage of exergy losses are summarized in **Table 5** for all components present in the plant. It is obviously that the exergy losses rate of the boiler is dominant over all other losses in the plant. The boilers accounts alone for 69.7% of losses in the plant, while the exergy destruction rate of the deodorizer is only 10.47%. The inefficiency in the boilers unit was an indication of significant losses due to high entropy generation at conditions at which the unit operates. At these conditions, the irreversibility of the combustion reaction occurring in the combustion cham-

Table 5. Energy intensity data from various literatures.

Process	Energy intensities (MJ/tonne)	Reference
Pelletized organic fertilizer production	350.0000	[9]
Powdered organic fertilizer production	280.0000	[9]
Fruit juice production	0.0011	[5]
Olive oil production	10028.6000	[10]
Sunflower oil production	7795.4000	[10]
Soya bean oil production	7619.1000	[10]
Palm kennel oil	487.0379	Present work

bers of the boilers increases, hence relatively low exergy efficiencies of the boilers is exhibited. Although, the bleaching column has a low exergy efficiency and a relatively low exergy lose, this situation arise because exergy efficiency values are quantitative measurement derived as a ratio of two numbers with the constraint that the ratio is not greater than one, whereas exergy losses are quantitative measurements derived as the difference between two numbers. Exergy efficiency was not defined for the condensers [11] and filters this is because the purpose of these devices is to reject waste heat rather than generate product. The calculated exergy efficiency of the palm kennel oil plant was found to be 38.6%, which is low. This indicates that tremendous opportunities are available for improvement. However, part of this irreversibility cannot be avoided due to physical, technological, and economic constraints.

The avoidable losses of the plant can be reduced by increasing the capacity of the plant which will result in the reduction of the load on the boiler following similar suggestion made by Dalsgard [12]. Also, the utilization of cooling and heating utilities can be reduced if the appropriate heat integration method is in place. The purpose of heat integration in this case is to identify all unmatched existing hot and cold streams. A composite line are drawn which provide a way to match them in an optimum fashion with respect to energy. The matching lines has precise counterpart in terms of exergy balance [13]. This will enable a longer production time and thus reduce avoidable energy wastage and the corresponding exergy destruction that will occur by plant start-up, shutdown, cleaning and sterilization. If this suggestion is taken, it may help the company to reduce its high expenditure on energy and thus improve the profit margin.

3.3. Comparison of the Energy Intensities and Exergy Inefficiencies in the Production of Edible Vegetable Oil from Palm Kennel Oil

The trend in energy utilization for the processing of edible vegetable oil from palm kernel oil has not been reported however effort is made by comparing this study

Table 4. Exergy efficiencies and percentage losses of the plant component.

Component	Exergy efficiency %	Percentage losses
Neutralization	53.5	2.36
Bleaching column	4.3	9.69
Bleaching condenser	-	1.59
Filtration	-	0.80
Deodorizer column	57.9	5.40
Deodorizer condenser		10.47
Boiler 1	23.9	30.12
Boiler 2	30.0	39.58

with edible vegetable oil from soybean oil, sunflower, olive oil and other non-vegetable oil production processes found in literature.

The energy intensities reported in this study is lower as compared to the production of vegetable oil from soybean oil, sunflower oil and olive oil [10] (**Table 5**). This can be attributed to varying factors such as the differences in inherent energy requirement of the production steps of the different products reported and the extent of the system boundary of the analysis. However, the result was higher than values found in other production processes such as organic fertilizer [9] and fruit juice production [5]. An indication that the production of edible oil from palm kennel oil is more energy intensive than this processes.

The exergy inefficiency in the production of the aforementioned vegetables oil (soybean oil, sunflower oil and olive oil) was not reported therefore comparison was not possible. Although, comparison with other works was also difficult due to lack of similar process, however, the present work can further be illustrated by comparing the pasteurizer inefficiency found by Fadare *et al.* [2] as 59.75%, and evaporator inefficiency reported by Rotstein [13] as 68% with the deodorizer inefficiency presented in this study which is 52.4%. All the processes considered are the point of highest inefficiencies in the main production steps from the literature and present work.

4. Conclusions

The energy and exergy consumption for production of 100 tonne of edible vegetable oil were estimated for a Nigerian palm kennel oil refining industry. Four defined group operations were identified in the production steps: neutralizer, bleaching, filtration and deodorizer. The energy audit revealed that the types of energy input for the production were electrical (4.65%), thermal (95.23%) and manual (0.12%) of the net energy input. The average energy intensity of the process was estimated as 487.04 MJ/kg. The deodorization unit was the most energy intensive process with 13127.96 MJ accounting for 52.4% of the net energy input the production. The company depended mainly on the use diesel and de-oils cake to power the steam boilers and diesel for the company's power generating set for supply of electrical power. The use of diesel in Nigeria is not cost effective [3] therefore the use of filtered cake of spent bleaching earth (de-oil cake) should be encouraged for the generation of steam in oil production processes.

An exergetic insight into the different unit operations reveals the amount of entropy generated and effluent losses associated in the operations. It can be seen that each section or unit of the plant is characterized by a certain level of entropy production. Considering the main group operation, exergy loss is highest in the deodorizer accounting for over half of the exergy losses in the production line while boiler 2 is the most inefficient component accounting for 39.6% of the overall loses.

The exergy losses in the system can be reduced by increasing the capacity of the plant which will in turn reduce the load on the boilers. Process heat integration of the plant can also help to improve the energy utilization and profitability of the system. The analysis also illustrates the fact that exergy balance is a powerful diagnostic tool for energy use optimization.

REFERENCES

[1] M. Akpan, "The Oil War: Local Vegetable Oil Producers Take Their Case against Importation to the Presidency," 2000. www.newswatchngr.com

[2] L. F. Pellegrini and S. de Oliveira, "Exergy Analysis of Sugarcane Bagasse Gasification," *Energy*, Vol. 32, 2007, pp. 314-327.

[3] D. A. Fadare, D. O. Nkpubre, A. O. Oni, A. Falana, M. A. Waheed and O. A. Bamiro, "Energy and Exergy Analyses of Malt Drink Production in Nigeria," *Energy*, Vol. 35, No. 12, 2010, pp. 5336-5346. doi:10.1016/j.energy.2010.07.026

[4] E. Sorguven and M. Ozilgen, "Energy Utilization, Carbon Dioxide Emission, and Exergy Loss in Flavored Yogurt Production Process," *Energy*, Vol. 40, No. 1, 2012, pp. 214-225. doi:10.1016/j.energy.2012.02.003

[5] M. A. Waheed, S. O. Jekayinfa, J. O. Ojediran and O. E. Imeokparia, "Energetic Analysis of Fruit Juice Processing Operations in Nigeria," *Energy*, Vol. 33, 2008, pp. 35-45. doi:10.1016/j.energy.2007.09.001

[6] E. U. Odigboh, "Machines for Crop Production," In: B. A. Stout, Ed., *CIGR Hand-Book of Agricultural Engineering*, American Society of Agricultural Engineers, 1998.

[7] R. P. Singh, "Energy Accounting in Food Process Operations," *Food Technology*, Vol. 32, No. 4, 1978, pp. 40-46.

[8] M. A. Rosen, "Energy- and Exergy-Based Comparison of Coal-Fired and Nuclear Steam Power Plants," *Exergy, An International Journal*, Vol. 1, No. 3, 2001, pp.180-192.

[9] M. Ozilgen and E. Sorgüven, "Energy and Exergy Utilization, and Carbon Dioxide Emission in Vegetable Oil Production," *Energy*, Vol. 36, No. 10, 2011, pp. 5954-5967. doi:10.1016/j.energy.2011.08.020

[10] E. Rotstein, "Exergy Analysis: A Diagnosis and Heat Integration Tool," In: R. P. Singh, Ed., *Energy in Food Processing*, Elsevier, Amsterdam, 1986.

[11] D. A. Fadare, O. A. Bamiro and A. O. Oni, "Energy and Cost Analysis of Organic Fertilizer Production in Nigeria," *Energy*, Vol. 35 No. 1, 2010, pp. 332-340. doi:10.1016/j.energy.2009.09.030

[12] I. Dincer and M. A. Rosen, "Exergy, Energy, Environment and Sustainable Development," Elsevier, Amsterdam, 2007.

[13] I. Dalsgard, "Simplification of Process Integration in Medium-Size Industry," Ph.D. Thesis, University of Denmark, Aarhus, 2002.

22

Model Establishment of Whole Life Cycle for Energy Efficiency of Rural Residential Buildings in Northern China

Chenxia Suo[1], Yong Yang[1], Solvang Wei Deng[2]
[1]Beijing Institute of Petrochemical Technology, Beijing, China
[2]Narvik University College, Narvik, Norway
Email: suochenxia@bipt.edu.cn, yangyonghebei@126.com, wds@hin.no

ABSTRACT

The building energy efficiency is determined by the climatic region and the energy-saving measures. In this paper an assessment model for energy efficiency of the rural residential buildings in the northern China was established by the method of whole life cycle. The energy consumption of the rural residential buildings in different stages was analyzed through quantitative method in this model. At the same time, the corresponding energy efficiency assessment system was developed.

Keywords: Building Energy Efficiency; Whole Life Cycle; Assessment System

1. Introduction

The energy efficiency assessment of buildings in their whole life cycles is used for analyzing the energy consumption of buildings and realizing overall consideration of the energy consumptions of building projects based on a whole life cycle theory, mainly including the energy consumptions of building materials in their manufacturing and transportation and the energy consumptions of building projects in their construction, operation, dismantling and recycling.

2. Principles of the Establishment of the Assessment Model

1) Scientific principle: The design of the post assessment indicator system of building energy efficiency process should fully consider the characteristics of building energy efficiency.

2) Overall principle: The design of the indicators used in the energy efficiency assessment of buildings in their whole life cycles should reflect the energy efficiency performance.

3) Practical principle: In the design of the indicator system of the energy efficiency assessment of a building in its whole life cycle, the principle of practicability should be followed, the accuracy and availability of data should be considered and targeted indicators that can reflect the building project comprehensively and are mutually independent should be selected in the light of the purpose of the study and the characteristics of the object of study.

4) System integrity principle: The establishment of the indicators for environment friendly buildings is a complex systematic work that must reflect the buildings' economic characteristic and other basic characteristics regarding resources, energy sources and indoor environmental quality.

3. Influencing Factors

In order to ensure its practicability, an assessment model should reflect the characteristics of the assessed building. Thus, an assessment model is influenced by factors like the regional factor, the economic development level, the local customs and so on.

3.1. Scope and Climate Features of the Regions Needing Heating in the Northern China

1) In the regions needing heating in the northern China, the winters are long, cold and dry; both the annual temperature range and the daily temperature range are large; the days with an average daily temperature equal to or lower than 5°C account for 25 to 40 percent of the total days of a year and the days with a maximum temperature equal to or higher than 35°C account for 22 percent of total days of a year.

*Beijing Energy Management Evaluation & Innovation Team (PHR 201007136); The Ministry of Education of Philosophy and Social Science Planning Project (10YJA630139).

2) The regions needing heating in the northern China are usually drouthy and rainless regions. In these regions, the annual average relative humidity is 50 to 70 percent; the annual number of rainy days is 60 to 100 days; the annual precipitation is 300 mm to 1000 mm; the daily maximum precipitation is 200 mm to 300 mm for most of places or over 500 mm for some places; and the annual number of snowy days is less than 15 days.

3) The regions needing heating in the northern China has strong solar radiation: The annual solar radiation intensity is 150 W/m² to 190 W/m²; the annual sunshine duration is 2000 h to 2800 h; and the annual percentage of sunshine is 40 to 60 percent [1].

3.2. Life Behavior Characteristics of the Peasants in the Northern China

1) The activities of the peasants in the northern China mainly happen on heated brick beds. The peasants have meals, study, do housework, receive visitors and sleep on heated brick beds, so they have special furniture different from the furniture used in cities, including short-legged table and stools for used on heated brick beds.

2) The peasants in the northern China have the habits of paying visits to their relatives and friends and doing some activities in front of or behind their houses, so the interiors of their houses have frequent ventilation. In their opinions, they will be easy to catch cold if the indoor temperatures of their houses are too high. Based on the analyses above, it is suggested to make the indoor calculating temperature 16˚C in the thermal performance design and energy efficiency design [2].

3.3. Characteristics of the Rural Residential Buildings in the Northern China

The rural residential buildings in the severe cold region is limited by the production and life behaviors and living patterns of the peasants there. the residential buildings mainly have a brick-concrete structure or a brick-wood structure and these two kinds of structures account for about 71.64 percent in all the structures; the severe cold region has 95,054,900 permanent resident households that account for 46.48 percent of the total permanent resident households of the country, with a total heating area that is 39 percent of the total area of the rural residential buildings of the country.

3.4. Characteristics of the Heating of the Residential Buildings of the Peasants in the Northern China

Viewed from the heating facilities, the heated brick beds (also called kangs for short) and suspended kangs (also called adobe kangs) used a lot in the rural areas in the northern China are common heating equipment in the residential buildings in the northern China. Suspended kangs are based on a new type of kang-linked stoves that are seasonally designed in the light of the scientific principles of combustion and heat transfer. Viewed from the form of heating, heating may be combined with cooking or separated from cooking. In the form of heating combined with cooking, heating in winter is combined with cooking, *i.e.* heating can be realized at the same time of cooking. The heating separated from cooking means that heating is separated from cooking [2].

3.5. Characteristics of the Materials of the Residential Buildings of the Peasants in the Northern China

Most of the building materials are locally obtained. Their houses have a brick-concrete structure and a brick-wood structure.

1) In a brick-concrete structure, the vertical load-bearing walls and posts of buildings are made of bricks or blocks and the transverse load-bearing beams, floor-slabs and roof panels have a reinforced concrete structure. In other words, a brick-concrete structure is a structure with a minority of parts made of reinforced concrete and a majority of parts made of brick walls for load bearing.

2) In a brick-wood structure that is a kind of architectural structure, the vertical load-bearing walls and posts of buildings are made of bricks or blocks and floorslabs and roof frames are made of wood. Due to the restrictions of mechanics engineering and engineering strength, a building with a brick-wood structure usually has one to three storeys [3].

4. Energy Efficiency Assessment System

4.1. Determination of Assessment Indicators

According to the function goal and hierarchical assessment indicator system structure of the energy efficiency assessment system of rural buildings in their whole life cycles, this paper has done level-by-level analysis and decomposition of the structure and logic level relationship of the assessment of energy-efficient residential buildings, determined the mutual subordinate relationships of the factors and established a hierarchical assessment system for the assessment model for the energy efficiency of the rural buildings in the regions needing heating in the northern China.

Overall indicator layer A: Overall goal of the energy consumption assessment of the rural buildings in their whole life cycles.

Classified indicator layer B: The compositions of the energy consumption assessment of the rural buildings in their whole life cycles.

Sub-classified indicator layer C: The classified indicators are made more detailed in this layer. The assessment content factors of the sub-classified indicators are more detailed contents and explanations of the classified indicators (See **Table 1**). The indicator system structure of the energy efficiency assessment system of rural buildings is as shown in **Table 1** and **Figure 1**.

4.2. Grade Standard of the Assessment Result

Grade standard is a quantitative standard for rating of Excellent, Good, Fair, Pass and Rejected. In an energy-efficient building assessment indicator system, dif-

ferent indicators have different dimensions, meanings and forms of expression. The grade standard is then determined based on the total score. The grade standard is as shown in **Table 2**.

4.3. Calculation of Building Energy Consumption

The energy efficiency assessment of buildings in their whole life cycles mainly include the assessment of the energy consumptions in five stages, including the manufacturing of building materials, the transportation of building materials, the construction of building projects, the operation of building projects and the dismantling

Table 1. Indicators for the energy efficiency assessment of the rural buildings in their whole life cycles.

Overall indicator A	Classified indicator B	Sub-classified indicator C	
		Assessment indicator	Explanation of assessment indicator
Energy consumption assessment system of buildings in their whole life cycles (A)	Energy consumption in manufacturing of building materials (B1)	Unit energy consumption of building materials (C11)	Calculate the energy consumptions in the exploration and manufacturing of different building materials.
		Consumption of building materials (C12)	Calculate the consumption of each building material.
	Energy consumption in transportation of building materials (B2)	Energy consumption of unit mileage in transportation (C21)	Calculate the energy consumption of unit mileage based on different modes of transportation.
		Distance of transportation (C22)	The total mileage of each mode of transportation
	Energy consumption in construction of building projects (B3)	Energy consumption in the use of building equipment (C31)	Sum up the power energy sources, rated powers and machine-teams of the main building machinery.
		Energy consumption in the management of building projects (C32)	The energy consumptions of other aspects except for main building machinery, e.g. the energy consumptions of construction quality, construction management, etc.
	Energy consumption in operation of building projects (B4)	Operation energy consumption of unit area (C41)	It refers to the standard coal consumed by unit building are in a heating period to keep the indoor calculating temperature under the average outdoor temperature in the heating period.
		Total building area (C42)	Total building area of a building
	Energy consumption in dismantling and recycling (B5)	Energy consumption in equipment dismantling (C51)	Sum up the power energy sources, rated powers and machine-teams of the main dismantling machinery.
		Recycling and re-manufacturing energy consumption of building materials (C52)	Recycled and reuse broken concrete steel, reinforcement, glass and concrete and bury other wastes and the energy saved by the recycling and reuse compared with the use of completely new building materials.

Table 2. Grade of the result of the building energy consumption assessment.

Grade	Score of the assessment	Energy-saving effect
Excellent	1 - 0.8	The energy-saving effect is national leading.
Good	0.8 - 0.6	The expected energy-saving effect is realized and the actual energy-saving effect is very obvious.
Pass	0.6 - 0.4	The expected energy-saving effect is realized.
Rejected	Lower than 0.4	The expected energy-saving effect is not realized.

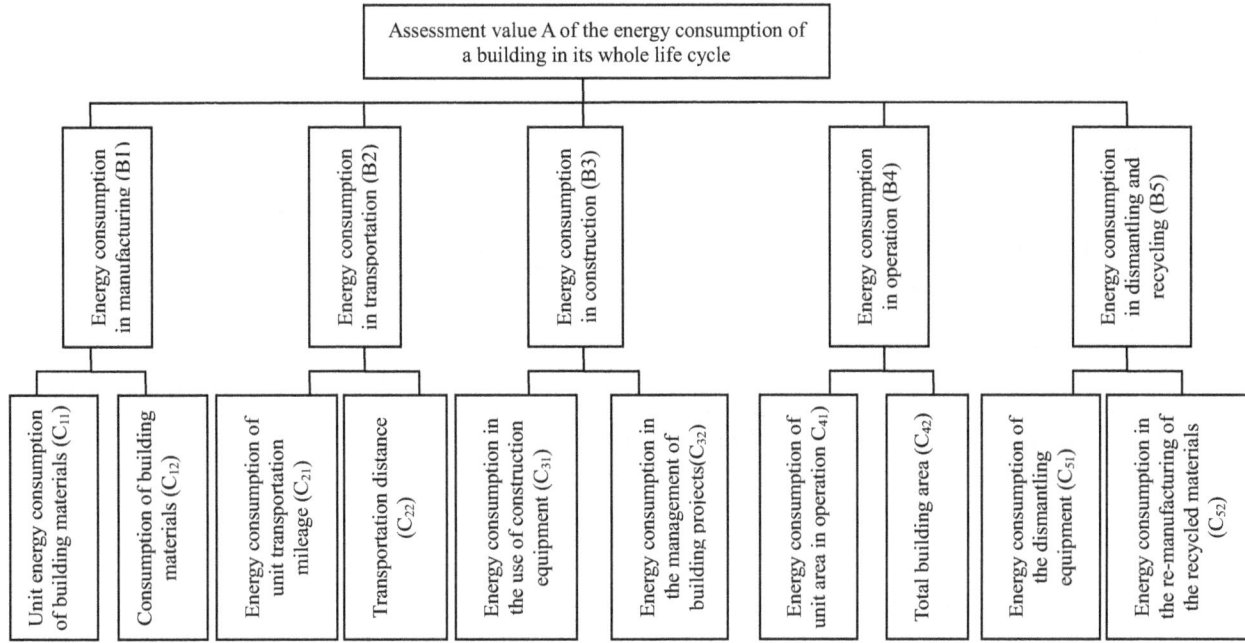

Figure 1. Assessment model for the energy efficiency of the rural residential buildings in the northern china in their whole life cycles.

and recycling of building projects [4].

$$E_{tot} = E_{manu} + E_{erect} + E_{occup} + E_{demo} + E_{dis} \quad (1)$$

In the formula:

E_{tot} refers to the total energy consumption of the buildings in their whole life cycles (unit: KJ);

E_{manu} refers to the energy consumption in the manufacturing of building materials (unit: KJ);

E_{erect} refers to the energy consumption in the transportation of building materials (unit: KJ);

E_{occup} refers to the energy consumption in the construction of building projects (unit: KJ);

E_{demo} refers to the energy consumption in the operation of building projects (unit: KJ); and

E_{dis} refers to the energy consumption in the dismantling and recycling of building projects (unit: KJ).

Hereunder are the calculation formulas for the energy consumptions in different stages.

1) Energy consumption in the manufacturing of building materials

$$E_{manu} = E_{manu, \, p_{rod}} + E_{manu, \, r_{enow}} \quad (2)$$

$E_{manu, \, p_{rod}} = \sum_{i=1}^{n} a_i * (1 + P) * A_i$

E_{manu} refers to the total energy consumption of building materials in the manufacturing stage (unit: KJ);

$E_{manu, \, p_{rod}}$ refers to the total energy consumption in the manufacturing of the building materials used in the construction of buildings (unit: KJ);

n refers to the number of the types of the building materials used for buildings;

a_i refers to the consumption of building material i and

is usually expressed in weight (Ton or Kg) or volume (m_3);

p refers to the ratio of the building material i abandoned in construction and is expressed in percentage (%); and

A_i refers to the unit energy consumption in the manufacturing of building material i (unit: KJ).

Different kinds of building materials have different service lives, so the energy consumption of building materials of building projects needs to be calculated repeatedly according to the service lives of building materials.

$$E_{manu, \, r_{enow}} = E_{manu} + p_{rod} * \left[\frac{Y_{bui}}{Y_{mat}} - 1 \right] \quad (3)$$

$E_{manu, \, r_{enow}}$ refers to the energy consumption in the manufacturing of the material buildings used for refurbishing buildings (unit: KJ);

Y_{bui} refers to the service life of a building (mainly the service life of the main structure of the building) (unit: year); and

Y_{mat} refers to the service lives of various building materials (unit: year).

2) Energy consumption in the transportation of building materials

The energy consumption in the transportation of building materials used in construction in building projects may be calculated based on the consumptions, transportation distances and unit transportation energy consumption of building materials of the buildings. If calculated based on the types and transportation distances of building materials, calculation may be done through the formula below.

$$E_{trans} = \sum_{i=1}^{n} a_i * (1 + P) B_i * A_i \qquad (4)$$

In the formula:

n refers to the types of the building materials used for buildings;

a_i refers to the consumption of building material i and is usually expressed in weight (Ton or Kg) or volume (m3);

p refers to the percentage of the building materials abandoned in construction;

B_i refers to the unit energy consumption of building materials transported in different means (unit: MJT/·Kln).

C_i refers to the transportation distance.

3) Energy consumption in the construction of building projects

The energy consumption in the transportation of building materials differs with the specific transportation means. If the building materials are basically local ones, the transportation means will be basically the same and the main factor affecting the energy consumptions are the transportation distances of the building materials.

$$E_{erect} = \sum_{i=1}^{n} m * a_i * p_i \qquad (5)$$

n refers to the number of the types of the construction methods used;

m_i refers to the unit construction area (m$_2$), weight (Ton) or volume (m$_3$) of each construction method; and

p_i refers to the unit energy consumption of the construction method (unit: KJ).

4) Energy consumption in the operation of building projects

For an ordinary reinforced concrete building, the total energy consumption in its service stage used accounts for about 70 to 80 percent of its whole life cycle. Even for a most energy-efficient building at present, the total energy consumption in its service stage accounts for 50 to 60 percent. The total energy consumption in the service stage of a building includes the energy consumed by air-conditioning, heating, water heating, cooking, lighting, electrical equipment [6].

$$E_{occup} = E_{occup, heat} + E_{occup, vent} + E_{occup, light} + E_{occup, sub} \qquad (6)$$

$E_{occup,heat}$ refers to the energy consumption of the air-conditioning system (unit: KJ);

$E_{occup,vent}$ refers to the energy consumption of the ventilation system (unit: KJ);

$E_{occup,light}$ refers to the energy consumption of the lighting system (unit: KJ); and

$E_{occup,sub}$ refers to the energy consumption of the auxiliary systems (unit: KJ).

5) Energy consumption in the dismantling and recycling of building projects

The energy consumption in this stage includes the energy consumption in dismantling and energy consumption in the transportation of covering soil and filling materials. The energy consumption in dismantling is regarded as 90 percent of the energy consumption in construction and the energy consumption in the transportation of covering soil and filling materials is calculated based on the construction area, the average soil covering and material filling, the average specific gravities of the covering soil and filling materials and the average transportation distance [5].

Waste and old building materials include recyclable ones and unrecyclable ones. Generally speaking, glass, wood, aluminum material, steel material and other building materials are recyclable.

$$E_{dis} = E_{dis, demo} - E_{dis, recycle} + E_{dis, unrecycle} \qquad (7)$$

$$E_{dis, demo} = 0.9 * E_{erect} + M * 1.5 * 2 * 1.836 \qquad (8)$$

$$E_{dis, recycle} = \sum_{i=1}^{n} a_i * q_i * A_i \qquad (9)$$

$E_{dis, demo}$ refers to the energy consumption in the dismantling stage (unit: KJ);

$E_{dis, recycle}$ refers to the energy consumption in the recycling of building materials;

$E_{dis, unrecycle}$ refers to the energy consumption in the disposal of building wastes;

E_{erect} refers to the total energy consumption in the construction stage (unit: KJ);

M refers to the construction area needing soil covering and material filling after the dismantling stage; and q refers to the recycling rate of building materials.

5. Assessment Model for Energy Efficiency of the Rural Residential Buildings in Northern China in Their Whole Life Cycles

Based on a full life cycle theory and an AHP, this paper establishes an assessment model for the energy efficiency of the rural residential buildings in the northern China in their whole life cycles.

1) Establishment of a hierarchical model

2) Establishment of a judgment matrix

The judgment matrix of the indicators on the first layer is a comparison matrix of the importance of the five indicators under the dominance of the overall goal assessment value A of the energy consumption of a building in its whole life cycle. The grade standard of judgment matrix is as shown in **Table 3**.

3) Determination of the weights of the judgment matrix

Determination of weights: It is realized through interview with experts. Several experts are invited to fill out a consultation form to form a judgment matrix between the

Table 3. Sadi scale for multiple comparison.

Scale	Meaning
1	Two factors have the equal importance.
3	The former factor is slightly more important than the latter one.
5	The former factor is obviously more important than the latter one.
7	The former factor is very important compared with the latter one.
9	The former factor is extremely important compared with the latter one.
2, 4, 6, 8	The means of the judgments above in comparison of two factors
Reciprocals of the figures above	The results above when the order of comparison above it opposite.

factor layer and the assessment value A of the energy consumption of a building in its whole life cycle.

4) Consistency check

The process of doing a multiple comparison between the factors of a layer and the factors of the higher layer to obtain the weights of the indicators of the lower layer is called single hierarchical sequencing.

$$CR = \frac{CI}{RI} \qquad (10)$$

$$CI = \frac{\lambda_{max} - n}{n-1} \qquad (11)$$

RI refers to the average random consistency indicator;

CI refers to the consistency indicator;

λ_{max} refers to the maximum of the characteristic equation; and n refers to the order of the judgment matrix.

5) General hierarchical sequencing and its consistency check

In the process, on the basis of the relative weights of the sub-systems, the weights of the factors on the last layer are multiplied by the relative weights of the controlled factors on the higher layer in turn to form the absolute weights of the factors for the overall goal.

$$CR = \frac{\sum_{i=1}^{n} B_i CI_i}{\sum_{i=1}^{n} B_i RI_i} \qquad (22)$$

If the CR is smaller than 0.1, it will be regarded that the general hierarchical sequencing has satisfying consistency.

Supposing the weights of the indicators of the indicator layer obtained through the AHP are w_1, w_2 and so on (ending with w_n) and the indicator values obtained after dimensionless processing are r_1, r_2 and so on (ending

with r_n), then the economic indicator A of an environment friendly building will be obtained through the formula below:

$$A = r * w^T \qquad (13)$$

$$r = (r_1, r_2, \cdots, r_n) \qquad (14)$$

$$w^T = (w_1, w_2, \cdots, w_n) \qquad (15)$$

6. Application of the Assessment Model

1) From the application of the model in the energy-saving reconstruction project of the residential b total energy consumption of other aspects took a small proportion (less than 1 percent) in the total energy consumption of the building project. In a word, the operation stage of the buildings is key for the energy efficiency of the buildings, and, after the reconstruction, the proportion of the energy consumption in operation of the buildings was decreased while that of the energy consumption in construction was increased.

2) With the increase of the service lives of the buildings, the energy consumption in the operation of the buildings keeps increasingly. The energy consumption in the operation of the buildings after the reconstruction is lower than that before the reconstruction and the energy consumption in the construction of the buildings after the reconstruction is higher than that before the reconstruction, so the total energy consumption of the buildings after the reconstruction will be increasingly lower than that before the reconstruction and the energy-saving effect of the buildings will be more and more obvious with time on.

3) Viewed from the total energy consumption, the buildings have shown an obvious reconstruction effect: The reconstruction of the single buildings has resulted in an energy consumption reduction of 11,620,037 Kj or 24.31 percent. Furthermore, the energy-saving effect is more and more obvious with increase of the building area.

4) Viewed from overall assessment, the grade of the assessment result is "Pass" before the reconstruction and "Good" after the reconstruction, *i.e.* the grade of the assessment result has been improved obviously.

REFERENCES

[1] W.-H. Tsai, S.-J. Lin, J.-Y. Liu, W.-R. Lin and K.-C. Lee, "Incorporating Life Cycle Assessments into Building Project Decision-Making: An Energy Consumption and CO2 Emission Perspective," *Energy*, Vol. 36, No. 5, 2011, pp. 3022-3029. doi:10.1016/j.energy.2011.02.046

[2] C. Y. Zhang, C. X. Suo and D. S. Wei, "Application of Energy-Saving Technologies in the Rural Buildings in China and Its Benefit Assessment," Economic Science

Press, Beijing, 2010

[3]   O. Bozdag and M. Secer, "Energy Consumption of RC Buildings during Their Life Cycle," *Sustainable Construction, Materials and Practices*: *Challenge of the Industry for the New Millennium*, Minho, 12-14 September 2007, pp. 480-487.

[4]   L. Uzsilaityte and V. Martinaitis, "Impact of the Implementation of Energy Saving Measures on the Life Cycle Energy Consumption of the Building," *7th International Conference Environmental Engineering*, Vol. 1-3, 2008, pp. 875-881,

[5]   Japan Sustainable Building Consortium, "Comprehensive Assessment System for Building Environmental Efficiency—Green Design Tool," China Architecture & Building Press, Beijing, 2005.

23

Research on Complementary of New Energy for Generation

Dan Li, Haiming Zhou, Fumin Qu
China Electric Power Research Institute, Beijing, China
Email: woandyu@126.com

ABSTRACT

Limited conventional energy and environmental issues have become increasingly prominent, so it has been more national attention to environmental protection and renewable new energy. The world's growing demand for energy, and the limited reserves of conventional non-renewable resources, mankind is facing a serious energy crisis. Coupled with the use of fossil fuels has brought serious environmental pollution problems, so the transition energy development way imminent, the need to be constantly developed and developing green renewable energy generation technologies.

Keywords: Energy; Environmental Pollution; Green Renewable Energy

1. Introduction

The world foot of the main power of the growing demand for energy, consume large amounts of coal and oil thermal power, while the contribution for the development of the national economy and the improvement of people's living standards, a lot of power, a lot of dust and smoke emissions, sulfur dioxide and other atmospheric pollutants, causing huge damage to our ecological environment. On the other hand, coal and oil reserves are limited, and the tighter supply. The limited reserves of conventional non-renewable resources, mankind is facing a serious energy crisis. From the protection of the ecological environment and energy consumption, we need to find new energy sources to replace conventional energy. Imminent transformation of energy development, need to be constantly developed. China has a wealth of new energy and renewable energy resources, mainly solar, wind, hydro.

2. Research Status at Home and Abroad

Solar new energy as a clean and pollution-free one, and has broad prospects for development in China's solar power photovoltaic-based. By the end of 2010, China's cumulative PV capacity of 860,000 kilowatts, including newly added 580,000 kilowatts, solar power projects currently under construction total size of up to 1 million kilowatts, the cumulative solar power capacity will reach 10 million kilowatts by 2015. However, the polycrystalline silicon solar cell is only in a small number of trial production stage. China's PV production and research

and development, there is a great gap compared with foreign countries, solar power is still in the small and medium-scale use of stage. Solar applications in Japan, a shortage of resources, Japan has been actively developing solar, wind, nuclear energy and other new energy sources, the use of biomass power generation, waste power generation, geothermal power and the production of fuel cells as a new energy, especially high hopes for the development and utilization of solar energy. Since 2000, solar photovoltaic, solar cell production for many years ranked first in the world, accounting for about half of the world's total output of. Wind energy is currently more mature a technology developed rapidly, the market value has been accepted by the people of clean energy. Wind energy has become an important part of the new energy plan for the U.S. government.

American Wind Energy Association, said New 8.35 GW wind power capacity in 2008, total production capacity has reached 25.1GW, accounting for 1/5 of global wind power, become the first in the world wind power. Wind energy resources have characteristics of randomness [1], intermittent, unpredictable nature and can not be stored, etc., tend to access the power system will bring greater harm [2-4] (such as the instability of the system is running, etc.), resulting in more wind power in the actual operation is limited to the Internet, has not been to maximize the use of wind power resources and social benefits [5-7]. The solution to this problem lies in how to control the power characteristics of wind power in the power system access, effective way of performance in wind power storage or complementary with energy joint operation and grid. The feng shui complementary power generation system is the organic combination of wind power generation system with the hydroelectric system with scheduling [8], when random fluctuations in the output of the wind farm to the grid, hydropower can quickly adjust the output of the generator output to compensate for the wind farm [9]. In recent years, the existing literature has studied wind water complementary joint optimization run, the use of pumping energy storage power plants, wind power and run complementary example optimized computing literature [10-12], especially more, but the pumping energy storage power station-building restrictions of natural conditions and requires a lot of investment. General regulation of hydropower and wind power complementary the literature relatively small and subject to geographical restrictions, [13] proposed using wind power / utilities complementary solve the reliability of power supply problems in the cold northern areas, and increase the economic efficiency of the system. [14] Proposed to solve the problem of power peaking and winter stable supply build wind, water and solar systems in Xinjiang. [15] With a specific example reveals the complementary characteristics of hydropower hydropower its power to support its capacity to support wind power and wind power wind power. If wind power configuration corresponding energy storage device (battery), with water and electricity the complementary to run (referred to as the wind build a hydropower complementary), the effective control of the power characteristics of the complementary system more secure and stable access to the grid, to improve the delivery of wind powercapacity, will make green energy to create more value.

Cold regions in northern China at the same time there is rich in hydropower resources and wind energy resources, from a seasonal point of view, winter and spring the water level of the reservoir is low, insufficient output of hydropower, when the wind speed of the wind farm, able to assume moreload. Wind speed in summer and autumn, the lower output of the wind farm, at this time it is the abundant rainfall, hydropower can bear the load, so precisely wind power hydropower season on complementary. In addition, small hydropower short-term volatility, the runoff in a day and night is basically uniform, while the short-term volatility of wind power is great, so that can reflect the complementary nature of wind power and hydropower [15]. In order to make effective use of wind energy, an energy storage device needs to be configured to reduce the supply of the electricity grid in the hydropower complementary supply of wind power, increase the effectiveness of the complementary system. Storage of wind energy [16], and the battery has a high efficiency, simple, reliable, high discharge power, quick charging, long cycle life, light weight, etc., to complete

the storage of excess wind power to supply the load. Select the battery to store excess wind energy on the feng shui complementary system and provide power to the load when necessary.

Hydro power generation will be the impact of the flood season and non-flood season. Flood season and non-flood hydropower optimization run around the nature of hydropower peaking hydropower generating capacity of wind power consumptive ability to maximize the optimization model to carry out comprehensive consideration. Therefore, the standard water and electricity to run the merits of large-scale wind power access grid case consists of two parts:

$$E = E_H + E_W$$

Formula: E is Target power, E_H is Hydropower generating capacity, E_W is Hydropower peaking power to the assimilative capacity of wind power.

Judging from the different periods of reservoir inflow characteristics, hydropower annual run can be divided into the flood season and non-flood season run.

3. Research on the Non-flood Season Run

Non-flood season have when hydropower plays a role in peak shaving, and wind power for the grid consumptive conditions. Run water and electricity in the grid load chart position, the peak load hydropower assumed more and more flat reserved for other power assumed power grid load on the grid clean energy consumptive conditions more favorable, but non-flood season by available water and hydropower installed capacity constraints, limited water and electricity in the grid during peak hours assumed power, hydropower peaking to eliminate the ability of wind power and hydropower generating capacity, hydropower installed capacity, power generation head, reservoir utilization requirements, as well as the power grid operation requirements related. Each charge relationship is calculated as follows:

$$E_{H,t} = KQ_t h_t \Delta t$$

$$E_{CH,t} = E_{H,t} - P_{b,t} \Delta t$$

$$E_{PL,t} = (P_{max} - P_{b,t}) \Delta t$$

Formula: $E_{H,t}$ is Generating capacity of hydropower t periods, K for hydropower output coefficient, Q_t is for t periods generation flow, h_t is the period t Hydropower average power head, Δt is The number of hours of the time period t, $E_{CH,t}$ is period t hydropower generating capacity net of minimum power after electricity, $P_{b,t}$ is the t periods hydropower lowest output, $E_{PL,t}$ is The hydropower t time adjustable power, $P_{max,t}$ is the period t hydropower expected to contribute, Size and average hydropower power head units available capacity and maximum contact line.

3.1. Non-the Flood Season Hydroelectric Peaking Elimination of Wind Power Capacity

According to the size of $E_{CH,t}$, Calculate the hydropower t periods peaking consumptive wind power capacity charge $E_{W,t}$ Can be divided into two kinds:

$$E_{W,t} = \begin{cases} E_{CH,t}, & E_{CH,t} < \dfrac{E_{PL,t}}{2} \\ E_{PL,t} - E_{CH,t}, & E_{CH,t} \ge \dfrac{E_{PL,t}}{2} \end{cases}$$

Visible, more non-flood hydropower generating capacity of peaking to consumptive, more wind power more favorable. Inevitably must bear trough power generation capacity over peak demand for electricity adjustable grid load when the hydropower t periods. The hydropower generating capacity is more and more unfavorable to the contrary, the power peaking. This was:

$$E_{W,t} = \min(E_{CH,t}, E_{PL,t} - E_{CH,t})$$

3.2. The Hutchison and Massive New Energy Access Non-flood Hydropower Optimal Operation Mode

As is shown in **Figure 1**, when the generating capacity to reach $E_{H,t}$, Hydropower peaking reaches a maximum contribution of electricity by wind power consumptive. Theoretically, when $E_{H,t} / E'_{H,t}$, and $E_{H,t} = E_{W,t}$ grid clean energy dissipation Granada to the best point. However, due to the water for reservoir operation with a lag, the best point of dynamic optimal point, need to be placed throughout the scheduling period to the consolidated balance.

Therefore, the non-flood season (in T_L) annual run model for hydropower:

$$T_L = \max \sum_{t \in TL} (E_{H,t} + \min(E_{CH,t}, E_{PL,t} - E_{CH,t}))$$

3.3. The Flood Season of Hydropower Optimal Operation

Since the flood season more runoff, and subject to the constraints of limited water level of reservoir flood season easily lead to disposable water, hydro mandatory participation peaking requirements to reflect the power grid, reservoir the annual operation mode making the introduction of daily load rate power peaking hydropower to participate describe, Hydropower t periods daily load expressed as:

$$\gamma_t = \frac{P_t}{P_{\max,t}}$$

where: P_t is the average output of the hydropower t periods. The visible daily load rates reflect the size of the contribution of hydropower flood season to participate peaking. To increase in terms of hydropower, flood season peaking, divided into the following two situations:

1) To allow a regulating storage reservoir ahead of pre-vent some water, will inevitably lead to reduced hydropower generation capacity.

2) Reservoir the abandoned water peaking, and the inevitable loss of some water resulting in reduced hydropower generation capacity, the measures equivalent consumptive wind power to give up part of the hydropower generating capacity.

The flood season of hydropower generating capacity should be:

$$E_{H,t} = \min(P(Q_t)\Delta t, P_{\max,t}\gamma_t \Delta t)$$

where: $P(Q_t)$, Q_t corresponds to the hydropower generation water flow output.

Flood season hydroelectric peaking assimilative capacity of wind power electricity:

$$E_{W,t} = (P_{\max,t}(h_t)\gamma_t - P_{b,t})\Delta t - E_{CH,t}$$

Therefore, the flood season (in T_H) build the target function is as follows:

$$T_H = \max \sum_{t \in TL} (E_{H,t} + E_{W,t})$$

4. Hydropower Peaking Contribution to the Power of the New Energy Consumptive Best Balance Analysis

Overall, the flood season to consider seeking the best balance of hydropower generation and hydropower peaking assimilative capacity of wind power electricity daily load rate constraints premise. Under normal circumstances, is to deal with the impact of wind power randomness of the grid, the grid is usually configured according to the same capacity with wind power conventional energy. Similarly, through the model theory analysis shows that, net power, hydropower generating capacity and adjustable power equal contribution of wind power consumer is satisfied most favorable. As is shown in **Figure 1**, $E_{H,t} = E_{W,t}$, the grid is the best balance of clean energy dissipation Granada. But this time is not the best clean energy generating capacity to achieve hydropower the largest contribution to the entire power grid clean energy consumptive target.

The objective function must also meet the following constraint conditions,

1) The water balance constraints:

$$W_{t+1} = W_t + (Q_{in,t} - Q_{out,t} - \Delta Q_t)\Delta t$$

$$Q_{out,t} = Q_{g,t} + Q_{e,t} + Q_{s,t}$$

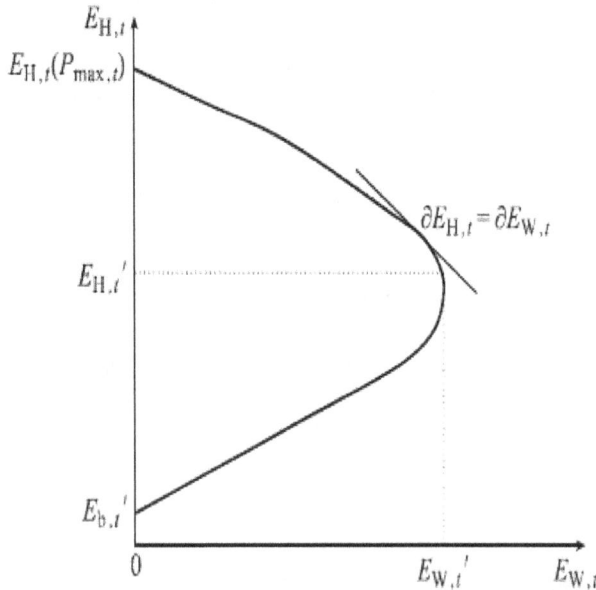

Figure 1. Non-flood hydropower generating capacity and peaking power capacity of wind power to dissolve the relationship.

where: W_t and W_{t+1}, respectively, for the the t periods early and late-period reservoir storage, $Q_{in,t}$ and $Q_{out,t}$, respectively, for the period t average inbound and outbound flow, ΔQ_t period t, the average loss (evaporation, leakage, etc.) flow, $Q_{g,t}$, $Q_{e,t}$, $Q_{s,t}$ respectively the time t average gate sluicing flow, generating traffic, comprehensive flow.

2) Water availability constraints

$$\sum_{i=1}^{T} Q_{in,t}\Delta t + \Delta W = W$$

where: ΔW is the storage capacity of the reservoir operation period; W scheduler period water availability, specific to a given year fixed runoff is a constant

3) The water level on the lower limit constraints

$$Z_{min,t} \leq Z_{t+1} \leq Z_{max,t}$$

wherein: $Z_{min,t}$, Z_{t+1}, $Z_{max,t}$ respectively, for the end of the period t, the minimum water level, period level, the highest water level of the period.

4) Comprehensive utilization of the flow limit constraints

$$Q_{min,t} \leq Q_{out,t} \leq Q_{max,t}$$

where: $Q_{min,t}$ and $Q_{max,t}$, respectively, for the period t minimum library traffic and outbound flow.

5. Conclusions

China's future sustainable energy development needs to be strong and smart grid support. Mostly with the conditions for the development of large-scale hydro, wind, solar and other clean energy and new energy, But far from load centers, the need for large-scale, long-distance transmission, the implementation of a wide range of energy optimal allocation of resources. So building a strong grid structure is the focus of China's smart grid development. In long-term point of view, the development of smart grid not only enhances the level of energy security, but also guides and changes the user's energy consumption habits, improve the efficiency of energy utilization. Rely on a strong and smart grid intelligent allocation of resources to become a key factor to enhance the sustainable development of China's energy.

REFERENCES

[1] H. H. Zhou and B. Tran, "Composite Energy Storage System with Flexible Energy Management Capability for Micro-grid Applications," *Energy Conversion Congress and Exposition (ECCE)*, 2010, pp. 2558-2563.

[2] P. Juang and Kollmeyer, "System Identification Based Lead Acid Battery Online Monitoring System," *Energy Conversion Congress and Exposition(ECCE)*, 2010, pp. 3903-3910.

[3] C. Stummer and K. Heidenberger, "Interactive R & D Portfolio Analysis with Project Interdependencies and Time Profiles of Multiple Objectives," *IEEE Transactions on Engineering Management*, Vol. 50, No. 2, 2003, pp. 175-183. doi:10.1109/TEM.2003.810819

[4] Y. P. Yu, M. Yong, L. Chen, *et al.*, "Analysis of Forced Power Oscillation Caused by Continuous Cyclical Load Disturbances," *Automation of Electric Power Systems*, Vol. 34, No. 6, 2010, pp. 7-11.

[5] Z. Y. Han, R. M. He, J. Ma, *et al.*, "Comparative Analysis of Disturbance Source Inducing Power System Forced Power Oscillation," *Automation of Electric Power Systems*, Vol. 33, No. 3, 2009, pp. 16-19.

[6] Y. P. Yu, Y. Min, L. Chen, *et al.*, "Disturbance Source Location of Forced Power Oscillation Using Energy Functions," *Automation of Electric Power Systems*, Vol. 34, No. 5, 2010, pp. 1-6.

[7] D. G. Yang, J. Y. Ding, H. Zhou, *et al.*, "Mechanism Analysis of Low — Frequency Oscillation Based on WAMS Measured Data," *Automation of Electric Power Systems*, Vol. 33, No. 23, 2009, pp. 24-28.

[8] D. J. Yang, J. Y. Ding, J. Y. Li, *et al.*, "Analysis of Power System Forced Oscillation Caused by Asynchronous Parallelizing of Synchronous Generators," *Automation of Electric Power Systems*, Vol. 35, No. 10, 2011, pp. 99-103.

[9] Y. Yang, F. Wen, L. Li, *et al.*, "Coordinated Model for Available Transfer Capability Decision-making Employ-

ing Multi-objective Chance Constrained Programming," *Automation of Electric Power Systems,* Vol. 35, No. 13, 2011, pp. 37-43.

[10] M. Wang and M. Ding, "Probabilistic Calculation of Total Transfer Capability Including Large Scale Solar Park," *Automation of Electric Power Systems*, Vol. 34, No. 7, 2010, pp. 31-35.

[11] S. K. Chung, "Phase-Locked Loop for Grid-Connected Three-Phase Power Conversion Systems," *IEE Proc eedings–Electric Power Applications*, Vol. 147, No. 3, 2000, pp. 213-219.doi:10.1049/ip-epa:20000328

[12] K. D. Zhu, Y. H. Song, Z. F. Tan, *et al.*, "China Wind Power Integration Status Quo and Its Benefit to Energy Saving and Emission Reduction," *Electric Power*, Vol. 44, No. 6, 2011, pp. 67-70.

[13] Z. H. Liu, L. Chen and Y. Min, "A Practical Energy-saving Generation Dispatching Model Integrated with Wind Power ," *Electric Power*, Vol. 44, No. 6, 2011, pp. 52-57.

[14] X. Z. Dai, Y. H. Shi, Z. X. Lu, *et al.*, "Impact of Large-scale Wind Power on Operation Stability and Security of Connected Jiangsu Grid," *Electric Power,* Vol. 44, No. 6, 2011, pp. 42-47

[15] F. Zhang, M. X. Li, G. W. Fan, *et al.*, "Transient Voltage Stability Study about a Regional Grid Integrated with Wind Power," *Electric Power*, Vol. 44, No. 9, 2011, pp. 17-21.

[16] G. L. Liu, J. J. Cai and W. Q. Wang, "Impacts of Grid Connection and Disconnection of Wind Turbines on System Operation and CopingSstrategies," *Electric Power*, Vol. 44, No. 4, 2011, pp. 7-10.

24

Analysis of Renewable Energy Utilization Potential in Buildings of China

Min-hua Cai, Lan Tang[*]

Guangdong Provincial Key Laboratory of Building Energy Efficiency and Application Technologies,
Guangzhou University, Guangzhou, China
Email: 316543111@qq.com, tanglan@gzhu.edu.cn

ABSTRACT

As a country of great population, China has increasing building energy consumption continuously. It not only threatens the lack of total energy but also hardens the progress of protecting environment. Therefore, it forces the country to accelerate finding substitution application of conventional energy in building, renewable energy building utilization. In base of 2010, this study explores the potential of the renewable energy building utilization by using energy consumption analysis until 2030 and predicts annual alternative quantity of renewable energy in different situations.

Keywords: Renewable Energy; Building Energy Consumption; Potential Analysis; Scenario Analysis Method

1. Introduction

Balancing every influence factors plays an important role in national prosperity and sustainable development, so energy balance has been a long-term and urgent task. The three global traditional energy, oil, coal and natural gas, accounted for more than 90% of the energy consumption. Implementing the scheme of traditional energy reduction is so necessary. Although our country is a rich coal country, because of the large population and causes personal average energy quantity is lower than the world average energy quantity. According to the analysis of journal academic paper "Thinking about the Problems of Chinese Energy", published in "Journal of Shanghai Jiaotong University" .China's average coal quantity is only 32.8% of the world's average coal quantity, demonstrating clearly our national energy tension.

Moreover, with the improvement of social life quality, China's building terminal energy consumption has accounted for nearly 27.6% of the total primary energy consumption. "2012 Annual Report on China Building Energy Efficiency" [1] issued that personal building energy consumption of China's urban people is only one-fifth of the total building energy consumption, However, due to the large population base, the total building energy consumption quantity is very large.

In the face of such serious situation, the state has adopted a series of measures:

On the one hand, main mandatory measures adopted by the national building field are as follow:

1) Strictly controlling the size of the general public buildings and the growth of building energy consumption;

2) Advocating saving the energy consumption model and way of life;

3) Establishing the supervision system of building energy efficiency;

4). Ensuring the smooth implementation of building energy efficiency;

5) Strengthening the incentive policy and giving priority to financial subsidies.

On the other hand, there is a large-scale application of renewable energy replacing conventional fossil energy. At the same time, people's consciousness of energy conservation and emission reduction has been improved. With the global appeal of "low-carbon development", so it's the tendency to limit fossil energy consumption and implement clean or renewable energy consumption nowadays. Utilization of renewable energy in building field has a great large of potential.

2. Forecast Analysis of Energy Saving Potential of the Renewable Energy Utilization in Building

2.1. Situation Forecast of Building Energy Consumption (2010-2030)

According to "2008 Annual Report on China Building Energy Efficiency" [2], we can organize the specific data of building energy consumption from 2002 to 2010 as following **Table 1** [3]. Taking the statistics data into account, the average annual growth rate from 2002 to 2010 comes out to be 12.64% .This is the reality state of building energy consumption growth recently.

According to Chinese long-term planning of energy [4], the total energy consumption should be controlled bellow 4.2 billion tce until 2020, with 20% ~ 25% occupied by building energy consumption. So the average annual growth rate of building energy consumption from 2010 to 2030 can come out to be 0.9%, while assuming 1 billion tce in 2020. This is the ideal state of building energy consumption growth.

During 2010 to 2030, with the development of the social mechanism and social progress, the state begins to pay more and more attention to building energy saving and introduces some corresponding laws or regulations. It improves the construction equipment efficiency by means of the improved science and technology. The society has formed a mature saving energy consumption lifestyle and so on .These lowers average annual growth rate. Taking a comprehensive consideration of the influence, we can set an average annual growth rate of 4.00% and forecast the development of building energy consumption from 2010 to 2030 (as **Table 2**)

2.2. The Alternative Quantity of Renewable Energy in Building Energy Consumption (2010-2030) (Baseline Scenario)

According to the published alternative quantity of renewable energy in 2010 and the index achieved by "the Medium and Long-term Development Plan of Renewable Energy" [4] until 2020, we can calculate the alternative quantity growth trend of renewable energy in building energy consumption and predict alternative data of 2010 to 2030.

By the end of 2010 [5] ，the installation of solar energy water heater amounts to 168 million square meters and its alternative fossil energy is about 20 million tons of standard coal; Roof photovoltaic power generation project is up to 4.4 billion kilowatt hour, which reduces 1.42 million tons of standard coal; Ground source heat

pump can save 4.6 million tons of standard coal; Methane is about 14 billion cubic meters and its alternative energy is 11.14 million tons of standard coal. The total alternative quantity is about 37.16 million tons of standard coal (shown as **Table 3**), accounting for 4.44% of building energy consumption.

In 2010, the solar energy water heater area is 168 million square meters and planning area of 2020 [5] is 800 million square meters. Refer to the growth speed of renewable energy production from 2010 to 2020, the average annual growth rate comes out to be 16.89%. Therefore we can predict the alternative quantity of solar energy

Table 2. The prediction of building energy consumption from 2010 to 2030.

Year	Population (hundred million)	Building area (hundred Million m^2)	Building energy consumption (hundred million tce)	Personal building energy consumption (kgce/per)
2010	13.40	453	8.36	623.88
2015	13.70	513.17	10.17	742.57
2020	14.00	569.36	18.32	1308.28
2030	14.50	632.05	27.11	1870.08

Table 3. the alternative quantity prediction of renewable energy in building energy consumption from 2010 to 2030 (converted into ten thousand tce/year).

year		2010	2015	2020	2030
Solar water heater	ten thousand m^2	16800	34661	80000	230000
	ten thousand tce/year	2000	3943	9100	26163
Roof photovoltaic power generation	hundred million kW·h	44	62	188	804
	ten thousand tce/year	142	201	608	2605
Shallow Geothermal energy	ten thousand tce/year	460	1000	2200	10522
Methane	hundred million m^3	140	200	440	1383
	ten thousand tce/year	1114	1591	3500	11000
Total	ten thousand tce/year	3716	6735	15408	50289
Proportion in building energy consumption	%	4.44	6.62	10.06	22.18

Table 1. The development of building energy consumption from 2002 to 2010 (unit: ten thousand tce).

Year	2002	2004	2006	2008	2010
North town heating	8584	12925	16205	17985	16330
Residential (except the northern heating)	11228	12787	14623	16290	47910
Public buildings (except the northern heating)	12229	26018	18716	30595	17370
total	32041	51730	49544	64870	83620
annual rate of growth (%)	6.29	35.32	8.77	16.90	14.23

Table 4. Prediction of renewable energy in building energy consumption under different situations from 2010 to 2030 (converted into ten thousand tce/year).

N	year	2010	2015	2020	2030
1	baseline scenario	3716	6735	15408	50289
	Proportion（%）	4.44	6.62	10.06	22.18
2	People know deeply	3901	7071	16179	52804
	Proportion（%）	4.67	6.95	10.56	23.29
3	Well-established policy	4087	7408	16949	55318
	Proportion（%）	4.89	7.28	11.06	24.40
4	Mature technology	4087	7408	16949	55318
	Proportion（%）	4.89	7.28	11.06	24.40
5	People know deeply + Well-established policy	4273	7745	17719	57832
	Proportion（%）	5.11	7.61	11.57	25.51
6	People know deeply+Mature technology	4273	7745	17719	57832
	Proportion（%）	5.11	7.61	11.57	25.51
N	year	2010	2015	2020	2030
7	Well-establish-ed policy + Mature technology	4459	8081	18490	60347
	Proportion（%）	5.33	7.95	12.07	26.61
8	Best scene	4645	8418	19260	62861
	Proportion（%）	5.56	8.28	12.57	27.72

Proportion: the proportion of renewable energy consumption in total building energy consumption.

water heater in 2015 and 2030. According to the same calculation methods, the alternative quantity of roof photovoltaic power generation, shallow geothermal energy and methane can be respectively predicted from 2010 to 2030 and the related data converted into unity unit: ten thousand tce/year, shown in **Table 3**. Until 2030, renewable energy in building applications can save 502.89 million tce and is 13.5 times that of 2010, accounting for 22.18% of the total energy consumption. It can ease the shortage of conventional energy sources which can be used in any other field.

2.3. The Alternative Quantity of Renewable Energy in Building Energy Consumption under Different Situations (2010-2030)

In the situational analysis, because of the popular utilization technology of the renewable energy in building, we can assume 5% of increase rate. Moreover, due to the standard system and incentive policy improvement, it makes the alternative quantity increased by 10%. Besides, it will be increased by 10% due to the mature utilization technology of renewable energy. In these ways, this research explores the alternative quantity of renewable energy in building under various situations (shown as **Table 4**).

When it comes to be the best scene, which means people know deeply, standard system and incentive policy of renewable energy construction become more perfect and utilization technology of renewable energy is mature, the alternative quantity is up to 628.61 million tce, accounting for 27.72% of building energy consumption. It's 125.72 million tce more than baseline scenario.

3. Conclusions

Based on 2010, this study explores the potential of renewable energy development in building energy consumption in 2030 and uses scenario analysis method to predict the alternative quantity of renewable energy in different situations. So the relevant departments led to realizing energy saving direction and key areas of renewable energy in the future. It will be bound to improve the building integration technology and perfect building design specifications.

4. Acknowledgements

We thank the support from the National Natural Science Foundation of China (51078092), "Yangcheng scholar" project of Education Bureau of Guangzhou (10A039G) and Guangzhou city-belonged university research project of Education Bureau of Guangzhou (10A020).

REFERENCES

[1] Qinghua University Building Energy Research Center, 2012 Annual Report on China Building Energy Efficiency, 2012.

[2] Qinghua University Building Energy Research Center, 2008 Annual Report on China Building Energy Efficiency, 2008

[3] National Bureau of Statistics of the People's Republic of China, 2001-2011China Statistical Yearbook, 2001-2011.

[4] National Development and Reform Commission of the People's Republic of China, Medium and Long-term Development Plan of Renewable Energy of the People's Republic of China, 2007.

[5] National Energy Bureau of the People's Republic of China, Renewable Energy Development 1025 planning, 2012.

Permissions

List of Contributors

Mohamed Talaat
Electrical Power and Machines, Faculty of Engineering, Zagazig University, Zagazig, Egypt

Reda Edris, Naglaa Ibrahim, Fatma Omar and Mohamed Ibrahim
Electrical and Computer Engineering, Higher Technological Institute, 10th of Ramadan City, Egypt

Mohamed Darwish and Rabi Mohtar
Qatar Environment and Energy Research Institute, Qatar Foundation, Doha, Qatar

Joanna Burger
Division of Life Sciences, Consortium for Risk Evaluation with Stakeholder Participation, Environmental and Occupational Health Sciences Institute, Rutgers University, Piscataway, USA

Michael Gochfeld
Consortium for Risk Evaluation with Stakeholder Participation, Environmental and Occupational Health Sciences Institute, Environmental and Occupational Medicine, Rutgers Medical School, Rutgers University, Piscataway, USA

Jose M. "Chema" Martinez-Val Piera, Alfonso Maldonado-Zamora and Ramon Rodríguez Pons-Esparver
ETSI Minas, Universidad Politécnica de Madrid, Madrid, Spain

Olugbenga Olanrewaju Noah
Department of Mechanical Engineering, University of Lagos, Lagos, Nigeria

Albert Imuentinyan Obanor
Department of Mechanical Engineering, University of Benin, Benin, Nigeria

Mohammed Luqman Audu
Department of Mechanical Engineering, Auchi Polytechnic, Benin, Nigeria

Ruiyuan Kong and Tiande Guo
School of Mathematical Sciences, University of Chinese Academy of Sciences, Beijing, China

Congying Han
College of Humanities and Social Sciences, University of Chinese Academy of Sciences, Beijing, China

Wei Pei
Institute of Electrical Engineering, Chinese Academy of Sciences, Beijing, China

Tarcisio Oliveira de Moraes Júnior, Yuri Percy Molina Rodriguez, Ewerton Cleudson de Sousa Melo, Maraiza Prescila dos Santos and Cleonilson Protásio de Souza
Federal Institute of Paraíba - IFPB, Cajazeiras, Brazil
Department of Electrical Engineering, Federal University of Paraíba - UFPB, João Pessoa, Brazil

Wilson Mungwena
Department of Mechanical Engineering, University of Zimbabwe, Harare, Zimbabwe

Cosmas Rashama
Department of Electrical Engineering, University of Zimbabwe, Harare, Zimbabwe

Joanna Burger
Division of Life Sciences, Rutgers University, Piscataway, USA
Environmental and Occupational Health Sciences Institute, UMDNJ and Rutgers University, Piscataway,USA

Michael Gochfeld
Consortium for Risk Evaluation with Stakeholder Participation, Vanderbilt University and Rutgers University, Piscataway, USA
Environmental and Occupational Medicine, UMDNJ-Robert Wood Johnson Medical School, Piscataway, USA

Ying Shen, Xintong Yang and Xiaoli Guo
Changchun University of Technology, Changchun, P.R. China

Jitiwat Yaungket and Tetsuo Tezuka
Graduated School of Energy Science, Kyoto University, Kyoto, Japan

Boonyang Plangklang
Faculty of Engineering, Rajamangala University of Technology, Thanyaburi, Thailand

Wasiu Oyediran Adedeji and Ismaila Badmus
Department of Mechanical Engineering, Yaba College of Technology, Lagos, Nigeria

Ding Ma and Yi-bing Xue
Department of Architecture and Urban Planning. Shandong Jianzhu University, Jinan, China

Wahied G. Ali and Sutrisno W. Ibrahim
Electrical Engineering Department, King Saud University, Riyadh, KSA

Gregory E. T. Gamula and Liu Hui
School of Environmental Studies, China University of Geosciences, Wuhan, China

Wuyuan Peng
School of Economics and Management, China University of Geosciences, Wuhan, China

Yasir A. Alturki and Mohamed A. El-Kady
SEC Chair in Power System Reliability and Security, Electrical Engineering Department, King Saud University, Riyadh, KSA

Noureddine Khelifa
Ministry of Water and Electricity, Riyadh, KSA

Vasilios Zarikas
Department of Electrical Engineering, Theory Division, Academic Institute of Technology of Lamia, ATEI Lamias, Lamia, Greece

Nick Papanikolaou
Department of Electrical Engineering, Power Electronics Division, ATEI Lamias, Lamia, Greece

Michalis Loupis
Department of Electrical Engineering, Informatics Division, ATEI Lamias, Lamia, Greece

Nick Spyropoulos
Research and Development Division, Kleeman Hellas, Kilkis, Greece

Xian Ma, Jingtian Bi, Weili Chen and Tong Jiang
State Key Laboratory of Alternate Electrical Power, System with Renewable Energy Sources, North China Electric Power University, Beijing, China

Zhisen Li
Tai'an Power Supply Company, Shandong Electric Power Corporation

Liyang Fan
Department of Civil Engineering and Architecture, Zhejiang University, Hangzhou, China
Department of Architecture, The University of Kitakyushu, Kitakyushu, Japan

Weijun Gao
Department of Architecture, The University of Kitakyushu, Kitakyushu, Japan

Zhu Wang
Department of Civil Engineering and Architecture, Zhejiang University, Hangzhou, China

Musediq Adedoyin Sulaiman
Mechanical Engineering Department, Olabisi Onabanjo University, Ago-Iwoye, Nigeria

Abayomi Olufemi Oni
Mechanical Engineering Department, University of Agriculture Abeokuta, Abeokuta, Nigeria

David Abimbola Fadare
Mechanical Engineering Department, University of Ibadan, Ibadan, Nigeria

Chenxia Suo and Yong Yang
Beijing Institute of Petrochemical Technology, Beijing, China

Solvang Wei Deng
Narvik University College, Narvik, Norway

Dan Li, Haiming Zhou and Fumin Qu
China Electric Power Research Institute, Beijing, China

Min-hua Cai and Lan Tang
Guangdong Provincial Key Laboratory of Building Energy Efficiency and Application Technologies, Guangzhou University, Guangzhou, China